ACS SYMPOSIUM SERIES **424**

Sound and Vibration Damping with Polymers

Robert D. Corsaro, EDITOR
Naval Research Laboratory

L. H. Sperling, EDITOR
Lehigh University

Developed from a symposium sponsored
by the Division of Polymeric Materials: Science and Engineering
at the 197th National Meeting
of the American Chemical Society,
Dallas, Texas,
April 9–14, 1989

American Chemical Society, Washington, DC 1990

Library of Congress Cataloging-in-Publication Data

Sound and vibration damping with polymers / Robert B. Corsaro, editor; L. H. Sperling, editor

p.   cm.—(ACS symposium series, ISSN 0097–6156; 424)

"Developed from a symposium sponsored by the Division of Polymeric Materials: Science and Engineering at the 197th National Meeting of the American Chemical Society, Dallas, Texas, April 9–14, 1989."

Includes bibliographical references.

ISBN 0–8412–1778–5

1. Vibration—Congresses. 2. Damping (Mechanics)—Congresses. 3. Polymers—Acoustic properties—Congresses. 4. Sound waves—Damping—Congresses. I. Corsaro, Robert D. II. Sperling, L. H. (Leslie Howard), 1932–   . III. American Chemical Society. Division of Polymeric Materials: Science and Engineering. IV. American Chemical Society. Meeting (197th: 1989: Dallas, Tex.). V. Series

TA355.S69   1990
620.3'7—dc20

90–325
CIP

The paper used in this publication meets the minimum requirements of American National Standard for Information Sciences—Permanence of Paper for Printed Library Materials, ANSI Z39.48–1984.

Foreword

The ACS SYMPOSIUM SERIES was founded in 1974 to provide a medium for publishing symposia quickly in book form. The format of the Series parallels that of the continuing ADVANCES IN CHEMISTRY SERIES except that, in order to save time, the papers are not typeset but are reproduced as they are submitted by the authors in camera-ready form. Papers are reviewed under the supervision of the Editors with the assistance of the Series Advisory Board and are selected to maintain the integrity of the symposia; however, verbatim reproductions of previously published papers are not accepted. Both reviews and reports of research are acceptable, because symposia may embrace both types of presentation.

Contents

Preface

A SIGNIFICANT DICHOTOMY EXISTS in the damping field: There are those scientists concerned primarily with viscoelastic relationships and material properties that surround the glass transition; then there are the acoustical engineers who design better damped structures in a very practical world. This dichotomy has led to a situation in which polymer scientists are surprisingly unfamiliar with the physical mechanisms by which material properties affect damping efficiency in engineering structures, and in which engineers do not fully use the ability of polymer science to control features of glass transition, loss and storage modulus.

The symposium on which this book is based was held for the specific purpose of bringing these two specialties together. Each group has aggressively presented background material it felt the other was lacking, and each presented its views on what is needed to produce more effective damping materials and structures in the future.

We invite you to read this book, whatever your technical background, and to consider it a first step toward the integration of this interdisciplinary field. To this end, the book contains many introductory tutorials and review chapters grouped within several sections, each section addressing (sometimes only broadly) a given topic. We caution those already involved with polymers or damping that reading only those sections of interest to your specialty reinforces the dichotomy this book is attempting to remedy. We encourage you to read the book as a whole, with particular emphasis on those subjects outside your specialty.

We want to thank the many people and organizations who made possible both the symposium and the book. The ACS Division of Polymeric Materials: Science and Engineering provided significant financial support for several overseas speakers and contributed to the cost of organizing the symposium. Lehigh University's Office of Research provided Sperling with a special grant to defer publication costs. We wish to thank all the contributors for their time and effort;

special thanks go to Bruce Hartmann for his technical assistance and to Virginia Newhard, secretary to Lehigh University's Polymer Laboratory, for her support in computer work and organizational activities.

ROBERT D. CORSARO
Naval Research Laboratory
Washington, DC 20375–5000

L. H. SPERLING
Lehigh University
Bethlehem, PA 18015

January 24, 1990

Introduction

Damping for suppression of sound and vibration has become dramatically topical, and justifiably so. There are an increasing number of high-payoff applications, both military and civilian. The trend is toward energy conservation and the attendant lighter weight and higher speeds. It follows that acoustic and vibratory disturbances are greater and structural response is both potentially larger and less desirable. The population has been conditioned to expect quiet. We don't want to hear aircraft taking off. We don't want to hear the dishwasher. We don't want to hear road noise from the car. Damping technology has quieted all of these and more.

Successful damping applications require high levels of expertise. First, the basic problem must be investigated (i.e., the disturbance, the response, and the problem mode mechanics must be understood). The operational conditions, especially operating and survival temperatures, are extremely important. With the problem mode and operational temperature known, accurate and efficient design methods may be used to develop possible solutions. Damping polymers must be selected both for function at the operating temperature range and for adequate environmental resistance and stability. Experimental and verification testing are essential to establishing design practicality. Fabrication processes must be technically satisfactory and economical.

Damping performance is related to structure and molecular weights of moieties; optimized polymeric compositions may be specified for specific applications. Techniques for reducing sound include scattering by inhomogeneities, mode conversion at boundaries, redirection, and intrinsic absorption. The temperature of peak damping depends on chemical composition variables: backbone flexibility, steric effects, polarity, plasticizers, crystallinity, pendant groups, and cross-link density. Inclusions and voids have a significant effect on reflection, absorption, and velocity of sound; theories may predict effects of microscopic or macroscopic inhomogeneity. Mode conversion scattering can be very effective. Further work is needed not only in the dynamic mechanical properties of polymers and reduced sensitivity to temperature, but also in properties such as toxicity, flammability, outgassing, resistance to heat, moisture, hydraulic fluid, air, atomic oxygen, and radiation.

The development of successful applications depends on continued advancement of scientific understanding of basic principles. Clearly, greater understanding of this subject area will point the way to materials and design concepts that underlie a wide range of damping applications. The chapters in this volume contribute significantly.

Dr. Lynn Rogers
Flight Dynamics Laboratory
Wright–Patterson Air Force Base, Ohio 45433

November 1989

1

DEFINITIONS AND CONCEPTS

DEFINITIONS AND CONCEPTS

The two papers in this section are tutorials on general polymer characteristics and behavior, with special emphasis on damping.

Polymers are unique materials, particularly well-suited for absorbing certain types of mechanical motion and forces. These two papers provide essential introductory material for describing the mechanical properties of polymers, and understanding how structural features (on the molecular level) control their mechanical behavior and characteristics.

In this section, **Sperling** first presents introductory definitions and descriptions of polymeric materials and their unique mechanical behavior. He also briefly introduces the key topics addressed in the remainder of this book. **Hartman** then deals more specifically with the damping factor, the most important property of polymers in the present context. He particularly considers the relationship of damping factor to polymer structure and composition.

Note that the damping factor is also known in this book variously as $\tan\delta$, the loss factor, and η, all defined as the ratio of the loss modulus (E") to the storage modulus (E'). Here we use E" and E' which refer specifically to the loss (imaginary) and storage (real) parts of the Young's modulus. This is appropriate, since polymers in damping applications are usually subject to uniaxial extension or compression. Other types of less frequently encountered deformations, such as bulk-compression or lateral-shearing motions, will couple to the corresponding moduli (bulk and shear moduli), and each will have a corresponding damping factor defined in a similar manner.

Chapter 1

Sound and Vibration Damping with Polymers

Basic Viscoelastic Definitions and Concepts

L. H. Sperling

Department of Chemical Engineering, Lehigh University, Bethlehem, PA 18015

The glass transition of a polymer involves the onset of coordinated chain motion. Dynamic mechanical spectroscopy characterizes the storage modulus, E', the loss modulus, E'', and the loss tangent, tan δ, as functions of temperature and frequency. In the presence of mechanical vibrations, the vibrational energy is absorbed by the polymer in the form of heat, the basis for damping with polymers. The phenomenon resembles infrared absorption, where electromagnetic waves increase molecular motion, actually warming the sample. The mechanics of extensional and constrained layer damping are reviewed, including the use of platelet fillers. Methods of engineering the width, height, and position of the transition, such as by plasticization, statistical copolymerization, graft and block copolymerization, and interpenetrating polymer networks are described. Application of current polymer materials to sound and vibration problems are delineated.

Sound and vibration damping with polymers utilizes the professions of chemistry, chemical engineering, materials, mechanical engineering, and polymer science and engineering. Each contributes its genius to an improved understanding of the nature of damping. How best shall noisy aircraft, cars, ships, machinery, etc., be quieted? The objective of this book is to provide an up-to-date introduction to the science and technology of sound and vibration damping with polymers, as well as a selection of papers showing the current status of research programs, world wide. Thus, theory, instrumentation, polymer behavior, and engineering systems will be described.

The objective of this paper is to introduce those terms and concepts which will appear throughout this book. Basic definitions and concepts will be emphasized. Reviews of both the polymer characteristics (1,2) and the acoustic requirements (3,4) are available.

Polymer Structure

Polymers are long-chain molecules with molecular weights often measuring in the hundreds of thousands. For this reason, the term "macromolecules" is often

0097–6156/90/0424–0005$06.00/0

employed when referring to polymeric materials. Another term, frequently used in the trade literature is "resins," which goes back before the chemical structure of the long chains was understood. The term polymer itself means "many mers," and a polymer is synthesized from its monomers, which are little molecules. The structure of poly(vinyl acetate), widely used in damping compounds, is written:

$$- CH_2 - CH - \quad CH_2 - \quad CH - \dots$$
$$\qquad\quad | \qquad\qquad\qquad\quad |$$
$$\qquad O - C - CH_3 \qquad O - C - CH_3$$
$$\qquad\qquad \| \qquad\qquad\qquad\quad \|$$
$$\qquad\qquad O \qquad\qquad\qquad\quad O$$

Polymers can be organic or inorganic in structure. Besides poly(vinyl acetate), above, common organic polymers include polyethylene, cis-polyisoprene (natural rubber), cellulose (cotton and rayon), and poly(methyl methacrylate) (Plexiglas). Some inorganic polymers include poly(dimethyl siloxane) (silicone rubber), and ordinary window glass. The glass transition (see below) is named for the softening of glass (1,2). Often polymers are made as copolymers, which means that they contain two or more kinds of mers.

<u>Sound and Vibrations</u>

Sound is air or water bourn acoustic vibrations. It is a pressure wave transmitted though air, water, or other fluid media. Vibrations are similar waves being transmitted through solid objects. Most importantly, these pressure waves are a form of energy. It is the removal or reduction of this energy, when necessary, that this book is all about. When the sound pressure is high enough, it may cause deafness or reduced efficiency when people are exposed to it. Vibrations may generate sound, or cause fatigue, and hence mechanical failure.

All real bodies are naturally damped, albeit many bodies of interest damp only modestly. Polymers, especially near their glass transition temperatures, damp much more. Commercially, polymers may be applied to the surface of the vibrating substrate to increase damping. Both single-layer (extensional) and two-layer (constrained) layer systems are in use, albeit for somewhat different purposes. The damping increase that can be afforded by applying polymeric materials to a steel reed is illustrated in Figure 1 (5). The waveform generated by an undamped reed is shown in Figure 1a, and a damped reed is shown in Figure 1b.

Interestingly, the human ear is an integrating device, determining input over a short period of time as a single sound. Thus Figure 1b sounds quieter than Figure 1a, although initially they are nearly the same. The human ear is also approximately a logarithmic device; this permits hearing at both very low and very high sound pressures.

<u>Basic Damping Concepts and Definitions</u>

The complex modulus, E^*, can be expressed as

$$E^* = E' + iE'' \tag{1}$$

where E' is the storage modulus, and E'' is the loss modulus. These quantities are illustrated schematically in Figure 2. Here, a ball is dropped on a perfectly elastic floor. It recovers a distance equivalent to E', a measure of the energy stored elastically during the collision of the ball with the floor. The quantity E'' represents the equivalent energy lost (as heat) during the collision of the ball with the floor.

Thus, the ball actually heats up during the collision. The equation expressing the heat gained, H, is

$$H = \pi E'' \epsilon_0^2 \tag{2}$$

where ϵ_0 represents the maximum deformation of the ball.

A further quantity is the loss tangent,

$$\tan \delta = E''/E' \tag{3}$$

where the ratio of the two modulus quantities represents an extremely useful damping quantity.

The two quantities, E'' and $\tan \delta$ are usually the prime parameters of interest for damping. If these quantities are small at a given temperature and frequency, damping will be small, and vice-versa.

The several quantities are applied to crosslinked polystyrene in Figure 3. At the glass transition temperature, T_g, which is about 120°C at 110 Hz, the storage modulus begins to decrease, while both E'' and $\tan \delta$ go through maxima, respectively, as the temperature is increased.   For dynamic data of this kind, the glass transition temperature is frequently defined as either the temperature of the loss modulus peak, or the temperature of the $\tan \delta$ peak.   Since all of these quantities are frequency dependent, it is convenient to define the glass transition temperature as the temperature of the middle of the Young's modulus, E, elbow at ten seconds.   Since this is equivalent to 0.1 Hz, the standard value of T_g can be converted to any desired frequency.

Five Regions of Viscoelastic Behavior

The log modulus-temperature or log modulus-log frequency plots of amorphous polymers show five distinct regions. As shown in Figure 4, at low temperatures a polymer will be glassy with a Young's modulus of about 3×10^9 Pa.   As the temperature is increased, it will exhibit a glass-rubber transition.   Physically, the polymer softens in the glass transition region, going from a glassy, plastic material to either a rubbery or liquid material above T_g, depending on molecular weight and crosslinking. As shown already in Figure 3, the storage modulus drops about three orders of magnitude through about 30°C increase in temperature.   Semi-crystalline polymers exhibit two transitions: melting, which involves exactly the same concepts as ice or iron melting, and the glass-rubber transition, which behaves quite differently. The two should not be confused (1).   Those polymers which are not crystalline because of irregular structure or other reasons only exhibit the glass transition.

As the temperature is raised still further, the polymer becomes rubbery, marked by the rubbery plateau. If the polymer is crosslinked, it will behave like a rubber band in this region. Because of extensive chain entanglement, very high molecular weight polymers also exhibit a rubbery plateau and concomitant elastic behavior. As the temperature is increased still further, the rubbery flow and liquid flow regions are encountered. Since the glass transition region is the most important for sound and vibration damping, special effort will be expended on describing this phenomenon.

The glass transition region is marked by the onset of long-range coordinated molecular motion. Some 10-50 backbone atoms are involved. According to modern theories of polymer motion, the chains begin to reptate back and forth along their length at T_g, resembling the motions of a snake, see Figure 5. However, according to increasing amounts of recent data, reptation as such may not be the motion directly associated with T_g and damping. T_g involves shorter range motions; for example, branching suppresses reptation, but has little effect on T_g. Also, the area under the

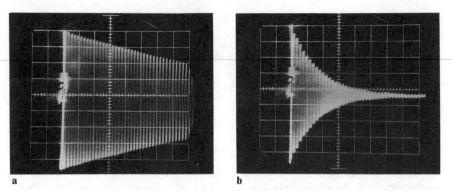

a   b

Fig. 1. Oscilloscope traces of vibration decay. Characteristics of an undamped reed(a), and a damped reed(b). Steel reed damped with an IPN constrained layer system. (Reprinted with permission from ref. 5. Copyright 1975 Wiley.)

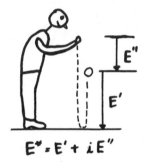

$$E^* = E' + iE''$$

Fig. 2. The ball actually gets warmer on bouncing. The critical time is the time the ball remains in contact with floor; this defines the equivalent frequency.

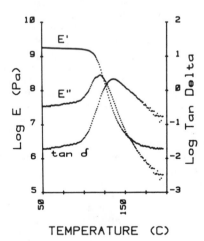

Fig. 3. Typical Rheovibron data for polystyrene. Note that tan δ always peaks at a higher temperature than E''.

E'' - T curve is independent of crosslink level up to very high levels, ca. 25%. Most importantly, E'' peaks when a natural frequency of chain motion equals the external vibrational frequency.

At temperatures both below and above the glass transition region, there may be still other transitions. The most famous of these is the β-transition, which involves an isoaxial motion of 4-8 backbone atoms, see Figure 6.

Electromagnetic Waves and Sound Waves

In fact, there is an analogy between the absorption of electromagnetic waves (such as infrared light or corresponding Raman spectra) and mechanical vibrational waves, see Figure 7. When the frequency of the waves (light or sound) equals the natural frequency of the chemical group or polymer chain, an absorption peak is found. The energy is converted into molecular motion, ie, heat.

In infrared spectra, absorption is almost always plotted against frequency (or wavelength). Often, absorption of acoustical waves (E'' or tan δ) is plotted against acoustical frequency. Equally often, measurements of mechanical behavior are made as a function of temperature. This is because temperature effects are usually more marked in polymer transitions than in infrared spectra. However, infrared absorption at a single frequency does vary with temperature in a manner somewhat analogous to the mechanical studies.

The onset of molecular motion associated with the glass transition can be related directly to damping. In fact, people speak about dynamic mechanical spectroscopy, in a manner analogous to infrared spectroscopy. A comparison of frequencies is in order. In the far infrared, there exists a region which is called the longitudinal acoustic mode, LAM, which goes down to 30-50 cm^{-1} wave number for polyethylene (5a). The LAM frequencies involve large sections of the chain, undergoing coordinated molecular motion. The low LAM frequencies correspond to the order of 10^{12} Hz. On the basis of the time-temperature superposition principle and concomitant master curves, a frequency of about 10^{12} Hz would be required to glassify polyethylene at room temperature. In other words, the lowest infrared frequencies, involving large segments of a polymer chain, are thought to correspond to those mechanical frequencies required for glassification.

KEY VISCOELASTIC PRINCIPLES

The WLF Equation. The WLF equation, derived in 1955, plays a key role in the theory of viscoelasticity. Space precludes its derivation here (1). The most widely used form of the equation may be written:

$$\log(t/t_g) = \frac{C_1(T - T_g)}{C_2 + (T - T_g)} \qquad (4)$$

where T is the temperature, T_g is the glass transition temperature, t is the time, and t_g is the duration of the experiment defining T_g. Often, C_1 is assigned the value - 17.44 and C_2 is assigned 51.6. These universal constants arise because C_1 and C_2 depend on free volume and expansion coefficient relationships; these values are nearly identical for many amorphous polymers near T_g. Importantly, any time-dependent variable can be substituted for time itself. Thus, viscosity, flow rates, etc. are excellent variables.

From the present point of view, it must be noted that by rule of thumb one decade of frequency increase is equivalent to an increase in temperature of approximately 6-7°C. Thus, T_g is higher at higher frequencies.

This can be explained by energies of activation, the requirements for free volume, or that time, in effect, moves faster for molecules at high temperature. All have a

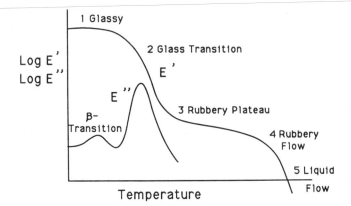

Fig. 4. The five regions of viscoelastic behavior. All polymers exhibit these five regions, but crosslinking, crystallinity, and varying molecular weight alter the appearance of this generalized curve. The loss modulus T_g peak appears just after the storage modulus enters the glass transition region.

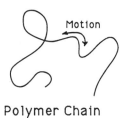

Fig. 5. The glass transition region involves the onset of long-range coordinated molecular motion. Usually, 10-50 backbone atoms are involved in this reptation-type motion.

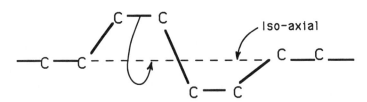

Fig. 6. The β-transition, illustrated in Fig. 4, involves the motion of four to eight carbon atoms about an iso-axial arrangement of the adjoining carbon atoms.

common basis in that when the experiment is conducted faster, the molecules must be moving faster, ie, be at a higher temperature, to match the imposed frequency.

It is interesting to observe the width of homopolymer transitions with respect to the WLF equation. As illustrated in Figure 3 for polystyrene, the transition is about 30 degrees wide, about average for such. Translated into frequencies, this means a range of 3-4 decades of frequency. Since human hearing spans the range of about 16 Hz to 20,000 Hz, or about three decades, it means that a polymer with a transition right in the middle of ambient temperature will just be able to damp all sound. If the temperature changes even 10 or 20°C, efficiency will be lost. Thus, a car door which is damped fine in the show room may not be damped at all in the July sun or in a January snow storm. The answer to this problem is to design polymer materials with broader damping peaks, hence blends, graft copolymers, and interpenetrating polymer networks.

The Time-Temperature Superposition Principle. For viscoelastic materials, the time-temperature superposition principle states that time and temperature are equivalent to the extent that data at one temperature can be superimposed upon data at another temperature by shifting the curves horizontally along the log time or log frequency axis. This is illustrated in Figure 8. While the relaxation modulus is illustrated (Young's modulus determined in the relaxation mode), any modulus or compliance measure may be substituted.

The data are determined over short ranges of time or frequency, as convenient for the investigator. With increasing time, these are shown as T_1, T_2, T_3, etc. These curves are then shifted either to the left, or to the right until they overlap with the next curve. A master curve at the the temperature of the unshifted data is formed. While the data can be shifted empirically, the shift can be shown to be controlled by the WLF equation, and the fit can be excellent in the range of T_g to $T_g + 50°C$. Thus, data for very long times, high frequencies, etc. can be easily estimated.

It must be noted that changes in density, as well as changes in temperature for higher temperatures, result in vertical shifts. Usually, these are modest in size, however, compared to the horizonal shifts. It should also be noted that the WLF equation, above, is a corollary of the time-temperature superposition principle.

Models for Analyzing Damping. A convenient method of modeling the damping behavior of polymeric materials involves combinations of springs and dashpots, see Figure 9. A spring is a perfect elastic element, following Hooke's law, and behaving like a metal spring. A dashpot is a perfect viscous element, following Newton's law. A spring is assigned a modulus E, and may be stretched or compressed at will, responding instantaneously. A dashpot may also be infinitely deformed, but will respond only slowly, depending on its assigned viscosity, η.

For a Maxwell element, which is a spring and a dashpot in series,

$$\sigma = \sigma_0 e^{-(t/\tau_1)} \tag{5}$$

$$\tau_1 = \eta/E \tag{6}$$

where τ_1 represents the relaxation time of the system. Interestingly, maximum damping occurs when $\tau_1 = t$; ie, one is in the glass transition region, see Figure 10.

In a limited way, the Maxwell element describes a liquid. Similarly, the Kelvin (Voight) element describes a solid. As the relaxation time, τ_1, is defined for the Maxwell element, the retardation time, τ_2, is defined for the Kelvin element. For the Kelvin element under stress,

$$\sigma = \eta \frac{d\epsilon}{dt} + E\epsilon \tag{5'}$$

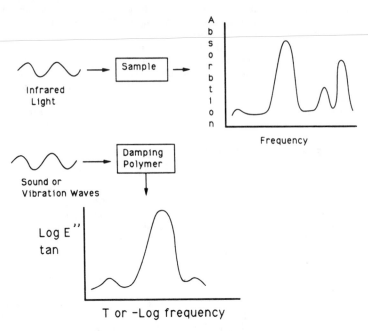

Fig. 7. Electromagnetic waves and sound waves. When the frequency of the waves (light or sound) equals the natural frequency of the chemical group or polymer, an absorption peak is found. The energy is converted into molecular motion, i.e., heat. Illustrated for the damping curve are the β-transition, the glass transition, and the liquid-liquid transition.

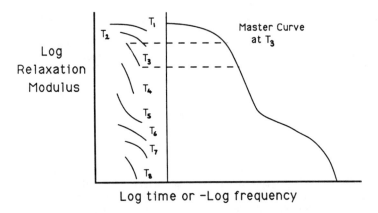

Fig. 8. Schematic representation of the construction of a master curve at T_3. Data are slid horizontally either left or right until they overlap data at the temperature of interest. A master curve can be used to predict results at very long or short times, or equivalently, very short or long frequencies.

where ϵ represents the strain. Equation $(5')$ can be integrated under conditions of constant stress,

$$\epsilon = \frac{\sigma}{E}\left(1 - e^{-(E/\eta)t}\right) \tag{6'}$$

In terms of the retardation time,

$$\epsilon = \frac{\sigma}{E}\left(1 - t/\tau_2\right) \tag{6''}$$

where τ_2 equals η/E.

Theories of Effective Damping. If one assumes that a range of temperatures and/or frequencies will be encountered, then the area under the damping curve (E'' or tan δ) determines the effectiveness of the polymer better than the hight of the transition alone, see Figure 11. For E'', the effective area is called the loss area, LA. This is determined after subtracting the background, as in any spectroscopic experiment. There are two theories to determine the quantity LA.

The first is the phenomenological mechanical theory. One form of the equation reads (6-9),

$$\int_{T_G}^{T_R} E''dT = (E'_G - E'_R)\,\frac{R}{(\Delta E_a)_{act}}\,\frac{\pi}{2}\,T_g^2 \tag{7}$$

where E'_G and E'_R represent Young's moduli in the glassy and rubbery states, respectively, just before and after the glass transition.

Another form of the equation reads (10),

$$2\int_{T_G}^{T_R} E''d(1/T) = (E'_G - E'_R)\,(R/\Delta E_{act})\,\pi^2 \tag{8}$$

These two equations are useful in providing energies of activation, ΔE_{act} (11), but require three parameters for useful predictions of actual LA values.

An alternate approach makes use of a group contribution analysis (7,9) approach. The group contribution analysis method has proved useful for the determination of solubility parameters (12,13), and has provided an approach for a host of other chemical properties (14). In a group contribution analysis, each moiety in the chemical or polymer contributes additively to the property in question.

By analogy with the solubility parameter approach, the loss area, LA, for the area under the loss modulus-temperature curve in the vicinity of the glass-rubber transition is given by (7,9)

$$LA = \sum_{i=1}^{n} \frac{(LA)_i M_i}{M} \tag{9}$$

where M_i represents the molecular weight of the ith moiety, and M is the molecular weight of the whole mer. The actual analysis of LA via this method will be examined further in a Chapter in this book by Fay, et al. There are several hidden assumptions in this theory. The first is that the modulus in the glassy state is constant. This is true for a wide range of polymers, with E'_G in equation (7) being about 3×10^9 Pa (3×10^{10} dynes/cm^2). Second, that E'_R is much smaller than E'_G, so its value can be ignored. Usually, E' drops about three orders of magnitude through T_g. Because the decline in E' with temperature is not known quantitatively, the equivalent theory for tan δ has yet to be worked out.

Damping and Dynamic Mechanical Spectroscopy. Dynamical mechanical spectroscopy means data taken with cyclical deformation of the sample. Often, the

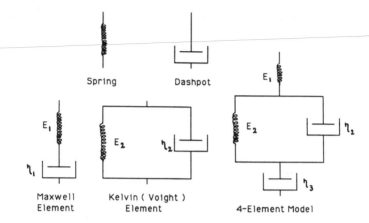

Fig. 9. Models for analyzing damping. The springs follow Hooke's law and contribute moduli, while the dashpots follow Newton's law and indicate viscous contributions.

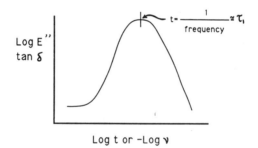

Fig. 10. The quantities E'' and tan δ peak when the natural frequency of chain motion equals the external vibrational frequency. At the peak, the relaxation time, τ_1, approximately equals the time of the experiment (or the inverse of the frequency).

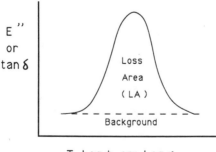

Fig. 11. Group contribution analysis leads to the determination of the effective area under the loss modulus-temperature curve. As in any spectroscopic experiment, background must be subtracted, and the instrument calibrated.

experiment involves a sinusoidal motion. There are several ways of representing this information, depending on the objective.

If an applied stress varies with time in a sinusoidal manner, the stress, σ, may be written,

$$\sigma = \sigma_0 \sin \omega t \tag{10}$$

where ω represents the angular frequency in radians, equal to 2πxfrequency. For Hookian solids, with no energy dissipated, the strain, ϵ, is given by

$$\epsilon = \epsilon_0 \sin \omega t \tag{11}$$

where the quantities with subscript zero represent the amplitude values.

For real materials which exhibit damping, the stress and strain are not in phase, the strain lagging behind the stress by the phase angle, δ, the same angle which appeared in tan δ, equation (3). The relationships among these parameters are illustrated in Figure 12. It must be noted that the phase angle defines an in-phase and out-of-phase component of the stress, σ' and σ'', respectively. The relationships among the in-phase component, out-of-phase component, and δ are given by

$$\sigma' \quad = \sigma_0 \cos \delta \tag{12}$$

$$\sigma'' \quad = \sigma_0 \sin \delta \tag{13}$$

$$E' \quad = \sigma'/\epsilon_0 = E^* \cos \delta \tag{14}$$

$$E'' \quad = \sigma''/\epsilon_0 = E^* \sin \delta \tag{15}$$

$$E^* \quad = \frac{\sigma_0}{\epsilon_0} = \left(E'^2 + E''^2 \right)^{1/2} \tag{16}$$

In terms of complex notation,

$$E^* = E' + iE'' \tag{17}$$

and

$$E = |E^*| = \sigma/\epsilon \tag{18}$$

again, $E''/E' = \tan \delta$.

There are several derivative functions of the above which are of interest. The first of these is the logarithmic decrement, Δ. This function is defined as the natural logarithm of the amplitude ratio between successive vibrations, see Figure 13. Thus,

$$\Delta = \ln(A_n/A_{n+1}) \tag{19}$$

where A_n is amplitude of the nth vibration. The quantity Δ is related to tan δ as follows,

$$\Delta = \pi \tan \delta \tag{20}$$

The percent critical damping, %C.D., is given by

$$\%C.D. = 100(2\pi\Delta) \tag{21}$$

and is a measure of the number of vibrations an object may undergo before decaying to substantially zero amplitude, a critically damped object decaying to zero amplitude in one vibration cycle.

Methods of Damping. Methods of damping tend to fall into two distinct groups, those that in some way isolate sound, or those that cause distructive interference of its waves. These methods include:

> Isolation
> Quarter-Wave Plates
> Honeycomb Structures
> Hollow Cavities

In addition, there are two general methods that are of great interest to the polymer scientist:

> Extensional Damping
> Constrained Layer Damping

Both of these methods make use of polymers, and in particular, their viscoelastic behavior, albeit somewhat differently.

Their arrangement in space is illustrated in Figure 14. An extensional damping system consists of two layers: The substrate to be damped, and the polymeric damping layer. The constrained layer system is a three layer affair, with the polymer in the middle, like jelly in a jelly sandwich. Either the constraining layer is part of the system, as in taking two thin sheets of metal and gluing them together with the damping layer, or the constraining layer can be applied afterwards. The best arrangements are when both the substrate and the constraining layer have high modulii, and the damping layer is soft.

Broadly speaking, extensional damping utilizes E'', while constrained layer damping is more closely related to tan δ. Actually, the acoustical engineering equations show that extensional damping works best when a polymer is at the top of its glass transition, and is stiff, while constrained layer damping works best when the sample is soft, at the bottom of its transition.

The two methods are compared in Table I. While extensional damping is clearly cheaper, constrained layer damping is more effective. Which one should be used in any given application clearly depends on the requirements. Both are in wide service.

Table I. Comparison of Extensional and Constrained
Layer Damping

Method	Advantages	Disadvantages	Utilizes
Extensional	Ease of Application Low Cost	Not very effective	E''
Constrained layer	Excellent damping	Requires three layers	tan δ

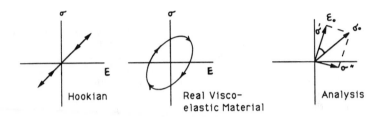

Fig. 12. Simple dynamic relationships between stress and strain, and the phase angle.

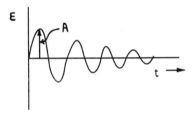

Fig. 13. Illustration of the amplitude for the calculation of the logarithmic decrement.

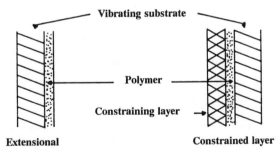

Fig. 14. Methods of applying damping polymers to vibrating substrates.

Use of Platelet Fillers. While the loss area tends to be constant for a given homopolymer or statistical copolymer system, there are important ways to augment the total damping. One of the more interesting methods makes use of platelet fillers, such as mica. Since the platelets tend to lie parallel in polymer dispersions, they behave like mini-constrained layer systems, see Figure 15. Thus, the area under both E'' and $\tan \delta$ will be augmented. Platelet fillers are mentioned several times in this monograph, particularly in a Chapter by Ting, et al.

Moving the Transition About. While the glass transition temperature is fixed for a given homopolymer and frequency, the scientist or engineer can still adjust the transition temperature within certain ranges. For example, plasticization lowers the glass transition, see Figure 16. The most common use of such plasticization is with poly(vinyl chloride). The neat homopolymer is a hard plastic with a glass transition temperature of about 80°C. An important use is as underground water pipes. The plasticized material, with a T_g just below room temperature, is as "vinyl" shower curtains.

When the polymer is filled, the filler tends to raise the glass transition slightly, as illustrated also in Figure 16. However, two effects must be distinguished. The filler will raise the modulus of the material, so that the beginning of a transition may be erroneously recorded as being higher using arbitrary methods. However, if the polymer binds to the filler, its free volume may be reduced, and the portion of polymer in direct contact with the filler (for about 50-100 Å interfacial thickness) may undergo real rises in T_g. Since this effect drifts off with distance from the filler surface, the total effect is to broaden the transition as well as raise its peak temperature.

A third general method of changing the glass transition temperature is through copolymerization. In this case, a monomer or monomers with either lower or higher glass transitions are incorporated directly into the polymer. If $\underline{A}$ is the homopolymer of interest, and $\underline{B}$ is the added monomer, the backbone of the copolymer will look like:

$$\text{AAAAABAABAAAAABABAAAAAAAAAAAB} \qquad (22)$$

where the B's are statistically arranged. The glass transition temperature of such copolymers follows the Fox equation,

$$\frac{1}{T_g} = \frac{w_A}{T_{gA}} + \frac{w_B}{T_{gB}} \qquad (23)$$

where w_A and w_B are the weight fractions of the two monomers, and T_{gA} and T_{gB} are the glass transition temperatures of the respective monomers. Thus, acrylics, styrenics, vinyls, dienes, and many other polymers can be synthesized with T_g's made to order.

Effect of a Multicomponent System. If one combines two immiscible polymers, two distinct glass transitions result, Figure 17. (Many people use incompatible to mean immiscible, and compatible to mean miscible.) If the thermodynamics of miscibility are such that the free energy of mixing is near zero, a microheterogeneous or more nearly homogeneous morphology may appear. In this case, there is extensive but incomplete mixing of the components and the phase domains may be of the order of 100 to 200 Å in size. Then, most of the material in the system is really interfacial "interphase," with variable composition on the scale of tens of Angstroms. This may result in a single broad transition which spans the range of the two original homopolymer transitions, see Figure 17. This result is highly desired if the application in mind covers a broad temperature range.

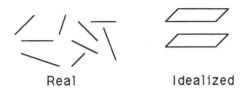

Fig. 15. Arrangements in space of platelet fillers.

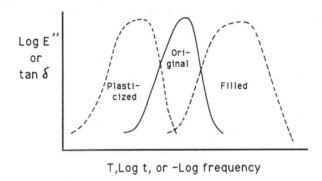

Fig. 16. Moving the glass transition about with plasticizer and filler.

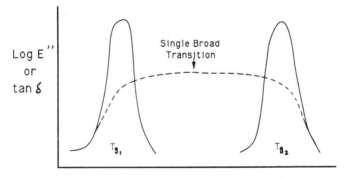

Fig. 17. The development of a single broad transition spanning the transition range of both polymers can be accomplished through extensive but incomplete mixing on the 100 Å scale, resulting in a microheterogeneous morphology.

A generally undesirable result is when the two polymers are less miscible in each other.   Residual mixing often causes a shifting of the two transitions inward, see Figure 18.   Generally in such cases, little damping with be obtained in the temperature range between the transitions.   Again, as illustrated in both Figures 17 and 18, variations in log time (of experiment) or minus log |frequency can be substituted for temperature.   All three quantities have the same general behavior. Equally to be avoided is the complete mixing of the two polymers, in which case a single, sharp transition may be observed.

Polymers Often Used for Damping

There are many polymers used commercially for damping, depending on the temperature and frequency range required.   Some of the more important materials include asphaltics, polyurethanes, poly(vinyl acetate) and copolymers, acrylics, natural rubber, and styrene-butadiene rubber (SBR), and silicone rubber.   For example, poly(vinyl acetate) has a glass transition temperature near 40°C.   However, slightly plasticized compositions exhibit rather sharp tan δ peaks at 25°C.   The asphaltics are low in price, and while highly irregular polymeric materials, make effective dampers in a number of cases.   Natural rubber is offered as a damper for uses under bridge bases, probably because its high compressive strength is important. The ether oxygen in polyether-urethanes is thought to contribute strongly to its damping capability.

Multicomponent polymeric materials with microheterogeneous mophologies include a number of polymer blends and block copolymers, however, an especially easy way to bring about the desired morphology is through interpenetrating polymer networks.   Several papers in the symposium are concerned with IPN's and related materials.

Mechanical Instrumentation

There are a large number of instruments which can be used to measure the glass transition characteristics, and hence their damping properties.   In addition to commercial equipment, a significant number of investigators have home-made instrumentation, often for special purposes.   Some of the better known commercial instrumentation is summarized in Table II.   While the Autovibron (Rheovibron) is useful for self-supporting samples, the Torsional Braid Analyzer is useful for non-self supporting samples.   The former will provide absolute values of the loss and storage moduli, while the latter can only provide relative values.   This is because the Torsional Braid Analyzyer utilizes a glass braid, within which is embedded the sample, usually by a dipping procedure.   The du Pont equipment is best for self-supporting samples, while the Weissenberg equipment actually measures dynamic viscosity best.   The Rheometrics is excellent for solids and liquids.   The Polymer Laboratory's are newer pieces of equipment.

The instruments also differ in frequency ranges.   The Autovibron, for example, has four frequencies, at 3, 11, 33, and 110 Hz.   The du Pont equipment operates between 0.001 and 10 Hz.

A Basic Engineering Question

Each individual damping situation must be treated separately.   There is no "panacea" polymer for damping.   A basic engineering question relates to the shape of the damping curve.   Which shape for loss modulus or tan δ will fit the particular needs best?   Note the differences between Figures 19 A and B.   The engineer needs to know the range of frequencies and temperatures to be encountered.   The polymer glass

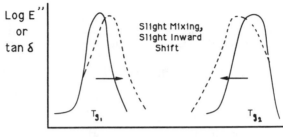

Fig. 18. Very slightly miscible polymer pairs often yield an inward shift of their respective glass transition.

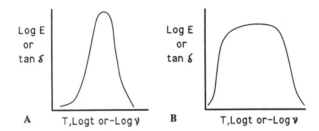

Fig. 19. The basic shape of the damping curve must fit the intended application.

Table II. Mechanical Instrumentation for Measuring the Glass Transition and Damping

Instrument	Manufacturer
Autovibron DDVIII	Toyo Instrument Co.
Torsional Braid Analyzer	Plastics Analysis Consultants
DuPont 983 Dynamic Mechanical Analyzer	DuPont
System 4 Mechanical Spectrometer	Rheometrics
Rheogoniometer	Weissenberg
Dynamic Mechanical Thermal Analyzer MKII	Polymer Laboratory
SDM5500 Dynamic Mechanical Spectrometer	Seiko
Viscoanalyzer	Metravib Instruments

transition should fit all expected temperature and frequency ranges, and no more. The engineer should invoke the time-temperature superposition principle, its corollary WLF equation, and the Fox equation to determine quantitatively changes in the glass transition from standard conditions to the particular needs of the moment.

Thus, whether the applications are for indoor or outdoor, for planes, ships, automobiles, industrial equipment, appliances, etc., polymer materials can be made to order for maximum damping.

Conclusions

The key variables needed for effective damping include T_g, E', E'', and tan δ. Important viscoelastic principles include the time-temperature superposition principle and its resultant WLF equation. These can be applied to understand the relationship between literature values of the glass transition temperature and actual needs. Thus, by using the growing amount of science now available in the field of damping, one can select that polymeric material which will damp most effectively.

Acknowledgment

The author is pleased to acknowledge the financial support of the Office of Naval Research.

Literature Cited

1.   Sperling, L. H.   Introduction to Physical Polymer Science; Wiley-Interscience: New York, 1986; Chapter 6.
2.   Ferry, J. D.   Viscoelastic Properties of Polymers, 3rd Ed.; John Wiley & Sons: New York, 1980.
3.   Hartman, B.   Acoustic Properties; Encyclopedia of Polymer Science and Engineering; Wiley: New York, 2nd Ed., 1984; Vol. 1, pp. 131.
4.   Trask, C. A.; Roland, C. M.   Macromolecules 1989, 22, 256.
5.   Grates, J. A.; Thomas, D. A.; Hickey, E. C.; Sperling, L. H.   J. Appl. Polym. Sci. 1975, 19, 1731.
5a. Painter, P. C.; Coleman, M. M.; Koenig, J. L   The Theory of Vibrational Spectroscopy and Its Application to Polymeric Materials; Wiley-Interscience: New York, 1982; pp. 336-344.
6.   Aklonis, J. J.; MacKnight, W. J.   Introduction to Polymer Viscoelasticity 2nd Ed.; Wiley-Interscience: New York, 1983.
7.   Chang, M. C. O.; Thomas, D. A.; Sperling, L. H.   J. Polym. Sci., Part B: Polym. Phys. 1988, 26, 1627.
8.   Staverman, A. J.; Schartzl, F.   in Die Physik der Hochpolymeren; Stuart, H. A., Ed.; Springer-Verlag: London, 1967; Chapter 8.
9.   Chang, M. C. O.; Thomas, D. A.; Sperling, L. H.   J. Appl. Polym. Sci. 1987, 34, 409.
10. McCrum, N. G.; Read, B. E.; Williams, G.   Anelastic and Dielectric Effects in Polymeric Solids; John Wiley & Sons: New York, 1976.
11. Hartmann, B.; Lee, G. F.   J. Appl. Polym. Sci. 1977, 21, 1341.
12. Small, P. A.   J. Appl. Chem. 1953, 3, 71.
13. Hoy, K. L.   J. Paint Technol. 1970, 46, 76.
14. Van Krevelen, D. W.; Hoftyzer, P. J.   Properties of Polymers, Their Estimation and Correlation with Chemical Structure, 2nd Ed.; North-Holland: Amsterdam, 1976.

RECEIVED January 24, 1990

Chapter 2

Relation of Polymer Chemical Composition to Acoustic Damping

Bruce Hartmann

Polymer Physics Group, Naval Surface Warfare Center, Silver Spring, MD 20903–5000

Acoustic damping in polymers is shown to be dominated by the glass transition occurring in the amorphous portions of the polymer. Thus to tailor damping characteristics for a particular application requires an understanding of the relation between chemical composition and glass transition temperature. Glass transition temperature depends on a number of variables, including backbone flexibility, steric effects, polarity, pendant groups, crystallinity, plasticizers, crosslink density, and co-polymerization. While the glass transition damping peak can be located at almost any desired temperature, there are limitations on the height and width that can be achieved in any real polymer.

The purpose of this chapter is to review the relation between the chemical composition of polymers and the acoustic damping produced by these polymers. After defining the necessary acoustic terms, the chapter will begin with a description of the experimental data for various polymeric systems, in particular the temperature and frequency dependence of the damping. It will be seen that acoustic damping properties are dominated by the glass transition in the polymer. The extensive literature on the relation between chemical composition and glass transition temperature will then be briefly reviewed, establishing the connection between chemical composition and the location of the damping peak. The discussion will then turn from the location of the damping peak to the height and width of the peak. It

will be pointed out that the height and width cannot be
chosen arbitrarily, even in principle. The material in
this chapter is based primarily on references 1-5.

  To understand what follows, some acoustic terms must
first be defined. The propagation of an acoustic wave
through a solid polymer can be characterized by two
parameters: sound speed and sound absorption. Sound
speed, c, in units of m/s, is the rate at which sound
waves travel through the solid. Sound absorption, α, in
units of dB/cm, is a measure of the loss in energy of the
sound wave as it travels through the solid. The energy
of the sound wave is converted into random thermal motion
or heat, usually with a negligible rise in temperature of
the solid. The units of α can be explained in the
following manner. The change in energy of an acoustic
signal is usually expressed in terms of the common log
(base ten) of the amplitude at two different positions.
This unit is called a Bel after Alexander Graham Bell.
Since this unit is rather small for most applications, it
is more common to use the deciBel, or dB, which is
obtained by multiplying the Bel by ten,

$$dB = 10 \; \log \; (I_2/I_1) \tag{1a}$$

$$= 10 \; \log \; (A_2/A_1)^2 \tag{1b}$$

$$= 20 \; \log \; (A_2/A_1) \tag{1c}$$

where I_1 and I_2 are the acoustic energies at the two
locations, A_1 and A_2 are the amplitudes of the sound
wave at these locations, and Equation 1b makes use of the
fact that the energy in a sound wave is proportional to
the square of the amplitude of the wave (6). With these
units for the change in energy between two positions,
absorption of acoustic energy per unit distance of
travel then has dimensions of dB/cm. In some cases,
rather than referring to absorption per unit distance in
dB/cm, it is more convenient to refer to the absorption
per wavelength or $\alpha\lambda$, where λ is the wavelength. The
units of $\alpha\lambda$ are then dB.

  In an unbounded, isotropic solid there are only two
independent modes of acoustic propagation: longitudinal
and shear. In the longitudinal mode, the particle motion
is parallel to the direction of propagation, while in the
shear mode, the particle motion is perpendicular to the
direction of propagation. Associated with each of these
modes of propagation there is an absorption. Thus, four
parameters are required to characterize the solid:
longitudinal sound speed, c_l, shear sound speed, c_s,
longitudinal absorption, α_l, and shear absorption, α_s.

  Some experimental techniques yield a modulus rather
than a sound speed, but the underlying physical
properties are the same. In a torsional pendulum, for
example, the shear modulus, G, is measured while in a
Rheovibron, Young's modulus, E, is measured. When there

is negligible absorption, the relations between sound speed and modulus are particularly simple,

$$c_s = (G/\rho)^{1/2} \tag{2}$$

$$c_l = (K + 4G/3)/\rho)^{1/2} \tag{3}$$

where K is bulk modulus and ρ is density. The moduli are related through the standard equations

$$E = 2G(1+\nu) = 3K(1-2\nu) \tag{4}$$

where ν is Poisson's ratio.

Absorption is taken into account by assuming that the modulus is complex

$$G^* = G' + iG'' \tag{5}$$

where G^* is the complex shear modulus, G' is the real part of the modulus, and G'' is the imaginary part of the modulus. Similar relations hold for Young's and bulk moduli. Since G' is usually much larger than G'', the absolute value of G^* is approximately equal to G', and G' is often taken to be the modulus G, written without the prime (7), as in Equation 2. The physical significance of G' is that it represents an elastic storage of mechanical energy and it is also called the storage modulus. G'' is a measure of absorption and is also called the loss modulus. When there is absorption, the strain lags (in time) behind the applied stress by a phase angle δ as shown in Figure 1. From Figure 1, it can be seen that

$$\tan \delta = G''/G' \tag{6}$$

where $\tan \delta$ is known as the loss factor and is an alternate way to express the absorption or loss of energy rather than G'' by itself.

In acoustic terms, absorption can be expressed as $\alpha\lambda$ while in modulus terms, absorption can be expressed as $\tan \delta$. Using an approximation (for small $\tan \delta$) that is almost always good to 1%, Ferry (2) points out that

$$\alpha\lambda = 8.686 \, \pi \, \tan \delta \tag{7}$$

in units of dB. Thus sound speed is related to modulus by Equation 2 and sound absorption is related to loss factor through Equation 7. The acoustic approach and the modulus approach are thus alternate and equivalent ways of describing the same physical phenomenon. We will consider the two approaches interchangeably in describing damping.

Before concluding this summary of acoustic property definitions, it is worth considering some related terms. The term acoustic is sometimes reserved for vibrations

that are in the audible range of frequencies, nominally
from 20 Hz to 20,000 Hz. Higher frequencies are referred
to as ultrasonic. Measurements of modulus made at low
frequency are often called dynamic mechanical
measurements. Examples include the torsional pendulum,
which operates at a nominal 1 Hz, and the Rheovibron,
which operates from about 1 to 100 Hz (8). We will adopt
the broader view that all of these measurements can be
considered as types of acoustic measurements.

Temperature and Frequency Dependence of Damping

For a given polymer, damping is found to depend on the
temperature and frequency of the measurements. A
schematic representation of the temperature dependence of
the damping and modulus for a typical polymer is shown in
Figure 2, adapted from Read and Dean (8). The major
feature of this plot is the glass transition, in the
vicinity of which the modulus drops by as much as three
orders of magnitude and the damping has a peak. It is
this peak that is usually relied upon in damping
applications. In addition to the glass transition, a
secondary transition is shown at lower temperature. The
loss peak for the secondary transition is much smaller
than that for the glass transition and secondary
transtions are not usually important for damping
applications. At temperatures above the glass
transition, the effect of melting is seen in Figure 2.
While the loss factor can be very high in this region
(9), the modulus is very low and polymers are not
generally used in the molten state for damping
applications. This paper will consider only the glass
transition. An example of experimental acoustic
measurements in the vicinity of the peak is shown in
Figure 3, where data for a poly(metacarborane siloxane)
(10) are displayed. The measurements were made in the
longitudinal mode as a function of temperature at a
frequency of 2 MHz. At the peak, the absorption per
wavelength has a value of $\alpha\lambda = 1.3$ dB so that tan δ =
0.05.
    When measurements such as those in Figure 2 are
performed at various frequencies, the peak shifts to
higher temperature at higher frequency, as shown in
Figure 4 for an epoxy polymer (11). Here the
measurements were made of Young's modulus at frequencies
from 0.33 to 30 Hz as a function of temperature. As a
rule of thumb, the peak shifts by about 7°C for every
decade increase in frequency (12). Specifically, it is
commonly assumed that the shift of the frequency, f, at
which damping has a peak varies with absolute
temperature, T, according to the Arrhenius rate equation

$$f = f_0 \exp(-\Delta H/RT) \tag{8}$$

where f_0 is a constant for each polymer, R is the gas

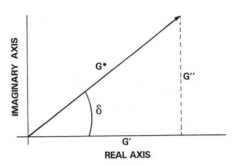

Figure 1.   Phase angle for complex shear modulus.

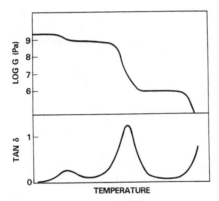

Figure 2.   Temperature dependence of damping and modulus for a typical polymer (schematic). (Reproduced with permission from Ref. 8.   Copyright 1978 Adam Hilger Ltd.)

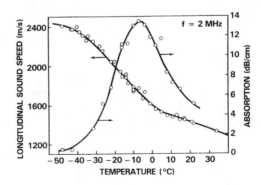

Figure 3.   Acoustic properties of a poly(metacarborane siloxane).   (Reproduced with permission from Ref. 10. Copyright 1971 John Wiley & Sons.)

constant, and ΔH is the activation energy of the process. This equation can be re-written as

$$\log f = \log f_0 - 2.303\Delta H/RT \qquad (9)$$

Thus an Arrhenius plot of log f vs 1/T has a slope of $-2.303\ \Delta H/R$, from which the activation energy can be determined. For the data in Figure 4, a value of 90 kcal/mol is found from the Arrhenius plot shown in Figure 5. For other polymers, the activation energy for the glass transition ranges from 20 to 100 kcal/mol ([13]). Arrhenius plots for various polymers, including some of interest in damping applications, are available ([1]).

Note that Equation 8 or 9 represents an equivalence between frequency and temperature, which can be expressed as a time-temperature equivalence. The Arrhenius equation is found to be most applicable at lower temperatures. At higher temperatures, a better representation of the equivalence between frequency and temperature is given by the WLF (Williams-Landel-Ferry) equation, which can be written as

$$\log f = \log f_0 + c_1(T-T_0)/\ (T-T_0+c_2) \qquad (10)$$

where c_1 and c_2 are constants for a given polymer and T_0 is a reference temperature. This relation is central to the understanding of the relation between frequency and temperature measurements and is the basis for extending the effective frequency range of the measurements. An in-depth treatment is given by Ferry ([2]).

Absorption is often observed to increase linearly with frequency, which is known in acoustics as hysteresis behavior. (Thus, the term hysteresis is restricted to linear frequency dependence of absorption and does not refer to absorption in general, as is common among polymer scientists.) An example is shown in Figure 6, for polyethylene ([9]). Both longitudinal and shear absorption increase linearly with frequency. We also note that, at any given frequency, shear absorption is much higher than longitudinal absorption. This behavior is found for polymers in general. Other absorption data for typical rubbers often used in damping applications are shown in Figure 7 ([14]). Hystersis behavior is common provided the measurements are not made too close to the damping peak, in which case the absorption approaches a frequency squared behavior. Since the absorption is directly proportional to frequency, the absorption per wavelength $\alpha\lambda$, which is directly proportional to the loss factor tan δ, is a constant.

Polymer Glass Transition

At the glass transition of an amorphous polymer, some 10 to 50 repeat units become free to move in cooperative thermal motions of individual chain segments, involving

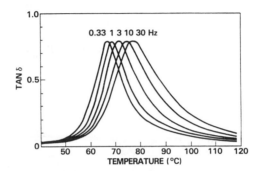

Figure 4.   Shift of damping peak with frequency for an epoxy polymer.   (Reproduced with permission from Ref. 11.   Copyright 1984 Elsevier Applied Science Co. Inc.)

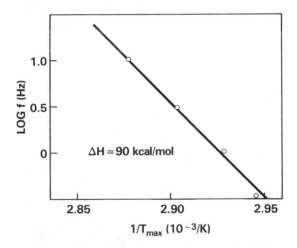

Figure 5.   Arrhenius plot for damping peak in an epoxy polymer.   (Reproduced with permission from Ref. 11. Copyright 1984 Elsevier Applied Science Co. Inc.)

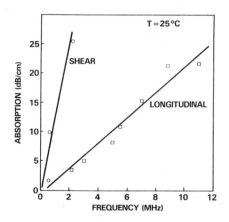

Figure 6.   Absorption vs frequency for polyethylene.
(Reproduced with permission from Ref. 9.   Copyright
1972 American Institute of Physics.)

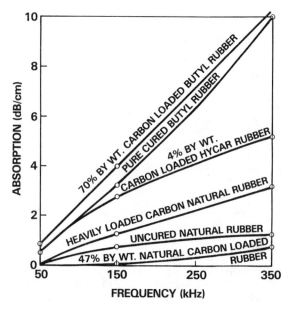

Figure 7.   Absorption vs frequency for various rubbers.
(Reproduced with permission from Ref. 14.   Copyright
1950 IOP Publishing Ltd.)

large scale conformational rearrangements of the chain
backbone (15). Below the glass transition, these large
scale motions become frozen and cannot occur. Major
changes in many physical properties, including acoustic
properties, take place at the glass transition.

The glass transition temperature, T_g, can be defined
as the temperature at which the specific volume vs
temperature plot has a change in slope. The measurement
is carried out in a dilatometer (16) at a slow heating
rate. In this apparatus, a sample is placed in a glass
bulb and a confining liquid, usually mercury, is
introduced into the bulb so that the liquid surrounds the
sample and extends partway up a narrow bore glass
capillary tube. A capillary tube is used so that
relatively small changes in polymer volume caused by
changing the temperature produce easily measured changes
in the height of the mercury in the capillary. Typical
results, for poly(vinyl acetate), are shown in Figure 8
(12). From Figure 8, we see that the slope changes from
a low value in the glassy state to a higher value in the
rubbery state. The transition is not sharp but occurs
over a range of temperatures. The intersection of the
two linear portions is taken to be the glass transition
temperature.

While the dilatometer method is the preferred method
of determining the glass transition temperature, it is a
rather tedious experimental procedure and measurements of
T_g are often made in a differential scanning calorimeter
(DSC). In this instrument (18), the heat flow into or
out of a small (10-20 mg) sample is measured as the
sample is subjected to a programmed linear temperature
increase (typically 10°C/min). The heat flow is
proportional to the specific heat of the sample. At the
glass transition, there is an increase in the heat flow
into the sample due to the increase in specific heat at
this point. Values obtained in this manner are only a
few degrees higher than the dilatometer values.

A convenient theoretical interpretation of the glass
transition is the free volume theory (2,3). In this
theory, the total macroscopic volume of a polymer is
considered to be the sum of the actual volume of the
polymer chains (the occupied volume) and the holes or
voids that constitute the free volume. The glass
transition occurs when there is enough free volume for
the large scale molecular motions associated with the
transition to take place. It is assumed that the
occupied volume increases linearly with temperature
throughout but the free volume undergoes a discontinuous
increase in expansion coefficient at the glass transition
when the number of holes increases. While there are some
objections to this theory (3), it leads to a correct
qualitative understanding of the process and will be
adequate for our purposes here.

It is important to keep in mind the difference
between the glass transition temperature, T_g, which is a

fixed number, and the glass transition, which is a
process that occurs at different temperatures depending
on the frequency of the measurement. In dynamic
mechanical measurements, the variation of tan δ peak
height with frequency is shown in Figure 4. This is a
general result: the glass transition is rate sensitive.
This rate sensitivity is one source of dissatisfaction
with the free volume theory of the glass transition,
which does not address the kinetic aspects of the
transition. We will take the view that there is a unique
T_g for each polymer but that, because of rate effects, we
need to make very slow measurements to measure it. For
practical purposes, the difference between the
dilatometer or DSC measurement and the infinitely slow
measurement theoretically required is negligible. On the
other hand, the difference between a dynamic mechanical
loss factor maximum and T_g cannot be ignored.
     There is a considerable literature on the effect of
chemical composition on T_g. A number of factors have
been shown to influence T_g, including backbone
flexibility, steric effects, polarity, pendant groups,
crystallinity, plasticizers, crosslink density, and
co-polymerization (13,5). After a brief qualitative
discussion of the first few factors, we will consider the
last two in more detail.
     A list of glass transition temperatures for common
polymers (12) of interest for damping applications is
given in Table I.

Table I. Glass Transition Temperatures

Polymer	Formula	T_g (°C)
polydimethyl siloxane	$-Si(CH_3)_2O-$	-123
poly(cis-1,4-butadiene)	$-CH_2CH=CHCH_2-$	-108
poly(cis-1,4-isoprene)	$-CH_2CH=C(CH_3)CH_2-$	-73
polychloroprene	$-CH_2CH=C(Cl)CH_2-$	-50
polystyrene	$-CH_2CH(C_6H_4)-$	100
polyacrylonitrile	$-CH_2CH(C\equiv N)-$	104

Note that a considerable range of T_g's is available,
covering more than 200°C in Table I. (Even higher T_g's
can be obtained than are listed in Table I, but such
polymers are not generally useful in damping
applications). The table also illustrates the effect of
some of the factors that influence the glass transition.
Poly(dimethyl siloxane) has the lowest T_g of any polymer
listed because of the two very flexible backbone
components. Comparing polybutadiene with polyisoprene
(natural rubber), we see that the substitution of a
methyl group for a hydrogen atom raises T_g as a result of
the steric hindrance caused by the larger size of the
methyl group. Steric effects are important because the
glass transition requires a certain free volume to occur.
Comparing natural rubber with polychloroprene (Neoprene),

the methyl group is comparable in size to the chlorine atom, but the T_g of Neoprene is higher because of the greater polarity of the chlorine atom. Polarity increases interchain attraction which decreases free volume and hence raises T_g. In polystyrene, the large, rigid pendant phenyl group raises T_g while the highly polar nitrile group raises the T_g of polyacrylonitrile.

The last two polymers in Table I are often co-polymerized with polybutadiene to form damping materials. The co-polymer of butadiene and styrene is the basis of styrene-butadiene-rubber, or SBR, which is the most commonly used type of rubber (19). One of the largest uses is in automobile tires. The co-polymer of acrylonitrile and butadiene is the basis of nitrile-butadiene-rubber, or NBR.

In small amounts, crystallinity raises T_g by limiting the motions of the amorphous regions. As the degree of crystallinity increases, the glass transition, which occurs only in the amorphous regions of the polymer, tends to be masked and may even be difficult to determine, as in the case of polyethylene. Crystallinity lowers the maximum value of the loss factor since this rigid filler has a lower loss factor than does the amorphous matrix. Similar effects are seen with rigid fillers that are sometimes added to polymers. However, there are cases where fillers increase the damping, probably by the introduction of new damping mechanisms which are not present in the pure polymer and which are outside the scope of this review (12).

Plasticizers are low molecular weight substances that can be mixed in with an acoustic material to lower the glass transition temperature (20). They may also serve another purpose in lowering the melt viscosity, making processing easier. Plasticizers must be soluble in the polymer and usually they dissolve it completely at high temperature. In some systems, there is a solubility limit so that higher concentrations of plasticizer separate out as a dispersed phase and are no longer effective. Plasticizers have very large free volumes so that even a small amount has a significant effect. Since no chemical reactions take place, plasticizers alter the glass transition by a physical change rather than a chemical composition change.

All of the above factors come into play in determining the effect of pendant groups on T_g. An aromatic pendant group raises T_g while an aliphatic pendant group can act as an internal plasticizer and lower T_g. However, if the pendant group is long enough, it may crystallize and raise T_g.

Crosslink Density

A chemical crosslink is a covalent bond linking two polymer chains together. At a crosslink, adjacent polymer chains are pulled closer together, and the free

volume is decreased (21). This reduction in free volume
raises T_g. While there are a number of theoretical
treatments of the effect of crosslinking on T_g, the
qualitative features can be expressed in the empirical
relation described by Nielsen (12)

$$T_g - T_{g0} = 3.9 \times 10^4 / M_c \qquad (11)$$

where T_g is the glass transition temperature of the
crosslinked polymer, T_{g0} is the glass transition
temperature of the uncrosslinked polymer having the same
chemical composition as the crosslinked polymer (both in
K), and M_c is the number-average molecular weight between
crosslink points.

The effect of crosslink density on acoustic
properties can be illustrated by some measurements on a
series of epoxy polymers (22). Polymers were formed by
reacting a resin with various curing agents. The resin
used was butanediol diglycidyl ether (BDGE), with the
chemical structure shown in Figure 9. Three aliphatic
curing agents were used: 1,3-propanediamine (PDA),
1,6-hexanediamine (HDA), and 1,12-dodecanediamine (DDA).
Crosslinking occurs at the diamine sites and these sites
are further apart as the chain length increases, hence
the crosslink density decreases as the chain length
increases. Calculated crosslink densities are 1.83,
1.70, and 1.49 mol/g for PDA, HDA, and DDA respectively.

Sound speed measurements as a function of temperature
at a frequency of 2 MHz for the different diamine curing
agents are shown in Figure 10. As can be seen, at any
temperature the greater the amount of the soft aliphatic
chain present, the lower the sound speed. Also, the
curves are similar in shape but shifted to different
temperature. Absorption curves for the same systems are
shown in Figure 11. Once again, the curves are similar
in shape, with approximately the same peak height, but
shifted in temperature. The behavior of all of the
systems is fundamentally the same but crosslinking has
shifted the curves by shifting the glass transition
temperature. Thus, the curves should be plotted as a
function not of temperature, T, but as $T-T_g$. In other
words, we should compare the polymers at the same
temperature relative to their glass transition
temperature. (The glass transition temperatures
determined in a DSC are 12, 5, and -10°C for PDA, HDA,
and DDA respectively.) The sound speed data can then all
be plotted on the same shifted temperature plot as shown
in Figure 12. All of the data fall on the same plot
within the experimental accuracy of the measurements. As
can be seen, most of the transition is mapped out except
for the glassy region. More of the transition is mapped
out using the shifted plot than can be seen in any one of
the individual curves. Similar behavior for the
absorption is seen in Figure 13. It is worth emphasizing
that the shifting was not performed arbitrarily in order

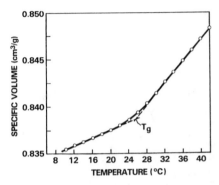

Figure 8. Specific volume vs temperature for poly(vinyl acetate). (Reproduced with permission from Ref. 17. Copyright 1957 Royal Society of Chemistry.)

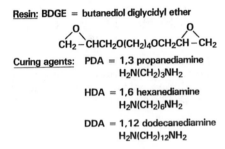

Resin: BDGE = butanediol diglycidyl ether

Curing agents: PDA = 1,3 propanediamine
$H_2N(CH_2)_3NH_2$

HDA = 1,6 hexanediamine
$H_2N(CH_2)_6NH_2$

DDA = 1,12 dodecanediamine
$H_2N(CH_2)_{12}NH_2$

Figure 9. Epoxy polymer chemical components used for varying crosslink density.

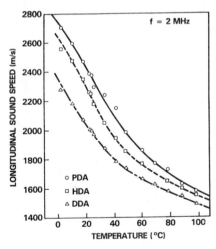

Figure 10. Longitudinal sound speed vs temperature for polyepoxides of varying crosslink density. (Reproduced with permission from Ref. 22. Copyright 1981 Butterworth & Co. Ltd.)

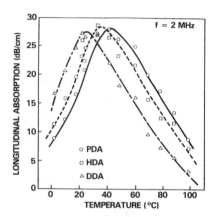

Figure 11. Longitudinal sound absorption vs temperature for polyepoxides of varying crosslink density. (Reproduced with permission from Ref. 22. Copyright 1981 Butterworth & Co. Ltd.)

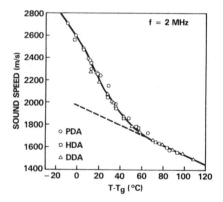

Figure 12. Shifted curve for longitudinal sound speed of polyepoxides of varying crosslink density. (Reproduced with permission from Ref. 22. Copyright 1981 Butterworth & Co. Ltd.)

to obtain superposition; the curves were shifted by the
glass transition temperature, which was determined
independently from the acoustic measurements. For an
aromatic curing agent (m-phenylene diamine) however, the
data is shifted by more than its glass transition
temperature and the absorption curve is lower and broader
than for the aliphatics. Thus, the substitution of an
aromatic curing agent for an aliphatic one changes more
than just the glass transition temperature

Co-Polymerization and Additive Properties

A random co-polymer or a blend of compatible polymers
will have a single glass transition temperature
intermediate between those of the two homopolymers. An
example is shown in Figure 14 for nitrile-butadiene-
rubber (23). The specific weight percents shown are
those of commercial interest for NBR. In contrast, most
polymer blends, graft and block copolymers, and
interpenetrating polymer networks (IPN's) are phase
separated (5) and exhibit two separate glass transitions
from the two separate phases. Phase separated systems
will not be considered here.

Various theories have been proposed for the
co-polymer equation (24). The simplest is a linear
combination of the T_g's of the homopolymers in the form

$$T_g = a \, T_{g1} + b \, T_{g2} \tag{12}$$

where a and b are constants that vary with the theory and
the polymer. This equation arises from the simple rule
of mixtures and has also been derived on the basis of the
free volume theory of the glass transition. In some
theories, the constants a and b are the weight fractions
of the two components while in other approaches the
number of main chain atoms is used (17).

A second commonly used theory involves the reciprocal
temperatures

$$1/T_g = a/T_{g1} + b/T_{g2} \tag{13}$$

where now a and b are different constants than above.
This equation can be derived from an energy theory of the
glass transition.

Finally, a logarithmic form follows from the entropy
theory of the glass transition (26) in the form

$$\ln T_g = a \, \ln T_{g1} + b \, \ln T_{g2} \tag{14}$$

In this case, the constants are related to the heat
capacity changes at the glass transition of the
homopolymers. While the above three equations have
rather different origins and appear to be quite
different, over the range of many measurements, the
results do not differ significantly. Couchman (26) has

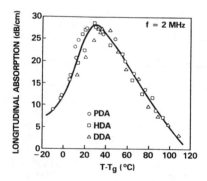

Figure 13. Shifted curve for longitudinal absorption
of polyepoxides of varying crosslink density.
(Reproduced with permission from Ref. 22. Copyright
1981 Butterworth & Co. Ltd.)

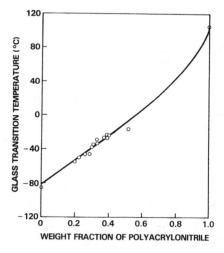

Figure 14. Glass transition temperature vs nitrile
content for acrylonitrile-butadiene co-polymers.

shown that the linear and reciprocal relations are both
approximations of the log equation. While more accurate
and detailed co-polymer equations are available, for the
qualitative understanding desired here, the simple linear
equation is sufficient.

It should be pointed out that the above co-polymer
equations are written as the sum of two terms indicating
that two components are involved, but more terms can
simply be added to the equations if there are more
components. Ter-polymers are sometimes encountered and
there is no reason in principle that even more components
could not be added.

The effect of co-polymerization on T_g is of
significant practical importance and is also a starting
point for an empirical procedure for predicting T_g known
as the method of additive properties. In this procedure,
T_g is estimated knowing only the molecular structure of
the polymer. Such estimates are useful in providing
guidance as to what new polymers should be synthesized in
order to achieve desired acoustic properties and in
providing insight into the competing factors that are
important in the properties of existing polymers.

The method of additive properties has been applied to
density, glass transition temperature, and many other
polymer properties by Van Krevelen (25). The basic idea
is that the properties of each chemical group in the
polymer are nearly independent of the other groups.
Because of this, each group can be assigned a
contribution to the glass transition temperature, for
example, and the T_g of a polymer is the sum of the
contributions of all the groups,

$$T_g = a\ T_{g1} + b\ T_{g2} + c\ T_{g3} + \ . \ . \ . \tag{15}$$

where T_{gi} is the group contribution to the glass
transition temperature and a, b, c, ... are the number of
atom distances along the main chain of the group divided
by the total number of atoms along the main chain of the
polymer repeat unit. For example, the polymer repeat
unit for polydimethylsiloxane consists of two components:
$-Si(CH_3)_2-$ and $-O-$. The number of atom distances along
the main chain is one in both cases. Thus, a = 1/2 and
b = 1/2.

In order to make use of Equation 15, one must first
determine the group contributions to T_g. This is done by
analyzing the measured T_g of polymers with known
molecular structure. Some typical values are listed in
Table II, taken from Van Krevelen (25). Two values are
listed for the methylene group contribution ($-CH_2-$)
depending on whether hydrogen bonding is present (for
polyamides, polyureas, and polyurethane the higher value
is used while for all other polymers, the lower value is
used). Likewise, the p-phenylene group ($-pC_6H_4-$) varies
depending on the type of polymer.

Table II. Group Contributions to T_g

Group	T_{gi}, K
$-CH_2-$	170, 270
$-CH(CH_3)-$	336
$-C(CH_3)_2-$	226
$-CH(C_6H_4)-$	576
$-pC_6H_4-$	300 to 550
$-CHCl-$	538
$-O-$	280
$-Si(CH_3)_2-$	20

Using results such as these, one can predict the T_g of
any polymer for which the group contributions are known.
The glass transition temperature for
polydimethylsiloxane, for example, is T_g = (1/2) 20 +
(1/2) 280 = 150 K = -123°C. The approach was originally
applied to linear polymers but has been extended to
cross-linked polymers as well (24).

Note the somewhat different interpretation of the
values in Equation 15 as compared with Equation 12. In
Equation 12, the homopolymer glass transition
temperatures are known and the multiplying coefficients
are determined by fitting to experimental data. In
Equation 15, the multiplying coefficients are known from
the structure of the polymer and the group contributions
are determined by fitting to experimental data. Also,
the group contribution values do not exist independently.
For example, the group contribution for oxygen (-O-) is
280 K, but this does not imply that oxygen polymer
exists, only that this is the contribution that oxygen
makes when it is a component of a polymer.

Height-Width Limitations

The discussion thus far has been confined to the location
of the damping peak, but not the height or width of the
peak. In many cases, an acoustic designer wants to
provide high damping over a wide range of frequencies.
This may not be possible since the height and width of
the damping peak cannot be adjusted independently. No
real polymer can have an arbitrarily high and broad loss
factor. In general, acoustic design relying on the glass
transition of a polymer involves a trade-off between
height and width. While it is difficult to be specific,
some generalities can be pointed out.

Experimentally, it is generally observed that when
the frequency range of the transition is broad, the
damping peak is low and when the transition is sharp, the
damping peak is high. One explanation for this
observation is that the integrated area under the damping
vs temperature curve has been shown (1) to be
proportional to the activation energy of the transition

$$\Delta H = (G_\infty - G_0) R \pi^2 [\int G'' d(1/T)]^{-1} \qquad (16)$$

where G_∞ is the high frequency (glassy) modulus, G_0 is the low frequency (rubbery) modulus, R is the gas constant, G'' is the imaginary part of the shear modulus, and T is absolute temperature. For transitions whose activation energies are not too different, broadening the transition comes at the expense of lowering the height.

Another approach to relating the width and height of the damping peak is through the use of an analytical model of the transition (27). Assuming, as commonly observed, that the glassy modulus of most polymers is fairly similar but the rubbery modulus varies by two or more decades, one can determine the allowable height and width combinations. In this manner, the trade-offs between height and width can be examined. One useful analytical model used to describe the data is based on the Cole-Cole equation originally proposed for dielectric relaxation (28) but which can also be used for dynamic mechanical relaxation (1,29,30). The Cole-Cole equation applied to shear modulus is given by

$$(G_0 - G_\infty) / (G^* - G_\infty) = 1 + (i\omega\tau)^\alpha \qquad (17)$$

where α is a dimensionless constant for a given polymer with a value between zero and one. (This parameter should not be confused with sound absorption, which unfortunately has the same symbol.) It can be shown that the value of α is a measure of the width of the absorption: the larger the value of α, the sharper the transition. For $\alpha = 1$, we have the sharpest possible transition, that for a single relaxation time. A plot of the Cole-Cole equation with parameters typical for a polymer is shown in Figure 15. The specific values used are: $G_0 = 2 \times 10^6$ Pa, $G_\infty = 6 \times 10^8$ Pa, $\tau = 0.1$ μs, and $\alpha = 0.6$. The shape of the curve shown in Figure 15 is a reasonably good fit to experimental data for many polymers and can be used to illustrate the trade-offs between height and width.

Defining height as the maximum value of the loss factor and width as the number of decades of frequency between the half height values, we can calculate from Equation 17 the relation between height and width for given values of G_0 and G_∞. The results are shown in Figure 16. Here we have fixed the value of G_∞ at 1 GPa, which is typical for many polymers, and considered two constant values for G_0: 10^6 Pa and 10^7 Pa. For fixed values of G_0 and G_∞, there is a different height and width for each value of α. The value of τ does not enter the calculation since it only determines the location of the transition along the frequency axis and not the height or width. While the two curves shown in Figure 16 show all the values mathematically possible, actual values tend to have lower G_0 when there is a narrower

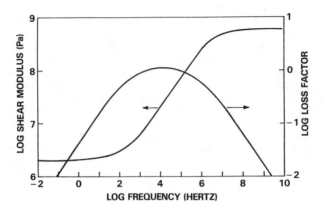

Figure 15.   Complex shear modulus in the modified Cole-Cole model.

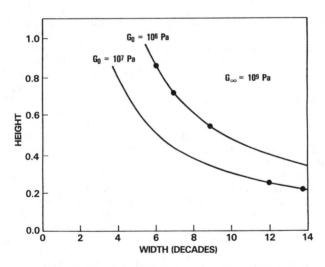

Figure 16.   Height vs width using the Cole-Cole model.

transition and to have higher G_0 values when the
transition is broader. Some actual values for several
polyurethanes are shown in Figure 16 as solid dots.

Modulus-Loss Factor Limitations

Another limitation on acoustic properties is expressed by
the Kramers-Kronig (KK) relations, which are general
relations between the real and imaginary parts of a
complex function. These relations were originally
derived for optics but can be applied in many other areas
as well. The essence of the relations is that the real
and imaginary parts of the function are not independent
of each other but one may be calculated from an integral
of the other. As applied to complex modulus, the
specific form of the relations is given elsewhere in this
book (J. Jarzynski, A Review of the Mechanisms of Sound
Attenuation in Materials).

The KK relations express the real part of the modulus
as an integral over all frequencies of the imaginary part
of the modulus, and the imaginary part of the modulus as
an integral over all frequencies of the real part. This
is a general mathematical result that follows from
causality (the effect cannot precede the cause). An
alternate way of viewing these relations that is more
familiar to polymer scientists is based on expressing G'
as an integral over a distribution of relaxation times.
Since G'' can also be expressed as an integral over the
same distribution of relaxation times, it is apparent
that G' and G'' are not independent. Equivalently, G'
and tan δ are not independent. Knowing G' in terms of
the distribution of relaxation times, there is no freedom
left to choose the value of tan δ.

The significance of this relation to the acoustic
designer is that there are limitations on the combination
of modulus and loss factor that can be achieved. It is
not possible to select the modulus and loss factor of a
material independently. As an example, if it is desired
to have a material whose modulus is independent of
frequency, then the loss factor must be identically zero.
Conversely, if a non-zero loss factor is required, there
will be frequency dependence in the modulus.

While the KK relation is exact, it requires knowledge
of one of the properties (G' or tan δ) for all
frequencies in order to predict the other. A useful
local approximation has been given (31) that is expected
to be valid in regions away from any resonances. Using
this local approximation, it has been shown that when
hysteresis absorption exists (absorption linearly
proportional to frequency), then the modulus will depend
on the log of the frequency (32).

The KK relation can also be used to check the
consistency of experimental data. Lack of agreement
between the measured loss factor and that calculated from
the modulus is an indication of some experimental error.

Likewise, an analytical model for complex modulus must obey KK in order to be a physically meaningful relation. The Cole-Cole model used above, for example, was shown (28) to obey KK. Consistency of experimental data is then demonstrated by showing that the same Cole-Cole fitting parameters describe the modulus and loss factor.

Conclusions

Measurements of acoustic damping have been reported in the literature for numerous polymers under a variety of temperature and frequency conditions. Based on a study of these results, a number of conclusions have been reached:
  • acoustic damping plotted over a wide temperature or frequency range is dominated by the glass transition
  • the location of the damping peak for a given polymer depends on the measurement frequency and temperature
  • glass transition temperature depends on chemical composition: backbone flexibility, steric effects, polarity, plasticizers, crystallinity, pendant groups, crosslink density, and co-polymerization
  • the location of the damping peak can be adjusted through chemical composition changes
  • the height and width of the damping peak cannot be chosen arbitrarily
  • the modulus and loss factor cannot be chosen independently
  Because the glass transition dominates the damping characteristics of polymers, it is necessary to the understanding of any given polymer to know the complete transition behavior, even though this means that the properties are plotted over a wide range of frequency that goes beyond a particular application. Only by knowing the overall picture can one see how variations in molecular weight, crosslink density, and the like will affect the damping properties.

Acknowledgments

This work was sponsored by the NSWC Independent Research Program and the Office of Naval Research.

Literature Cited

  1. N. G. McCrum, B. E. Read, and G. Williams, Anelastic and Dielectric Effects in Polymeric Solids; Wiley: New York, 1967.
  2. J. D. Ferry, Viscoelastic Properties of Polymers; Wiley: New York, 1980; 3rd ed.
  3. I. M. Ward, Mechanical Properties of Solid Polymers; Wiley: New York, 1983, 2nd ed.
  4. B. Hartmann, Acoustic Properties, Encyclopedia of Polymer Science and Engineering; H. Mark, Ed.; Wiley: New York, 1984; Vol. 1, 2nd ed., pp 131-160.

5. L. H. Sperling, Introduction to Physical Polymer Science; Wiley: New York, 1986.
6. L. E. Kinsler and A. R. Frey, Fundamentals of Acoustics; Wiley: New York, 1962, 2nd ed.
7. Ref. 3, p 96.
8. B. E. Read and G. D. Dean, The Determination of Dynamic Properties of Polymers and Composites; Wiley: New York, 1978.
9. B. Hartmann and J. Jarzynski, J. Appl. Phys. 1972, 43, 4304.
10. B. Hartmann and J. Jarzynski, J. Polym. Sci. A-2 1971, 9, 763.
11. R. E. Wetton, in Measurement Techniques for Polymer Solids; R. P. Brown and B. E. Read, Eds.; Elsevier: London, 1984.
12. L. E. Nielsen, Mechanical Properties of Polymers and Composites; Marcel Dekker: New York, 1974.
13. R. F. Boyer, Rubber Chem. Tech. 1963, 36, 1303.
14. P. Hatfield, Br. J. Appl. Phys. 1950, 1, 252.
15. Ref. 5, p 232.
16. N. Bekkedahl, J. Res. Natl. Bur. Stds. 1949, 42, 145.
17. P. Meares, Trans. Farad. Soc. 1957, 53, 31.
18. E. A. Turi, ed., Thermal Characterization of Polymeric Materials; Academic Press: New York, 1981.
19. F. W. Billmeyer, Jr., Textbook of Polymer Science; Wiley: New York, 1984, 3rd ed, p 507.
20. Ref. 2, p 582; ref. 3, pp 174-5.
21. Ref. 3, p 171.
22. B. Hartmann, Polymer 1981, 22, 736.
23. R. H. Wiley and G. M. Brauer, J. Polym. Sci. 1948, 3, 704.
24. G. Lee and B. Hartmann, J. Appl. Polym. Sci. 1983, 28, 823.
25. D. W. Van Krevelen, Properties of Polymers: Correlations with Chemical Structure; Elsevier: Amsterdam, 1972.
26. P. R. Couchman, Macromolecules 1978, 11, 1156.
27. B. Hartmann and G. F. Lee, Proc. Xth Internatl. Cong. Rheol., 1988, p 392.
28. K. S. Cole and R. H. Cole, J. Chem. Phys. 1941, 9, 341.
29. C. R. Taylor, C. J. Aloisio, and S. Matsuoka, Polym. Eng. Sci. 1985, 25, 105.
30. B. E. Read, Polymer 1989, 30, 1439.
31. M. O'Donnell, E. T. Jaynes, and J. G. Miller, J. Acoust. Soc. Am. 1981, 69, 696.
32. N. Lagakos, J. Jarzynski, J. H. Cole, and J. A. Bucaro, J. Appl. Phys. 1986, 59, 4017.

RECEIVED January 24, 1990

DYNAMIC EVALUATION

DYNAMIC EVALUATION

Selecting a polymer for a particular application requires an accurate knowledge of its dynamic mechanical behavior. It is the response of a polymer to an applied force which defines its usefulness in damping applications. It must be remembered that all polymers are viscoelastic, exhibiting behavior which varies with time and temperature.

A typical dynamic mechanical evaluation involves either applying a constant stress to the sample (for example, hanging a mass on it) and measuring the resulting strain (the extent of elongation), or conversely applying a constant strain and measuring stress. Because the response of the polymer is time dependent, these studies must be dynamic. Quasi-static techniques such as creep or stress relaxation are often useful, however such techniques are usually limited to conditions where the relaxation will proceed sufficiently slowly that accurate data can be collected on-the-fly.

For dynamic polymer property evaluations, a periodic or cyclic disturbance is more typically used. The sample is then disturbed at some selected frequency (typically 0.1 to 10,000 Hz) and some physical property is monitored as sample temperature is varied. Conversely the frequency can be varied at a fixed temperature.

In this section, **Dlubac** et. al. first describe three techniques for measuring the dynamic mechanical properties of polymers. **Capps and Beumel** then describe one transfer-function technique, and provide examples of measurements on a materials system intended for use in a vibration isolator. **Sattinger** describes a technique which directly measures the dynamic shear properties of polymers, rather than the more usual Young's or longitudinal properties. **Ganeriwala and Hartung** present a new variation on dynamic/mechanical testing techniques - FTMA. They discuss the limitations and potential of this technique, particularly regarding its ability to more rapidly obtain a detailed material characterization. **Weissman and Chartoff** concentrate on the data analysis procedure typically used to reduce the data, namely the creation of Master Curves using the time-temperature supperposition principle. **Sircar and Drake** describe a frequently used alternate polymer characterization technique, differential scanning calorimetry (DSC). They compare measurements made with DSC to those of dynamic-mechanical testing, and show that DSC results can be used in place of mechanical properties for many engineering applications. Finally, **Urban and McDonald** describe a novel new technique using FT-IR to study polymers undergoing mechanical stress.

Chapter 3

Comparison of the Complex Dynamic Modulus as Measured by Three Apparatus

James J. Dlubac[1], Gilbert F. Lee[2], James V. Duffy[2], Richard J. Deigan[1], and John D. Lee[2]

[1]Ship Acoustics Department, David Taylor Research Center, Bethesda, MD 20084–5000
[2]Nonmetallic Materials Branch, Naval Surface Warfare Center, Silver Spring, MD 20903–5000

This chapter compares complex dynamic modulus data on two viscoelastic materials obtained with three apparatus: (1) a forced torsional apparatus that determines shear modulus from the relative amplitude and phase at the ends of a harmonically torqued right circular cylinder, (2) a resonance apparatus that measures the Young's modulus by analyzing the response of a bar sample at extensional resonance, and (3) a cantilever beam bending apparatus that is used to determine the Young's modulus. In all of these apparatus, the complex dynamic modulus is measured over a limited frequency range at a number of fixed temperatures. Time-temperature superposition is then employed to generate master curves of modulus and loss factor at a fixed temperature over a very broad frequency range. The data presented here spans the glass transition of the two polyurethane materials. One material has a low, broad transition and the other has a high, sharp transition. Good agreement was found among the different apparatus for both materials, but more care must be taken when measuring a sharp transition. It is concluded that the three apparatus give reasonably consistent data, a sharp transition is more sensitive to the testing procedure, and the shifting algorithm used in the time-temperature superposition must be considered as part of the test technique when comparing data.

The design of effective sound and vibration damping
materials assumes an understanding of the mechanisms
controlling the dissipation process and knowledge of
candidate material properties. The use of viscoelastic
materials as sound and vibration absorbers is wide-spread
and well-known. Accurate measurement of the complex
dynamic moduli of these materials is therefore vital to
the control of acoustic and vibrational energy. This
chapter discusses and compares three apparatus used to
measure the dynamic modulus of viscoelastic materials.

DESCRIPTION OF APPARATUS

Various methods (1-3) have used to determine the dynamic
mechanical properties of polymers. Many of the
instruments described are well known and are widely used
(torsional pendulum, rheovibron, vibrating reed, and
Oberst beam ASTM D4065-82). Newer instruments like the
torqued cylinder apparatus (4), resonant bar apparatus
(5) and Polymer Laboratories Dynamic Mechanical Thermal
Analyzer (6) are becoming more popular in recent times.
It is of interest of this chapter to show that these
newer instruments are accurate and easy to use.
    Each device considered in this chapter determines the
complex dynamic modulus from a theoretical description of
the measurement. These descriptions or solutions are
derived by making assumptions about the experiment. The
degree to which these assumptions are realized determines
the accuracy of the measurement. The most important and
restrictive assumptions are those concerning the boundary
conditions, sample geometry and stress state.

TORSION OF A CYLINDER. The complex dynamic shear modulus
can be determined through the relative motion, amplitude
and phase, of the ends of a dynamically torqued
cylindrical sample (4). Figure 1 is a setup sketch of
this non-resonant experiment. The sample to be tested is
bonded to rigid discs to which are attached
accelerometers. An oscillatory torque is applied to the
bottom of the sample through a force couple created by
driving two shakers in phase. The accelerometers'
signals are amplified and filtered before being measured
by a phase angle voltmeter. A computer uses the
acceleration amplitude and phase to invert the
theoretical solution of the torqued cylinder to infer the
real part of the shear modulus, G', and the loss factor =
G''/G' where G'' is the imaginary part of the modulus.
The sample, accelerometers, torsion spring, and shakers
are all in an environmental chamber.
    Torsional waves are propagated through the sample and
are resisted at the top plate by a torsion spring mounted
to a rigid frame. The torsion spring at the top serves
two functions. First, the spring prevents significant

bending of the sample due to a chance misalignment of the
shakers. This is possible since the "X" shape of the
spring has a bending stiffness much greater than its
torsional stiffness. Secondly, the torsion spring forces
the sample to distort while allowing a measurable motion
at the top plate.

The frequency range of the torqued cylinder apparatus
is 50 Hz to about 1500 Hz. The temperature range of the
experiment is -40°C to 70°C. The maximum temperature is
limited by the durability of the shaker diaphragms.
Though a thorough study of the modulus and loss factor
measurement ranges has not been conducted, current
experience indicates the range of the real part of the
shear modulus, G', is 10^7 to 10^{11} dyn/cm^2; the range of
loss factor is from 0.05 to 1.2.

The instrument is capable of handling samples from 20
to 90 mm in diameter by 30 to 150 mm in height. For
these tests,the sample dimensions were 50 mm diameter by
50 mm in height. An advantage of such large samples is
that measurements can be made of the effective shear
modulus of materials with large inhomogeneities. A
disadvantage is that thermal equilibrium takes longer to
achieve. This lengthens the test time required.

Measurements on the torqued cylinder apparatus are
made isothermally, from 50 to 1500 Hz, in 5°C intervals
starting at -40°C. Thermal equilibrium time between
temperature changes is about 1.5 hours. Typically, a
material can be evaluated in about 20 hours using this
method.

For the worst case, calculation of temperature rise
within the sample due to mechanical energy dissipation is
about 1°C. With heat loss from the sample, this value
should be lower.

EXTENSION OF A BAR. The complex dynamic Young's modulus
can be inferred through the response of a sample bar at
extensional resonance. The resonance apparatus (5) is
shown schematically in Figure 2. An electromagnetic
shaker is used to drive a test sample (6.35 by 6.35 by
100 mm) at one end while the other end is allowed to move
freely. Miniature accelerometers are adhesively bonded
on each end to measure the driving point acceleration and
the acceleration of the free end. The weight of the
accelerometer and mounting block is about 3 grams. The
output signals from the accelerometers are amplified by
charge amplifiers. The output from the charge amplifiers
are routed to a dual channel Fast Fourier Transform (FFT)
spectrum analyzer. The analyzer digitizes and displays
the measured signals as the amplitude and phase of the
acceleration ratio. The analyzer also provides a random
noise source to drive the shaker and is effective over a
frequency range of three decades (25 Hz to 25,000 Hz).
The data are always sampled and rms averaged at least 8
times, for low noise data, and up to 256 times, for noisy
data. A minicomputer is used to collect and store the
data from the analyzer for later calculations.

The spectrum analyzer is used to identify the resonant frequencies of the sample. The number of resonant modes that can be measured is dependent on the loss factor of the material. At low loss, on either the glassy side or the rubbery side of the glass transition, four to five resonant modes are easily measured on the analyzer. As expected, the resonant modes appear at higher frequencies in the glassy state than in the rubbery state. At the glass transition of the material, where the loss is high, only three to four resonant modes can be measured. The higher frequency resonant modes (modes 4 and 5) are not detectable.

From the peak amplitude and frequency, the real part of the Young's modulus (E') and the loss factor are determined as functions of frequency and temperature. The resonant apparatus can measure E' from 10^5 to 10^{13} dyn/cm^2 and loss factor over the range of 0.01 to 5.0.

In making the measurements, the following thermal cycle is used: cool the test sample (mounted in the apparatus) from room temperature to -60°C. The sample is allowed to soak at -60°C for at least 12 hours. Measurements are then made as the temperature is raised in 5°C intervals. Approximately 20 minutes are allowed after a temperature change to obtain thermal equilibrium in the sample. The operating temperature range is -60°C to 70°C. The time required to complete the measurements is about 24 hours. Since the system is automated, the instrument can run unattended overnight. The frequency range for a typical set of measurements is from 1 to 15 kHz.

Conservative calculation of temperature rise within the sample due to mechanical energy dissipation is very much less than 1°C.

BENDING OF A BEAM. The complex dynamic Young's modulus can be determined from the forced, non-resonant oscillations of a single or double cantilever beam. The apparatus considered in this paper is the Dynamic Mechanical Thermal Analyzer (DMTA) (6), manufactured by Polymer Laboratories, Inc. Figure 3 shows the experimental setup for the single cantilever measurement. A thin sample is clamped at both ends. One end is attached to a calibrated shaker through a drive shaft. The force and displacement are measured at the driven end for each fixed frequency. The low frequency/low mass bending solution is used together with the measured input impedance to infer the Young's modulus and loss factor.

The DMTA operates at fixed frequencies over a broad temperature range. Sixteen discrete frequencies from 0.01 Hz to 200 Hz are available. The very low frequencies, below about 0.1 Hz, require a long time to complete, while frequencies above 30 Hz are often near or above the system resonance and require special consideration. Though the system is capable of a

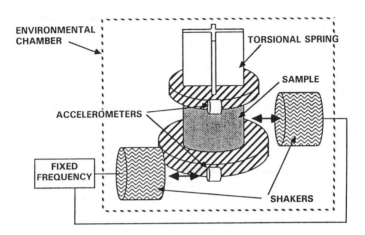

Figure 1.   Torqued cylinder apparatus.

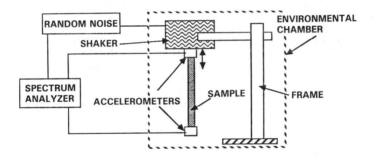

Figure 2.   Resonance apparatus.

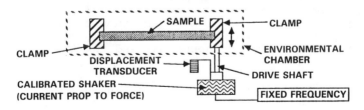

Figure 3.   Dynamic mechanical thermal analyzer (DMTA) apparatus.

temperature range from -150°C to 300°C, runs much above
100°C tend to overly soften the samples of interest and
in some cases the samples begin to melt. Only the
sample, its clamps and part of the drive shaft are in the
temperature-controlled chamber. The DMTA can measure E'
over the range 10^6 to 10^{12} dyn/cm^2 and loss factor over the
range of 10^{-4} to 9.99.

Length to thickness ratio from 4 to 6 mm are
encouraged in this experiment to optimize sample
stiffness through the entire range of modulus. Three
clamps are available with single cantilever beam spans
ranging from 5 mm to 18 mm. Sample widths are typically
about 10 mm while the thickness can range up 5 mm. For
the measurements reported here, a length about 12 mm and
a thickness of 3 mm was used. The dynamic modulus and
loss factor data was collected isothermally from 30 to
0.30 Hz. Material evaluation requires about 4 hours.

Special care must be exercised in clamping the DMTA
samples into place. Samples were prepared by bonding
aluminum blocks to each end with epoxy adhesive. The
epoxy was chosen such that, together with the bond
thickness, the stiffness of the adhesive is always much
greater than the sample. The advantage of using the
aluminum blocks is that the assumed sample boundary
conditions are obtained at all temperatures and
frequencies. The blocks also prevent sample pinching
when directly clamping to the sample.

Sample preparation techniques using aluminum blocks
and a shifting algorithm are not provided by the
manufacturer of the DMTA and were developed
independently.

Conservative calculation of temperature rise within
the sample due to mechanical energy dissipation is less
than 1°C.

PARAMETER RANGE COMPARISON. Table I summarizes the
parameter ranges of the torqued cylinder apparatus, the
resonance apparatus and the DMTA. Since the bulk moduli
of the materials under consideration in this paper are
much larger than the Young's or shear moduli, the
materials are considered incompressible. For
incompressible materials, the shear modulus is one third
of Young's modulus. Comparisons are then made by
converting Young's modulus to shear modulus for the data
measured by the resonance apparatus and the DMTA.

TIME-TEMPERATURE SUPERPOSITION PROCEDURE

Almost always the data from the apparatus above is
analyzed by using the time-temperature superposition
principle to form a master curve over a wide frequency
range at a selected reference temperature. The basis for
this procedure is that for thermorheologically simple
materials the effect of a change in temperature on

Table I.   Comparison of Dynamic Modulus Apparatus

DEVICE	FREQ (Hz)	TEMP (°C)	REAL MODULUS dyn/cm^2	LOSS FACTOR	SAMPLE SIZE (mm)	REMARKS
Torqued Cylinder	50 to 1500	-40 to 70	10^7 to 10^{11}	0.05 to 1.2	50 dia x 50 = 9.8 x 10^4 mm^3	Large samples
Resonant Bar	2.5 to 25,000	-60 to 70	10^5 to 10^{13}	0.01 to 5.0	6x6x 150 = 5.4 x 10^3 mm^3	Extension of a bar
DMTA Beam	0.01 to 200	-150 to 300	10^6 to 10^{12}	10^{-4} to 9.99	3x10 x12 = 3.6x 10^2 mm^3	Bending of a beam

complex modulus is indistinguishable from a change in frequency (7). Thus, making measurements over a range of temperatures is equivalent to making measurements over a range of frequencies. The advantage of this is that temperature measurements are much easier to make than frequency measurements. In most cases, the actual frequency measurements are not made at the frequency of interest but one can determine what the properties would be at the frequency of interest using time-temperature superposition. Specifically, in this study the three apparatus do not operate in the same frequency range and a direct comparison of results would not be possible without superposition. Thus the final comparison depends not only on the instruments but how the data is analyzed. While the principle of superposition is well established, significant differences can result if the implementation of the shifting is not done in a consistent manner. For this reason, the superposition procedure used will be described in some detail.

Figure 4 illustrates the mechanics of the principle. Data collected at various temperatures is shifted along the log frequency axis to form a modulus curve over an extended frequency range. The incremental shift along the log frequency axis, represented by the change in the shift factor (log a_T), is summed to form the shift curve as illustrated in Figure 5. The mechanics of shifting was performed by using an algorithm implemented on a computer. No attempt was made to fit the frequency shift versus temperature to the Williams-Landel-Ferry (WLF) equation (3), though usually the fit is good in the glass transition region.

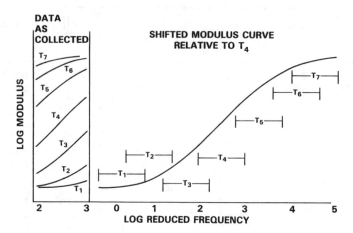

Figure 4.   Time-temperature superposition.

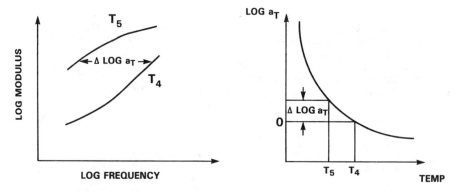

Figure 5.   Construction of a shift curve.

        The specific algorithm used was as follows. The log
modulus vs log frequency data at each temperature was
least square fitted to a linear equation to determine the
slope of the data. This slope is equal to the difference
in log modulus, at a given frequency, between two sets of
data divided by the difference in log frequency. The
difference in log frequency is the change in shift
factor, log a_T. When implementing this procedure, data
at the reference temperature is fixed, while data at the
other temperatures are shifted relative to the reference
temperature. Note that the loss factor data is not used
in determining the shift. For the apparatus discussed
here, the modulus measurements are more accurate than the
loss measurements and give more reliable shift. The loss

factor data are shifted using the same   log a_T's that are used for the modulus, and the smoothness of the resulting curve is an independent indication of the validity of the shift function.

MATERIALS

Polyurethanes were chosen for this study because these materials are becoming more widely used in sound and vibration damping and because they offer a wide range of material properties against which to compare the apparatus.   In particular, of the two materials chosen for this study, one has a broad glass transition and the other has a narrow glass transition.   Therefore, with these two samples, a more stringent test can be made on the apparatus.   The polyurethanes are prepared from a prepolymer of poly(tetramethylene ether)glycol (nominal molecular weight of 1000) and 4,4'-diphenylmethane diisocyanate in which the molar ratio of the two components is 1 to 3.   The prepolymer is chain extended with either 1,4-butanediol, forming a polymer designated as H01, or with a 50/50 mixture of 1,4-butanediol and 2,2-dimethyl-1,3-propanediol, which is designated as H14. The synthesis and details of the chemical components are discussed in another chapter of this book (J. V. Duffy, et al., Effects of Diol Chain Extenders Structure on the Dynamic Mechanical Properties of PTMG Polyurethanes).

A cautionary note when using polyurethanes.   There can be considerable variation in properties depending on the processing technique even when the chemical composition is nominally the same.   For this reason, it is important to do quality control checks to verify that the material evaluated in the three apparatus is in fact the same.   Two good ways to characterize polymeric materials are the density and glass transition temperature.   The glass transition temperature is particularly important since it governs the dynamic mechanical response of the material.   Density and glass transition temperatures are listed in Table II.   Glass transition temperature, T_g, values were determined in a differential scanning calorimeter and density values were obtained by water immersion.

Table II.   Material Properties

polymer	density, g/cm^3	T_g, °C
H01	1.139	−48
H14	1.107	0

One of the first attempts to compare the apparatus was not successful because the material used for the three test samples was found to have a significantly different glass transition temperature.   This material was then eliminated from further consideration.

RESULTS

Figures 6, 7 and 8 are shear modulus and loss factor
master plots for H01 as collected by each of the three
apparatus. Note that the left ordinate is log G', the
right ordinate is loss factor and the abscissa is log
frequency. In each case, the plotting reference
temperature is 10°C.

The shifted data collected by the torqued cylinder
apparatus is shown in Figure 6. The individual
temperature runs are apparent as groups of data,
especially in the glassy region. This is a direct
consequence of the limited frequency range of the
apparatus. Overlapping data could be obtained, if the
measurement was made at more closely spaced temperatures,
thus greatly increasing the time required for the
measurement. There is almost no scatter in the G' data,
but there is some scatter in the loss factor data near
the transition. These results are typical in that
modulus measurements generally show less scatter than
loss factor measurements and are considered to be more
accurate.

Figure 7 contains the shifted data of the resonant
bar apparatus for H01. The data obtained at each
temperature with this device covers a slightly broader
frequency range than the torqued cylinder apparatus,
resulting in overlapping of the data when shifted. The
G' data contains very little scatter, but there is
moderate scatter in the loss factor data.

Figure 8 is a plot of the H01 data obtained with the
DMTA. The data clearly overlap due to the relatively
broad frequency range of operation of this device. The
G' data contains very little scatter, but there is
moderate scatter in the loss factor data.

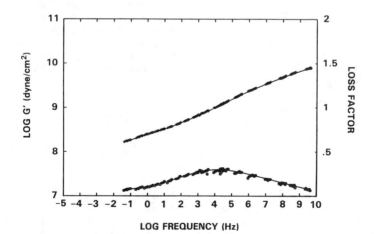

Figure 6.   Torqued cylinder data for H01.

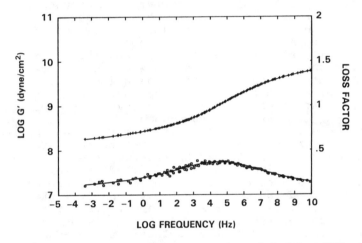

Figure 7. Resonance apparatus data for H01.

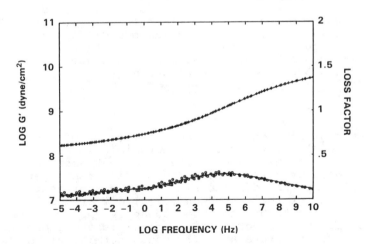

Figure 8. DMTA data for H01.

The shifted data for the three devices is compared
in Figure 9 for H01. For ease of comparison the actual
data points are not plotted. At the transition
frequency, G' is approximately 10^9 dyn/cm^2. The spread in
G' among the three curves is minimal at about 18%. At
the transition frequency the peak loss factor is
approximately 0.33, slightly higher for resonant bar with
a deviation among the three apparatus of about 24%. Thus
for H01, with its broad transition and low loss factor,
there is good agreement among the apparatus.
    For H14, a plotting reference temperature of 10°C
was first used. The agreement between the torque
cylinder and the beam apparatus was quite good. However,
the agreement with the resonant bar was poor. It was
reasoned that choosing a reference temperature close to
the T_g (0°C) may not have been a wise choice. Figure 10
demonstrates the effect of shifting errors and reference
temperature on the modulus curve. At temperatures near
T_g on a log a_T versus temperature curve, the shift factor
changes rapidly with temperature. Thus, a small change
in the temperatures can result in large differences in
the shifted modulus curve. Since the shift factor curves
are slightly different from each apparatus for the same
material due to errors in modulus measurements and
shifting, then it is best to select a temperature where
the shift factor is less dependent with temperature,
which occurs at higher temperatures. So, a plotting
reference temperature of 35°C was chosen. Figure 11
contains a overlay of the curves for the shifted data of
the cylinder, bar and beam devices for H14 at 35°C. The
agreement is fairly good. The maximum difference in the
modulus, about 36%, occurs at the transition frequency.
The peak loss factor of the resonant bar apparatus is
about 10% higher than those measured by the other
systems.

DISCUSSION

The inherent difficulty in the measurement of the complex
dynamic moduli of viscoelastic materials is emphasized by
the results of this paper. The agreement among the
shifted modulus data as measured by different systems is
limited by several difficulties: (1) measurement
inaccuracies of the instruments, (2) differences in the
data reduction techniques used to apply the
time-temperature superposition principle and propagation
of shift curve errors and, (3) nonuniformity of the test
samples.
    Though the measurement uncertainty of each device
has been checked, and care was taken to minimize
measurement errors, inaccuracies of measurement cannot be
ruled out. The higher peak loss factor for the resonant
apparatus for example may be due to vibrational energy
propagating into the dangling accelerometer cable.

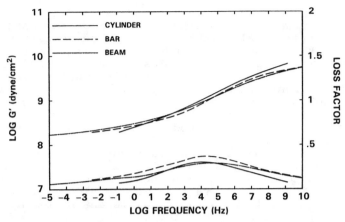

Figure 9.   Comparison of all three apparatus for H01.

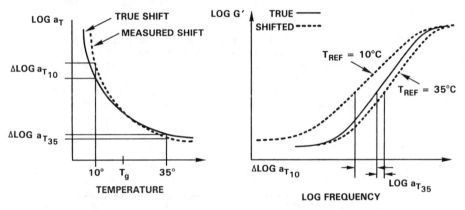

Figure 10.   Effect of errors on the modulus curve.

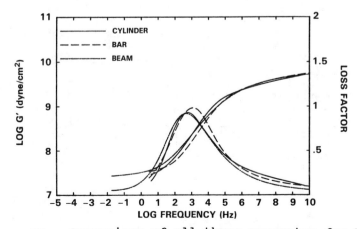

Figure 11.   Comparison of all three apparatus for H14.

     In this study all data was reduced and shifted using
the same software in order to eliminate differences
between different algorithms. It is also necessary to
use a shifting reference temperature well above T_g. Good
agreement was obtained on samples H0l at 10°C and Hl4 at
35°C, both reference temperatures being well above the
respective glass transition temperatures. Since the
final results are displayed in the form of master curves,
the shifting algorithm must be considered part of the
test procedure and can introduce large errors if not done
properly.
     An obvious and very important consideration in
dynamic modulus comparisons is the uniformity of test
samples. Candidate materials should be checked to be
stable in time. Sample fabrication should be meticulous.
Special care should be given to different sample
geometries, especially when the chemical reaction during
fabrication is exothermic. Finally, comparisons should
be made with materials that possess a relatively narrow
glass transition region and high loss factors. These
materials more readily display differences among the test
apparatus.
     In choosing among the three apparatus, it is seen
that the results are comparable when proper care is taken
so that all three can be considered equivalent data.
Also, because all three rather different apparatus give
fairly similar and reproducible results, one has greater
confidence that all three are measuring intrinsic
materials properties with acceptable accuracy.

LITERATURE CITED

  1. Ward, I. M., Mechanical Properties of Solid Polymers;
     John Wiley and Sons: New York, 1971.
  2. Read, B. E. and Dean, G. D., The Determination of
     Dynamic Properties of Polymers and Composite; John
     Wiley and Sons: New York, 1978.
  3. Murayama, T., Dynamic Mechanical Analysis of
     Polymeric Materials; Elsevier Scientific Publishing
     Company: New York, 1982.
  4. Magrab, E. B., J. Res. Natl. Bur. Stds. 1984, 89,
     193-207.
  5. Madigosky, W. M. and Lee, G. L., J. Acoust. Soc. Am.
     1983, 73, 1374-1377.
  6. Brown, R. P. and Read, B. E., Measurement Techniques
     for Polymeric Solids; Elsevier Applied Science
     Publishers: New York, 1984.
  7. Ferry, J. D., Viscoelastic Properties of Polymers;
     John Wiley and Sons: New York, 1980.

RECEIVED January 24, 1990

Chapter 4

Dynamic Mechanical Testing

Application of Polymer Development to Constrained-Layer Damping

Rodger N. Capps[1] and Linda L. Beumel[2]

[1]Naval Research Laboratory, P.O. Box 568337, Orlando, FL 32856–8337
[2]TRI/TESSCO, 9063 Bee Caves Road, Austin, TX 78733–6201

Viscoelastic materials are widely used for acoustic
attenuation, isolation of continuous vibration, and
shock mountings. The properties of these materials are
dependent upon temperature and frequency of excitation,
molecular structure of the base polymer, and chemical
cross-linking systems and fillers. This paper describes
a transfer function technique for the measurement of the
frequency-dependent Young's modulus and loss tangent.
Algorithms for time-temperature superposition are also
discussed. It is then shown how the results of such
measurements can be used in the selection of
viscoelastic materials and fillers in the design of
constrained-layer damping structures. Comparisons of
mathematical modeling and experimentally determined
damping are given for some of the chlorobutyl
formulations discussed.

Viscoelastic materials are widely used for acoustic attenuation,
isolation of continuous vibration, and shock mounting for damping of
transient disturbances. For example, elastomers are commonly found
in automotive engine and body mounts; load-bearing pads for
machinery, railroad rails, and bridges; and constrained-layer
damping treatments for decreasing structure-borne noise in airplanes
and ships. In particular, constrained-layer damping is becoming
increasingly important in naval applications.

The engineering property that is of interest for most of these
applications, the modulus of elasticity, is the ratio of unit stress
to corresponding unit strain in tension, compression, or shear. For
rigid engineering materials, unique values are characteristic over
the useful stress and temperature ranges of the material. This is
not true of natural and synthetic rubbers. In particular, for
sinusoidal deformations at small strains under essentially
isothermal conditions, elastomers approximate a linear viscoelastic

0097–6156/90/0424–0063$06.00/0

medium in mechanical behavior. The dynamic mechanical properties of elastomeric materials are dependent upon a number of factors, including the frequency of excitation, temperature, molecular structure of the base polymer, and chemical cross-linking systems. Additionally, reinforcing fillers such as the various types of carbon black will significantly alter both the physical properties and viscoelastic behavior of elastomeric materials. Under cyclic deformations, the strain will lag the stress with a phase angle, δ, that is between 0 and $\pi/2$ radians. The complex elastic modulus describing the behavior of the material consists of real and imaginary components. The ratio of imaginary to real is the tangent of the phase angle, δ, and is commonly denoted as the loss factor or loss tangent. It is a measure of the mechanical hysteresis or internal damping of the polymer. This quantity is of considerable interest for many vibration control applications.

The design of vibration control structures and selection of materials for use in them requires a accurate determination of the viscoelastic properties of the polymers used. The modulus that is appropriate for consideration will depend upon the geometry and boundary conditions found in the vibration control structure.

A number of commercial and "home brew" instruments have been devised for the measurement of viscoelastic properties of elastomers. Most of these measure either the Young's modulus and loss tangent, or shear modulus and loss tangent. Many of the commercially available instruments suffer the disadvantage that they operate only at relatively low frequencies which may be inappropriate for certain applications. This paper describes a transfer function technique for the measurement of the frequency-dependent Young's modulus and loss tangent. Techniques for time-temperature superposition are also discussed. It is then shown how the results of such measurements can be used in the selection of viscoelastic materials in constrained-layer damping treatments. Comparisons of mathematical modeling and experimental results are given for some constrained-layer damping assemblies.

MEASUREMENT METHOD

A block diagram of the automated measurement system (1) is shown in Figure 1. The principle of measurement is based upon measuring the transmissibility of a mass-loaded rod with high internal damping undergoing longitudinal sinusoidal excitation (2-4). The sample is harmonically excited using discrete frequency excitation and a lock-in analyzer is used to measure the transfer function (amplitude ratio and relative phase of the free and driven ends). It can be shown (2-5) that the displacement solution of the equation of motion in the bar can be separated into a pair of coupled transcendental equations relating the experimentally measured transfer function to the elastic modulus and loss tangent in the bar. They can be unambiguously solved by Newton's method at the $\pm$ 90 degree phase crossing points in the transfer function. The seed values of the modulus and loss tangent can then be used at frequencies other than resonance to obtain a family of frequency dependent modulus and loss

tangent curves at each experimental temperature of measurement
(4,5). These are then stored on floppy disks for further
processing.

For linear visoelastic behavior, the interrelated dependence
upon frequency and temperature of the behavior of polymers subjected
to a periodically varying stress can be expressed through the well-
known time-temperature superposition principle (6). The criteria
for its application have been discussed by Ferry (6). This
technique can be used to superimpose modulus curves covering a
limited frequency range at different temperatures into an extended
frequency curve at a single reference temperature. The amount of
horizontal frequency shift required to superimpose a storage or loss
modulus curve measured at some temperature T onto another curve at a
reference temperature T_0 is expressed in terms of a shift factor,
a_T, which may be viewed as either being equal to the ratio of the
shifted frequency to the reference frequency f_0, or the ratio of
relaxation times for an elastomer at some temperature T and a
temperature T_0 which is a characteristic temperature for the
material under consideration. The modulus, E', and loss tangent
that one would observe at a reference temperature, T_0, relative to
an experimentally measured modulus and loss tangent at some
temperature, T, are given by

$$E'(f,T_0) = (T_0 \cdot \rho_0 / T \cdot \rho) \cdot E(f a_T, T) \qquad (1a)$$

and

$$\tan \delta(f,T_0) = \tan \delta(f a_T, T) \qquad (1b)$$

The correction ratio involving temperatures in Eq. 1(a) is
included because the modulus is directly proportional to
temperature. Similarly, a correction for densities is included
because the volume of a polymer is a function of temperature, and
the modulus, defined per unit of cross-sectional area, will vary
with the amount of material contained in a unit volume. Under
normal circumstances, the correction for densities is a small one,
and is often neglected.

As a matter of convenience, the highest temperature within a
range of measurements is normally chosen as T_0 in our system. A
polynomial least squares curve fitting routine is then used to find
best fit through the experimental logarithm modulus-logarithm
frequency points at each temperature. These coefficients, along
with the beginning and ending x coordinates, are stored on disk with
a file containing the experimental temperatures. The program then
reads the first two temperatures, T_0 and T_1. The first x
coordinate in the curve at T_1 is then used in the equation for the
curve at T_0 to calculate the value of the modulus. The two values
are compared. If E_0 is less than E_1 the x coordinate representing
the logarithm of frequency is incremented a small amount and the
procedure is repeated until the two values match within a certain
tolerance. This procedure is repeated at four separate intervals
along the curve. The x increments required to have the two curves

match are used to find the average horizontal distance between the
pair of curves ($\delta \log a_T$). This procedure is repeated for each pair
of curves in the experimental temperature range. The selected
values of $\delta \log a_T$ are progressively added, beginning at T_0, to give
$\log a_T$ at each temperature.

In cases where the material follows the characteristic form (6)
of the William-Landel-Ferry (WLF) equation,

$$\log a_T = -c_1 \ (T-T_0) \ / \ (c_2 + T - T_0) \tag{2}$$

a least squares fit of $T-T_0$ vs $(T-T_0)/\log a_T$ will have a slope of
$-c_1$ and an intercept of $-c_2$, where c_1 and c_2 are the WLF constants.

The shift constants obtained by this procedure can be
transformed to another reference temperature through the relations

$$c_1 = c'_1 \ c'_2/(c'_2 + T_0 - T_1), \ c_2 = c'_2 + T_0 - T_1 \tag{3}$$

where T_1 is the original reference temperature, T_0 is the new
reference temperature, and c'_1 and c'_2 are the shift constants
obtained at the original reference temperature. The initial choice
of T_1 and its effect upon the values of c_1 and c_2 determined by this
procedure have been discussed in detail by Ferry[2] (6).

Over certain temperature ranges, the shift factor may exhibit a
behavior which agrees better with an Arrhenius temperature
dependence (6,7). This will normally be the case when a rather
narrow range of experimental temperatures is used or the reference
temperature is significantly greater than the glass transition
temperature. A least squares treatment of $\log a_T$ versus $1/T$ will
lead to a shift function of the form

$$\log a_T = (\Delta H_a/R) \ (1 \ / \ T - 1 \ / \ T_0) \tag{4}$$

where R is the gas constant, ΔH_a is an apparent energy of activation
for the molecular relaxation process, and T_0 is the reference
temperature. Other investigators (8,9) have proposed various
empirical forms for the shift function. These two forms have been
found to be adequate for the materials that we have studied.

When the proper form of the shift function has been determined,
a computer program is then used to shift the raw data with respect
to the chosen reference temperature. A polynomial least squares is
then used to find the best fit through the shifted modulus and loss
tangent points. The resulting curves and shifted points are then
stored on floppy disks for plotting or recall by various modeling
programs.

APPLICATIONS

Constrained Layer Damping. The concept of constrained layer damping
has been described by Ungar and others (10-12). The simplest type
of constrained layer structure is the three layer sandwich type,
shown schematically in Figure 2. It consists of a viscoelastic
layer coated onto a rigid substrate but constrained by a stiff top
plate. The addition of the top plate produces a shearing action as
the composite structure is deformed through an angle ϕ with respect
to a neutral plane in the composite. In contrast, damping in single
layer treatments is achieved primarily through energy dissipation
through the flexural and extensional motions of the damping layer.
The constrained layer system normally exhibits much better damping
on an equal weight basis than the single layer extensional system.

The damping of the composite structure will be affected by the
thicknesses of the various layers, stiffnesses of the base and top
plates, and the viscoelastic properties of the constrained layer
(12). In the present instance (13), it was desired to develop a a
broad-band material to damp a model composite structure consisting
of a 2.54 cm. base plate (H_1), 0.079 cm. polymer layer (H_2), and
0.159 cm. cover (H_3). The base and cover were composed of brass
with a modulus of 10^{11} Pa. In this instance, the only variable was
the viscoelastic behavior of the polymer layer. A temperature range
from 0 to 20 degrees Centigrade and a frequency range from 100 Hz to
10 kHz were desired.

Selection of a viscoelastic material for use in a constrained-
layer damping application will involve a practical compromise
between the maximum structural damping that can be obtained for a
particular temperature or frequency, and the extent of the
temperature or frequency range that the designer desires to cover.
Certain types of materials, such as nitrile rubbers, may exhibit
very high damping over a relatively narrow temperature and frequency
range. Other materials may not exhibit loss factors that are as
high, but show less temperature and frequency dependence. This is
depicted in Figures 3 and 4 for a nitrile-epichlorohydrin rubber
blend (ASTM designation NBR-ECO) and a chlorobutyl rubber (CIIR),
both compounded and cured to a shore A hardness of 55. The
formulations are given in Table I. Figures 3 and 4 show the
isochronal temperature response of the Young's storage modulus and
loss tangent of these two materials at 10 Hz as measured by a
Polymer Laboratories Dynamic Mechanical thermal analyzer. The
chlorobutyl elastomer was selected as a damping material because it
offered a good combination of desired viscoelastic behavior, aging
characteristics, and tensile and tear properties (14).

Laboratory batches of compounded rubber samples were obtained
from either Smithers Laboratories of Akron, Ohio, or Mare Island
Naval Shipyard. Proper cure conditions were determined for each
compound as per ASTM D 20845-79 through the use of a Monsanto 100S
oscillating disk rheometer. Samples used for experimental purposes
were cured to 90% of optimum cure. Density was determined by
Archimedes' method.

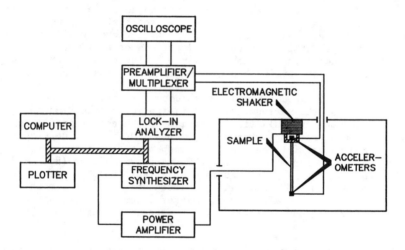

Fig. 1. Block diagram of measurement system for Young's modulus.

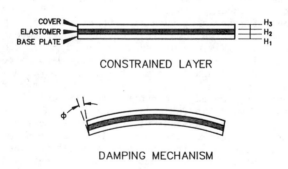

CONSTRAINED LAYER

DAMPING MECHANISM

Fig. 2. Schematic of three member constrained-layer structure
and mechanism of loss.

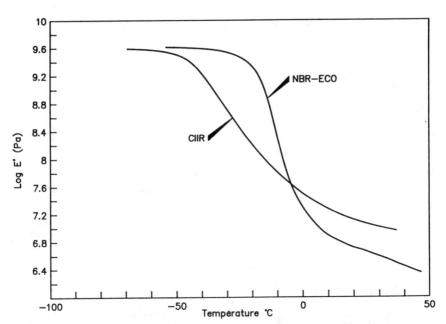

Fig. 3.   Plot of Young's storage modulus versus temperature in
degrees Centigrade at 10 Hz for nitrile-epichlorohydrin blend
and chlorobutyl rubber.

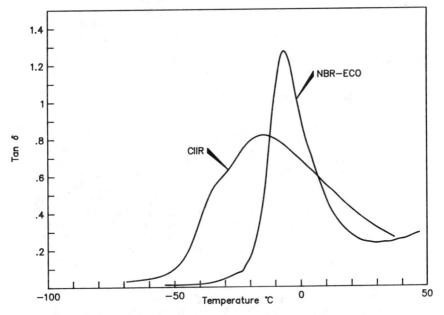

Fig. 4.   Plot of loss tangent versus temperature in degrees
Centigrade at 10 Hz for nitrile-epichlorohydrin blend and
chlorobutyl rubber.

TABLE I. FORMULATIONS FOR CIIR AND NBR-ECO RUBBERS

COMPONENT	(P.H.R.)	COMPONENT	(P.H.R)
Chlorobutyl HT1066	100.0	Hycar 1052	80.0
Zinc Oxide	5.0	Hydrin T	20
Schenectady SP1055	4.0	Agerite Resin D	2.0
MBTS	2.0	Saret 500	2.0
Antioxidant 2246	1.0	Dyphos	2.0
Stearic Acid	1.0	Stearic Acid	1.0
Diphenylguanidine	0.5	Dicup 40C	3.0
Poly DNB	0.5	Struktol WB 300	10.0
N347 Black	35.0	N990 Black	65.0
		Elast-O-Cal C75	2.5
Total:	144.0		187.5
Cure time (min.) @ 155 °C	32		32

The approach taken was to measure the viscoelastic properties
of the polymer as a function of temperature and frequency, using the
transfer function technique, and use these in the Ross-Kerwin-Ungar
(RKU) model (12, 15) to calculate the expected structural loss
factor in the composite structure. The Ross-Kerwin-Ungar model is a
generalized solution for a three-layer beam configuration. The
problem has been simplified by assuming that the extensional
stiffness of the viscoelastic layer is small and that shear
deformations of the base plate and constraining layer are
negligible. It is also assumed that the constraining material is
purely elastic and dissipates no energy. Linear behavior of the
viscoelastic layer is assumed, with uniform shear effects throughout
the viscoelastic layer. The loss factor of the three-layer beam is
calculated by taking the ratio of the imaginary part of the bending
rigidity to the real part.

In order to test the predictions of the model, plates were made
up from selected formulations and tested in air. The thicknesses of
the various layers were as stated previously, while the length was
one meter and the width was one-third meter. Damping measurements
were made by two different methods. In both cases, the plates were
suspended in a shock chord and accelerometers placed at different
locations on the plate. The first used a reverberation meter and
the half-power method (16). In the second, an impact hammer was
used to tap the structure and the outputs of the accelerometers fed
into a Fourier transform based spectrum analyzer to examine the
envelope of vibration (17). The results presented here are based
upon the second method.

An HAF carbon black, N347, was selected for use as the primary
filler. This was primarily because HAF blacks gave satisfactory
adhesion of vulcanized chlorobutyl rubbers to metal, and good
tensile and tear properties.

In order to obtain maximum damping from a given constrained layer configuration, it is desirable that the Young's or shear modulus vary approximately linearly with frequency, and that the loss tangent be as high as possible. In order to determine the loadings of fillers that should be used, the RKU model was used to predict the damping that would be obtained from polymers with different dynamic mechanical properties. Figure 5 shows these. Figures 6 and 7 show the effect of increased carbon black upon the storage modulus and loss tangent. Comparison of Figure 5 with Figures 6 and 7 indicates that the optimum damping will be obtained from a material which has a storage modulus varying from approximately 10^7-10^8 Pa over the 100 Hz to 10 kHz frequency, and a relatively high loss tangent. This implies that relatively low loadings of fillers (on the order of 35 phr or less) will give the best damping in this particular constrained-layer configuration.

For broad-band damping, the glass transition region should be as diffuse as possible. For many polymers, use of reinforcing fillers will broaden the transition region, but result in an overall decrease of the magnitude of the loss tangent. As shown in Figures 6 and 7, this has been found to be the case with chlorobutyl (14). Alternatively, use of platelet types of fillers such as mica and graphite may serve to introduce additional loss mechanisms (18-20) and increase the overall dampening, while simultaneously broadening the transition region.

Carbon black at various loadings, as well as in combination with mica, and mica and flake graphite, was used. The range of loadings used is shown in Table II. The graphite used was Cummings Moore 6894 graphite, while the mica used was Micro-Mica C1000. This has a platy particle shape, with particle size of 10 to 20 microns in diameter and a theoretical mesh size of 1000.

It was found that lower loadings (20-35 phr range) of carbon black alone did not give damping in the constrained layer structures that was as good as that obtained with combinations of mica and graphite. Figures 8 and 9 show the viscoelastic behavior of two different loadings of fillers at a total loading of 35 parts per hundred (phr) of rubber. Use of mica gives a material that is dynamically softer, but with a higher loss tangent than carbon black alone. Adding an additional 20 phr of graphite gives a material that becomes dynamically stiffer, although not as dynamically stiff as the carbon black alone, but with a higher loss tangent (Figures 10 and 11). Of all the formulations tested, these two gave the best overall damping and tensile performance. Use of higher loadings of graphite gave higher loss tangent values, but gave very poor cohesive strength in the rubber.

TABLE II.   TYPE AND RANGE OF LOADING OF FILLERS USED IN CIIR RUBBER

FILLER :	P.H.R.
Vulcan 3H (N347)	0-50
Micro-Mica C1000	0-45
Cummings-Moore #6894 Graphite	0-40

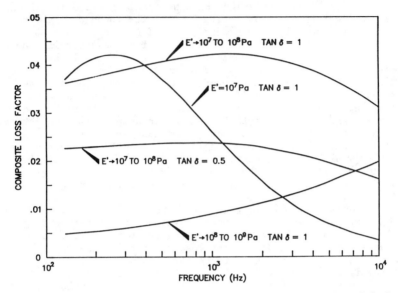

Fig. 5. Calculated structural loss factor based on RKU model for
three-layer composite with varying Young's storage modulus and
loss tangent of viscoelastic layer.

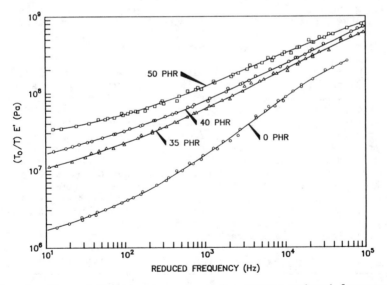

Fig. 6. Plot of Young's storage modulus versus reduced frequency
at 10 degrees Centigrade as a function of carbon black loading
for chlorobutyl rubber. (Reproduced from Ref. 14. Copyright
1986 ACS). (PHR = parts per hundred of rubber).

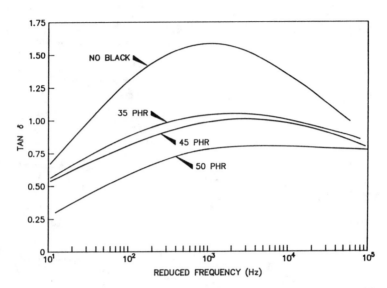

Fig. 7. Plot of loss tangent versus reduced frequency at 10 degrees Centigrade as a function of carbon black loading for chlorobutyl rubber. (Reproduced from Ref. 14. Copyright 1986 ACS). (PHR-- parts per hundred of rubber).

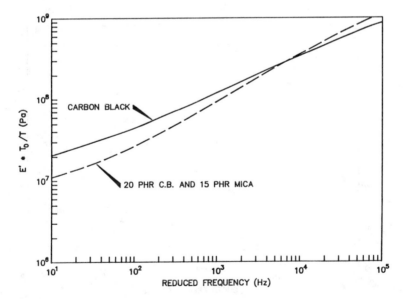

Fig. 8. Comparison of Young's storage moduli as a function of reduced frequency at 5 degrees Centigrade for chlorobutyl reinforced with 35 parts per hundred of rubber (PHR) of N347 carbon black, and 20 PHR carbon black and 15 PHR of mica.

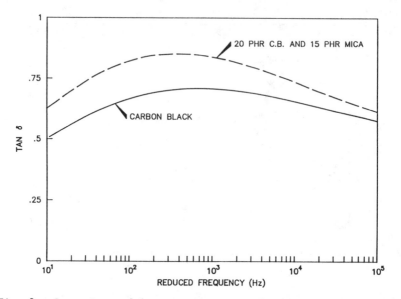

Fig. 9. Comparison of loss tangents as a function of reduced frequency at 5 degrees Centigrade for chlorobutyl reinforced with 35 parts per hundred of rubber (PHR) of N347 carbon black, and 20 PHR carbon black and 15 PHR of mica.

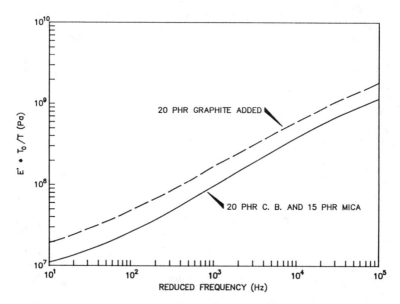

Fig. 10. Comparison of Young's storage moduli as a function of reduced frequency at 5 degrees Centigrade for chlorobutyl reinforced with 20 PHR carbon black and 15 PHR of mica, and with an additional 20 PHR of graphite added.

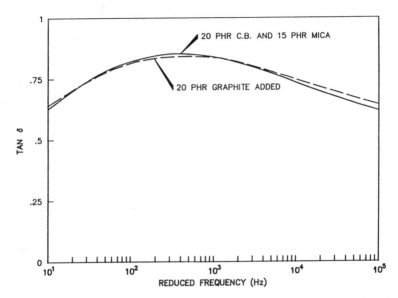

Fig. 11. Comparison of loss tangents as a function of reduced frequency at 5 degrees Centigrade for chlorobutyl reinforced with 20 PHR carbon black and 15 PHR of mica, and with an additional 20 PHR of graphite added.

In general, it was found that the RKU model gave only qualitative agreement with the experimental damping results. The model accurately predicted that the damping obtained from the formulation with carbon black and mica would be better at lower temperatures than when graphite is added. This is shown in Figures 12 and 13. This indicates that the formulation with graphite is too stiff dynamically at lower temperatures, even though the two formulations have the same loss tangent. The model predicted that the damping would be lower than the experimental values at both 20 and 5 degrees Centigrade, while the agreement was much better at 20 degrees Centigrade (Figures 14 and 15). In all instances the model was able to qualitatively predict the relative merits of the damping capability of different polymer layers. As the model predicted, relatively soft, lightly filled formulations gave the best damping performance.

The exact reasons for differences between the calculated and measured damping are not clear. There are several spectulations that can be offered. One of these is that the RKU model neglects any damping effects that are due to the adhesive and the adhesive interface. At certain temperatures, the adhesive may contribute a significant portion of the damping in the composite structure. The viscoelastic materials used in testing the Young's modulus and loss tangent were also compounded in diferent batches of rubber than those used in the composite structures. This might have resulted in significant differences in the viscoelastic properties of the two groups of materials.

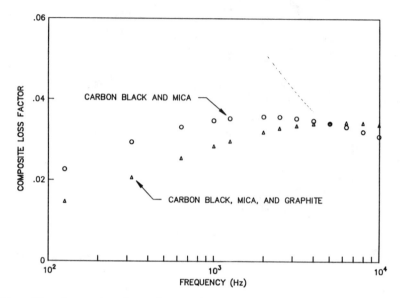

Fig. 12. Composite loss factor for three-layer constrained
layer assembly as a function of frequency at 5 degrees Centigrade
as predicted by the RKU model for the formulations listed in
Figs. 10 and 11.

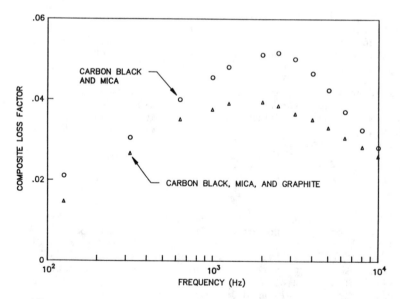

Fig. 13. Experimentally determined structural loss factor for
three-layer constrained layer assembly as a function of frequency
at 5 degrees Centigrade for the formulations listed in Figs. 10
and 11.

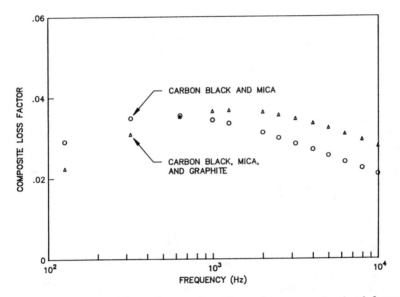

Fig. 14. Composite loss factor for three-layer constrained layer assembly as a function of frequency at 20 degrees Centigrade as predicted by the RKU model for the formulations listed in Figs. 10 and 11.

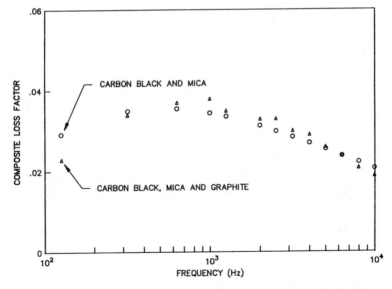

Fig. 15. Experimentally determined structural loss factor for three-layer constrained layer assembly as a function of frequency at 20 degrees Centigrade for the formulations listed in Figs. 10 and 11.

SUMMARY

A transfer function technique for the determination of the dynamic Young's modulus and loss tangent has been described. Algorithms for performing time-temperature superposition have been discussed.

　　The development of constrained-layer damping materials through the use of dynamic mechanical testing and mathematical modeling has been described. It has been shown how different types and loadings of fillers will affect the measured viscoelastic properties of chlorobutyl rubbers. It has then been shown how these changes will affect the damping performance of these materials in constrained layer structures.

ACKNOWLEDGMENTS

The authors would like to thank Mr. Douglas Noll of David Taylor Research Center for performing the plate damping measurements.

Literature Cited.

1.　Capps, R.N. J. Acoust. Soc. Am. **73**, 2000 (1983).
2.　Norris, Jr., D.M.; Young, W.C. Exp. Mech. **10**, 93 (1970).
3.　Snowden, J.C. "Vibration and Shock in Damped Mechanical Systems," Wiley, New York, 1968.
4.　Pritz, T, J. Sound and Vibration **81**, 359 (1982).
5.　Buchanan, J.L. J. Acoust. Soc. Am. **81**, 1775 (1987)
6.　Ferry, J.D. "Viscoelastic Properties of Polymers," Wiley, New York, 1980, 3rd. ed., Ch. 11.
7.　Yin, T.P.; Pariser, R. J. Appl. Polym. Sci. **8**, 2427 (1964).
8.　Jones, D.I.G. J. Sound and Vib. **33**, 451 (1974).
9.　Rogers, L. J. Rheology **27**, 351 (1983).
10.　Ungar, E.E. "Damping of Panels," in "Noise and Vibration Control," L. L. Beranek, Ed., McGraw-Hill, New York, 1971, Ch. 14.
11.　Kerwin, Jr., E.M. J. Acoust. Soc. Am. **31**, 952 (1959).
12.　Ross, D.; Ungar, E.E.; Kerwin, Jr., E.M. "Damping of Plate Flexural Vibrations by Means of Viscoelastic Laminae," Section III of "Structural Damping," J. E. Ruizcka, Ed., Am. Soc. Mech. Eng., New York, 1959.
13.　Capps, R.N.; Beumel, L.L. "Influence of Fillers on Constrained-Layer Vibration-Damping Capabilities of Chlorobutyl Elastomers", J. Acoust. Soc. Am. **83**, S82(A), 1988.
14.　Capps, R,N. Rubber Chem. Technol. **59**, 103 (1986).
15.　Noll, D.A.; Maga, L.J.; Palladino, D.J. "Computer Programs for Calculation of Structural Loss Factors Using the Ross, Kerwin, and Ungar Model", Research and Development Report DTNSRDC/SAD-87/69E-1944, September 1987, David. W. Taylor Research Center, Bethesda, MD, 20084-5000.
16.　Plunkett, R. "Measurement of Damping," Section V of "Structural Damping," E. V. Ruzicka, Ed., Am. Soc. Mech. Eng., New York, 1959.
17.　Schulze, K.D. "Investigation of Damping Characteristics of Constrained Layer Plates and Small Homogenous Specimens," Thesis, September 1985, Naval Postgraduate School, Monterey, CA 9343-5100.
18.　Ball, G.L.; Salyer, I.O. J. Acoust. Soc. Am. **36**, 386(1964).
19.　Wong, D.T.; Williams, H.L. J. Appl. Polym. Sci. **28**, 2187 (1983)
20.　Klempner, D.; Frisch, K.C., Eds. "Polymer Alloys III", Plenum Press, New York, 1983.

RECEIVED September 26, 1989

Chapter 5

Direct Method for Measuring the Dynamic Shear Properties of Damping Polymers

S. S. Sattinger

Mechanics Department, Westinghouse Science and Technology Center, 1310 Beulah Road, Pittsburgh, PA 15235

A non-resonance, direct-force method for dynamic shear properties measurements is described, and the results of tests on two commercially available damping polymers are presented. Novel aspects of this method include the means for supporting the sample and for measuring the imposed force and the resultant shear deformation. Addressed in this article are the test configuration, the principle of operation, the data reduction procedure, some typical measured properties, consistency checks on the data, and a brief description of an initial application of the data.

To generate accurate damping performance predictions, designers of constrained-layer damping treatments must have a good knowledge of the dynamic shear properties of the viscoelastic damping polymers they will use. The conventional characterization of these damping materials is the complex shear modulus,

$$G^* = G' + jG'' = G' \ (1 + j\eta) \tag{1}$$

where G' is the real part of this modulus (often designed as the storage modulus), G'' is the quadrature part (often designated as the loss modulus), and η is the material damping loss factor. The quantity η is frequently expressed as tan δ, where δ is the phase angle between the stress and strain phasors.

Dynamic properties tests on viscoelastic materials fall into the general categories of resonance tests and non-resonance tests (1). They can be further subdivided into tests using base motion excitation (2), (3) and direct force excitation (4), (5). Still another classification may be made according to whether the specimen is stressed in extension, shear, or dilatation. An important issue in the selection of test methods is whether it will

0097–6156/90/0424–0079$06.00/0

be necessary to obtain data well into the rubbery and glassy
regions of temperature and frequency for a given material.

The measurement method described in this article is an
embodiment of the non-resonance, direct-force-excitation approach
that subjects a double-lap shear sample of damping polymer to force
from a vibration shaker. In concept this approach can be applied
irrespective of whether the material is in a rubbery, glassy, or
intermediate state. Each material specimen is small in size and
behaves as a damped spring over the entire frequency range. The
small specimen size is in contrast with some alternate approaches
in which the specimens have sufficiently large dimensions to be
wave-bearing.

This method combines the advantages of simplicity, direct
control of frequency, and minimum reliance on mathematical modeling
assumptions. It differs from other embodiments of the direct-force
approach (4), (5) in the means for reacting the forces applied to
the sample and in the techniques for measuring the imposed force
and the resultant shear deformation. Described in this article are
the test configuration, the principle of operation, the method of
data reduction, typical measurements, consistency checks, and an
application of measured data.

Test Configuration

Figure 1 shows the entire test system in schematic form. The test
sample is comprised of two specimens of a viscoelastic damping
polymer loaded in shear by force from an electromagnetic shaker.
Each specimen is cemented between a centrally located, driven steel
sample block and one of two clamped reaction blocks. The
dimensions of each block in this apparatus are 25.4mm height,
38.1mm width, and 12.7mm thickness (1.00 in x 1.50 in x 0.50 in),
but the specimens do not necessarily cover the full areas of the
25.4mm x 38.1mm faces of the blocks.

Excitation force is transferred from the armature of the shaker
to the driven sample block through a piezoelectric force gage. The
reaction blocks of the sample are bolted to a steel fixture that
transfers the sample reaction force back to the shaker field
assembly through a set of thermally-insulating fiberglass/ epoxy
standoffs. The entire force path upward to the sample and then
downward into the field assembly is designed to be as stiff as
possible. Not shown in the figure is a pair of spacers that
preclude beam-mode lateral vibrations of the fixture legs
supporting the sample.

At frequencies in the initial range of interest (100 Hz to
1 kHz), accelerometers are the preferred means of motion
measurements. Because the stiff design of fixturing does not
ensure total immobility of the reaction blocks, signals from a pair
of accelerometers, one on the driven block and one on a reaction
block, are differenced electronically by an operational amplifier.
The resultant difference signal provides an accurate measure of
shear deflection in the specimens. The signal from a third
accelerometer, which is mounted on the specimen-side flange of the
force gage, is used to electronically compensate the force signal
for the effects of specimen-side force gage mass. Sinusoidal
dwells are used to maximize signal/noise ratios.

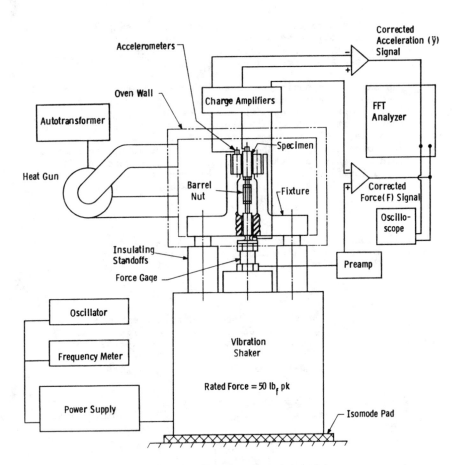

Figure 1.   Test system for dynamic shear property measurements on viscoelastic damping polymers.

Data Reduction

The complex shear modulus of the viscoelastic damping polymer is
obtained directly from measurements on the sample using the
relationship

$$G^* = \frac{k^* h}{2A}$$ (2)

where k^* = the complex dynamic stiffness of the sample
  h = the thickness of each of two damping polymer specimens
     included in the sample
  A = the shear area of each specimen.

The sample's dynamic stiffness is obtained from the measured
force and motion amplitudes. The formula for this dynamic
stiffness is derived below with reference to the Figure 2
measurement system model. The support impedance element in this
model represents the combined effects of the flexibility of the
downward force train into the shaker field assembly, the mass of
the field assembly, and the oscillating magnetic reaction force
exerted on the field assembly from the armature.
The sample block's differential equation of motion is

$$m\ddot{x} + k^*(x-x_o) = F$$ (3)

where the double dot superscript denotes the second derivative with
respect to time, and all quantities are defined in Figure 2.
Equation 3 can be rewritten as:

$$m\ddot{y} + k^* y = F - m\ddot{x}_o$$ (4)

where $y = x-x_o$ is the vertical (shear) displacement across the
thickness of each specimen. Under steady-state sinusoidal
excitation, the time varying quantities can be expressed as:

$$y = \bar{Y} e^{j\omega t}; \quad F = \bar{F} e^{j\omega t}; \quad x_o = \bar{X}_o e^{j\omega t}$$ (5)

where $\omega = 2\pi f$ is the excitation frequency in radians/sec and the
barred quantities are complex amplitudes. Substituting (5) into
(4) gives:

$$k^* = \frac{\bar{F}}{\bar{Y}} + \omega^2 m \left[1 + \frac{\bar{X}_o}{\bar{Y}}\right].$$ (6)

Observation of the accelerometer signals shows that the
support motion, x_o, is considerably smaller in amplitude than the
shear displacement, y. In addition, the driven block mass
correction term, $\omega^2 m$, is small in comparison with the ratio $\bar{F}/\bar{Y}$ for
tests at frequencies well below resonance of the sample.
Therefore, the approximately-corrected complex stiffness value can
be expressed as:

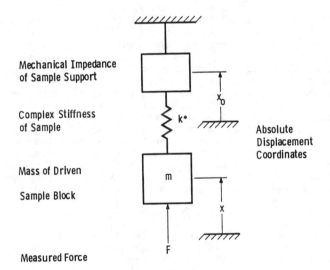

Figure 2.   Dynamic model of measurement system.

$$k^* \simeq \frac{\overline{F}}{\overline{Y}} + \omega^2 m \qquad (7)$$

Combining (1), (2), and (7) yields the results

$$G' = \frac{h}{2A}\left[Re\left(\frac{\overline{F}}{\overline{Y}}\right) + \omega^2 m\right]; \quad \eta = \frac{Im\left(\frac{\overline{F}}{\overline{Y}}\right)}{Re\left(\frac{\overline{F}}{\overline{Y}}\right) + \omega^2 m} \qquad (8)$$

where Re (z) denotes the real component of the complex quantity z
and Im (z) denotes the imaginary component. These components are
obtained as output data from the FFT analyzer, which is operated in
transfer-function mode. The analyzer applies a $(j\omega)^{-2}$ frequency-
domain weighting to convert the amplitude of the measured
acceleration difference, $\ddot{y}$, to a displacement amplitude, Y.

An alternate arrangement that would relocate the force gage
from the shaker armature to the driven sample block could be
advantageous, provided that the force gage sensitivity remains
constant over the required temperature range. This relocation
would eliminate the need to take account of the mass correction
term, $\omega^2 m$, in the reduction of data. Instead, this mass loading
could be electronically compensated by using the driven-block
accelerometer signal to correct the force signal, eliminating the
need for the the third accelerometer in the Figure 1 arrangement.
Another alternate would be to use an impedance head, but space
limitations could rule out this choice in some instances.

Examples of Measurements and Consistency Checks

Figures 3 and 4 show the results of tests completed on single lots
of the commercially available Soundcoat Dyad 606 and 3M ISD 112
damping polymer films. The measured results are plotted without
the application of any time or temperature shifts. For each
material, one sample was assembled using two specimens of the full
face dimensions of the sample block. The 1.27-mm (0.050-in.)-thick
Dyad specimens were cemented to the sample blocks using Soundcoat
B-Flex epoxy, whereas the 0.051-mm (0.002-in.)-thick 3M specimens
used 3M Type 1838 epoxy.

For comparison, Figures 3 and 4 include typical property
values reported by the polymer manufacturers for corresponding
conditions (6-8). Reference 8 is a recent update of 3M damping
polymer properties previously furnished in Reference 7.
Differences between the measurements and the manufacturers'
reported storage modulus values are large in some instances. The
causes for these differences are not known, nor are the extent of
lot-to-lot property variations for these materials.

Figures 5 and 6 examine the 3M material measurements for
consistency of temperature and frequency effects using plot formats

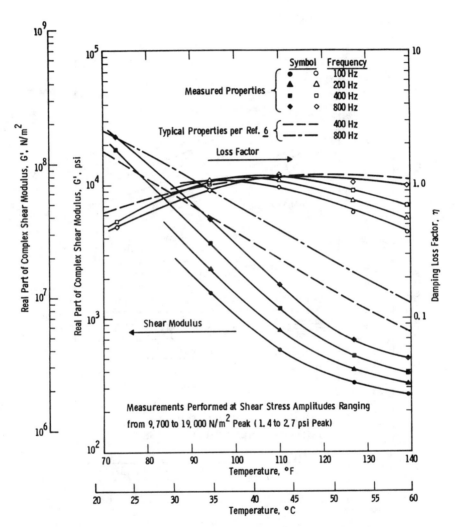

Figure 3. Dynamic shear properties for Soundcoat Dyad 606 damping polymer from tests with 1.27-mm (0.050-inch) nominal specimen thickness.

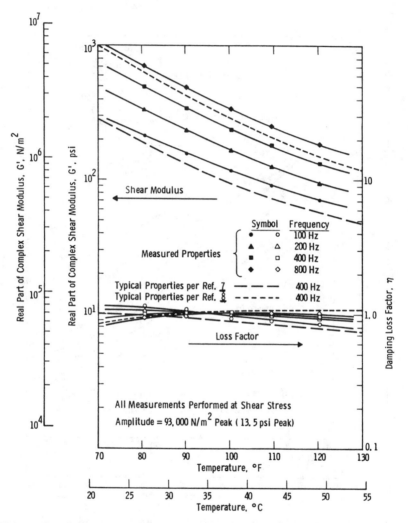

Figure 4.   Dynamic shear properties for 3M ISD 112 damping polymer
from tests with 0.051-mm (0.002-inch) nominal specimen thickness.

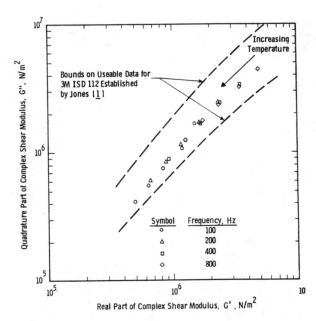

Figure 5.   Self-consistency check on measured shear properties data: quadrature part vs. real part of complex modulus for 3M ISD 112.

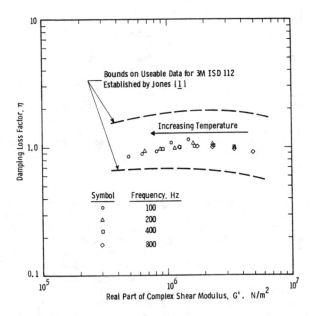

Figure 6.   Self-consistency check on measured shear properties data: damping loss factor vs. real part of complex modulus for 3M ISD 112.

recommended by Jones (1). In the absence of experimental errors, the loss modulus and the loss factor should be uniquely related to the storage modulus, irrespective of temperature and frequency, for low cyclic strain levels. Except for two data points for the lowest-frequency conditions in Figure 6, the present measurements satisfy these checks. Also shown in these figures are curves showing bounds on usable data for 3M ISD 112. These limits were established by Jones on the basis of a survey of measurements performed by a number of investigators employing a variety of different test methods. The present measurements fall well within these bounds.

Application of Data

The test apparatus was developed to aid in resolving discrepancies between calculations and measurements in a study of the extensional vibration damping performance of constrained-layer damping treatments (9). That study demonstrated that constrained-layer treatments can be designed to be effective in dissipating not only flexural but also extensional strain energy. The viscoelastic material layer in a finite-length segment of damping treatment can be made to sustain shear due to oscillatory extension, just as occurs in treatments applied to members in flexure. Figure 7 shows the distinction between these two mechanisms of structural damping. The extensional damping action of constrained-layer treatments is not to be associated with free-layer or unconstrained-layer treatments, which have occasionally been referred to as extensional damping treatments.
     In the Reference 9 study a number of barbell-shaped test assemblies of differing designs had been specially constructed using 3M ISD 112 damping polymer. Figure 8 shows a design typical of those evaluated. Figure 9 illustrates the sizeable differences between measured system-damping values and initial predictions (the dashed curve) generated for this design using property values from Reference 7. The 3M ISD Type 112 shear properties tests described in this article were subsequently performed on the same lot of material as used in the barbell test assemblies. When the system damping performance was recalculated using the measured properties, the revised calculations (the solid curve) matched the performance measurements very closely. This close agreement helped confirm the method that had been identified for generating extensional damping performance predictions.

Discussion

To date this method has been applied in measurements of polymer properties in their glass transition regions, where there is high damping. In concept these applications could be readily extended to measurements well into the glassy and rubbery regions. It could prove advisable to use specimens of more than one geometric configuration to accommodate variations in stiffness over wide ranges of temperature and frequency on a given polymer.
     The frequency range over which accurate data may be obtained is bounded, on the lower end, by the frequency response characteristics of the force gage and accelerometers, and on the

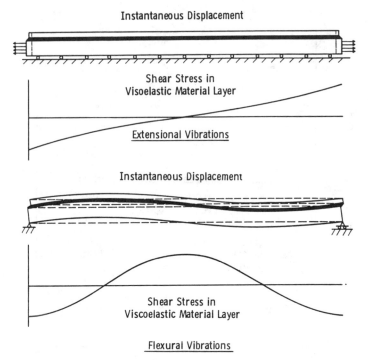

Figure 7. Constrained-layer damping of extensional and flexural vibrations of bars and plates.

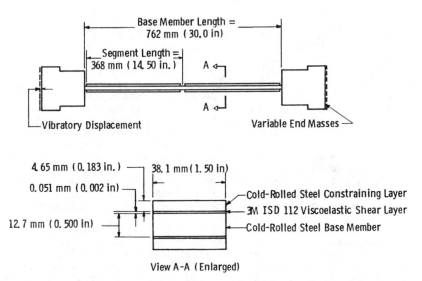

Figure 8. Design of a barbell test assembly used to evaluate the extensional damping performance of constrained-layer treatments.

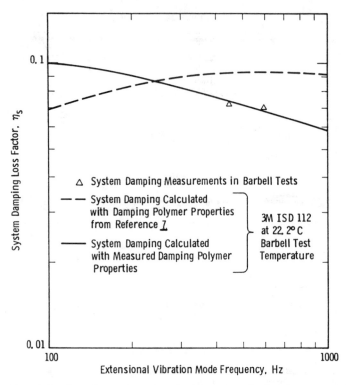

Figure 9. Calculated vs. measured extensional damping performance for the Figure 8 test assembly.

upper end, by the approach of sample resonance, at which the force signal becomes small. The resonant frequency of the sample was well above the excitation frequency at every data point shown. The 1-kHz upper frequency limit demonstrated here is well above the 100-Hz limited cited by Jones (1) for direct measurement methods. In principle the frequency range could be extended upward into sample resonance, but it would become necessary to add processing of the support motion signal due to the invalidation of Equation 7. The lower limit of the frequency range could also be extended further downward by the simultaneous use of a displacement transducer to sense shearing motion at frequencies where accelerometer data becomes unreliable. A substitution for the piezoelectric force gage might also be required for lowering this limit. The temperature sensitivity of both the force and the motion instrumentation is an issue of concern in any apparatus of this kind.

No universal standard appears to exist for the selection of specimen height in relation to its thickness. Adkins (2) attributes about a 1% decrease in shear stiffness to end effects at a height-to-thickness ratio of 24. Parin, et al (10) recommends a height-to-thickness ratio greater than 4. The values of this ratio were 20 and 500 in the present tests on the Soundcoat and the 3M materials, respectively.

Measured storage modulus values for individual samples can differ substantially from typical values reported by manufacturers of damping polymers. The causes of these deviations have not been identified.

Conclusions

The primary advantage of this method of measuring damping polymer properties is its simplicity. This apparatus generates accurate measurements by minimizing unwanted inertia forces and sample support motions, yet taking them into account in the process of data reduction. The need for mathematical modeling assumptions on deformation or other sample behavior is minimized. A wide range of material stiffnesses can be accommodated, and direct control of test frequency conditions is afforded.

Acknowledgments

The encouragement and support of Mr. P. C. Warner of the Westinghouse Marine Division, sponsor of this investigation, are gratefully acknowledged. K. B. Wilner, D. V. Wright, and J. M. Zomp of the Science and Technology Center are also thanked for their contributions. Correspondence with Dr. David Jones of the Air Force Materials Laboratory regarding the validation of data was very helpful.

Literature Cited

1. Jones, D.I.G. Damping 1986 Proceedings, AFWAL-TR-86-3059, U.S. Air Force Flight Dynamics Laboratory, 1986.
2. Adkins, R. L. Experimental Mechanics, 1966, 23, 362-7.
3. Smith, G. M.; Bierman, R. L.; Zitek, S. J. Experimental Mechanics, 1983, 40, 158-63.
4. Coote, C. T. Journal of Sound and Vibration, 1972, 21, 133-47.
5. Ganeriwala, S. N.; Rotz, C. A. Proc. Am. Chem. Soc. Polym. Mat. Sci. Eng., 1984, 50, 37.
6. Constrained-Layer Damping Materials for Control of Noise and Vibration, Soundcoat Company, Inc., Publication 118010MTP.
7. Product Information-Scotchdamp, 3M Company, Publication SP-SD-1 (50.25).
8. Product Information: Scotchdamp Vibration Control System, 3M Industrial Specialties Division, Bulletin 70-0702-0235-6 (18.05) CFD 257A.
9. Sattinger, S. S. in The Role of Damping in Vibration and Noise Control, American Society of Mechanical Engineers: Technical Publication H00405, 1987; pp. 33-40.
10. Parin, M. L.; Nashif, A. D.; Lewis, T. M. Damping 1986 Proceedings, AFWAL-TR-86-3059, U.S. Air Force Flight Dynamics Laboratory, 1986.

RECEIVED January 24, 1990

Chapter 6

Fourier Transform Mechanical Analysis and Phenomenological Representation of Viscoelastic Material Behavior

S. N. Ganeriwala and H. A. Hartung

Research Center, Philip Morris, USA, Richmond, VA 23261-6583

Fourier transform mechanical analysis (FTMA) measures material properties over broad frequency spectra by using random noise input. Modulus-frequency isotherms are obtained in just a few seconds. FTMA is superior for characterizing moisture and additive effects because it can be carried out with minimal temperature and moisture changes. Data are obtained over a series of temperatures and then regressed with an analytical sigmoid-shaped function for modulus-temperature responses. This yields parameters with clear, distinct physical meaning which can be correlated with material variables. The groundwork needed for comprehensive studies of moisture and plasticizers in viscoelastic behavior is provided.

The viscoelastic properties of polymers make them valuable for suppression of sound and vibration. A comprehensive, useful understanding of the viscoelastic damping inherent in these systems can come only from studies of mechanical properties over wide ranges of time (frequency) and temperature. If materials are moisture sensitive, the effects of water activity also should be determined. A similar rule holds for plasticizers and solvents.

Techniques for measuring the dynamic mechanical properties of polymers are tabulated in figure 1 (1-9). Although each category has advantages and disadvantages, the forced vibration methods are now preferred for basic dynamic studies and tests (6). A few of the commercially available instruments for measurements of dynamic properties are listed in figure 1. Each instrument has some advantages and limitations depending upon the material, temperature and frequency range, accuracy, resolving power, and the information sought (7).

Typically, these instruments measure dynamic mechanical responses to sinusoidal input. To characterize the viscoelastic properties of a material, these tests must be repeated over a range of temperatures and frequencies. This is sometimes done at a fixed frequency while the polymer specimen is heated or cooled and

0097–6156/90/0424–0092$06.00/0
© 1990 American Chemical Society

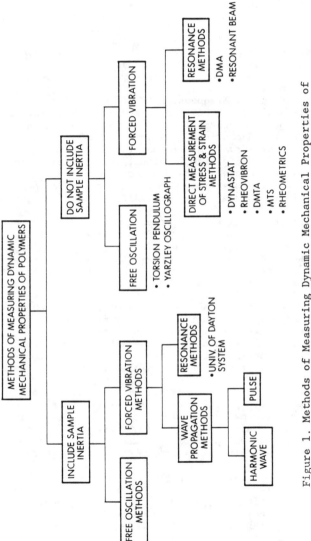

Figure 1. Methods of Measuring Dynamic Mechanical Properties of Polymers. Commercially available instruments are listed by schemes of measurement.

measurements are made periodically at different temperatures.
Another method utilizes frequency variations while the temperature
is held constant. In both procedures, the material is subjected to
cyclic deformation over a period of time with uncontrolled
temperature rise, loss of volatiles and other changes from energy
dissipated in the material. These effects may be compounded by lags
of sample from ambient temperature if heating is carried out at
constant frequency. Thus it is difficult to obtain truly isothermal
properties using most of the commercial instruments.

     Another problem with these instruments is determination of
mechanical properties of moisture sensitive materials (10-12). Such
materials have tendency to gain moisture when subjected to
mechanical excitation at constant relative humidity and temperature.
This makes single frequency tests impractical for isomoisture
studies over a range of frequencies.

     Spectral analysis techniques to study the behavior of polymers
subjected to dynamic mechanical loads and/or deformation is called
Fourier Transform Mechanical Analysis (FTMA). FTMA measures the
complex moduli over a range of frequencies in one test by exciting
the sample by a random signal (band limited white noise) (13,14).
FTMA overcomes or circumvents problems inherent in other test
methods because it measures dynamic mechanical properties over a
wide range of frequency with minimal temperature and moisture
changes within the sample.

     Often the range of frequency covered by an instrument is not
large enough to fully analyze the dynamic mechanical response of a
material system. The principle of time-temperature superposition is
then utilized to obtain master curves that presumably approximate a
mechanical response isotherm for a wide range of time (or frequency)
(15-20). The underlying assumption is that the mechanical response
of a material at all temperatures is governed by the same
viscoelastic mechanism. This assumption has been found unsuitable
in many cases (19,20). However, it is very much in use and it does
provide a limited, semi-empirical perspective on the effects of time
(or frequency) and temperature. It is not very suitable for
treating the obvious effects of moisture, plasticizers, molecular
weight, etc. Also, it does not provide a predictive, constitutive
model.

     With the long term objective of treating the effects of
moisture and other plasticizers on the mechanical properties of
materials, a new scheme that yields a complete constitutive model of
viscoelastic materials has been developed. The time-temperature
principle is an integral part of this modeling with a quantitative
description of the glass transition behavior of polymers.

<u>Theory</u>

<u>Linear Viscoelasticity Theory</u>. FTMA is based on linear
viscoelasticity theory. A one dimensional form of constitutive
equation for linear viscoelastic materials which are isotropic,
homogeneous, and hereditary (non-aging) is given by (21):

$$\sigma(t) = \int_{-\infty}^{\infty} G(t,t-t')\dot{\gamma}(t')dt' \qquad (1)$$

where the kernel $G(t-t')$ is a monotonic nonincreasing function of time known as the stress relaxation modulus, $\sigma(t)$ is current stress, and $\dot{\gamma}(t')$ is the strain rate history.

For the case of sinusoidal strain history Equation 1 can be transformed to yield an expression for the complex modulus, $G^*(j\omega)$:

$$G^*(j\omega) = G'(\omega) + jG''(\omega) \tag{2}$$

where $j = \sqrt{-1}$, ω is the frequency, and $G'(\omega)$ and $G''(\omega)$ are the storage modulus and loss modulus, respectively. G' is related to the amount of energy stored and released in a cyclic oscillation and G'' indicates the energy dissipated. In damping applications Equation 2 is expressed:

$$G^*(j\omega) = G'(1 + j\eta) \tag{3}$$

where η is designated as the material damping factor defined as the ratio of G'' over G'.

When the strain $\gamma(t)$ is sinusoidal in time with an amplitude of γ_0:

$$\gamma(t) = \gamma_0 \sin \omega t \tag{4}$$

and stress response $\sigma(t)$ will be

$$\sigma(t) = \sigma_0(\omega) \sin [\omega t + \delta(\omega)] \tag{5}$$

where $\sigma_0(\omega)$ is the stress amplitude and $\delta(\omega)$ is the phase angle between stress and strain. Note that $\eta = \tan \delta$. Then the storage and loss moduli are given by (13,21)

$$G'(\omega) = \frac{\sigma_0}{\gamma_0}\cos\delta \tag{6}$$

$$G''(\omega) = \frac{\sigma_0}{\gamma_0}\sin\delta \tag{7}$$

Thus, dynamic mechanical viscoelastic properties may be measured in tests with sinusoidal strain input at fixed frequency. Such tests have to be repeated at different frequencies over the range of interest to completely characterize the material.

FTMA Formulation. When $\gamma(t)$ is an arbitrary function of time such that its Fourier transform $\Gamma(\omega)$ exists. Then (22):

$$\gamma(t) = \frac{1}{2\Pi} \int_{-\infty}^{\infty} \Gamma(\omega)\, e^{j\omega t} d\omega \tag{8}$$

and

$$\Gamma(\omega) = \frac{1}{2\Pi} \int_{-\infty}^{\infty} \gamma(t)\, e^{-j\omega t} dt \tag{9}$$

and there are similar coupled relationships between $\sigma(t)$ and its Fourier transform $\Sigma(\omega)$. Then (13):

$$G^*(j\omega) = \frac{\Sigma(\omega)}{\Gamma(\omega)} \qquad (10)$$

Thus, FTMA determines complex modulus as the transfer function between input strain and output stress. A prerequisite is that the Fourier transform of $\gamma(t)$ must exist. White noise should suffice since it contains all frequencies. Note that $G^*(j\omega)$ in Equation 10 will be the complex Young's modulus if $\sigma(t)$ and $\gamma(t)$ are the normal stress and normal strain, respectively; and the complex shear modulus if they are the shear stress and shear strain.

Samples and Measurements. A scheme for shear measurements is shown in figure 2. Two identical polymer samples of length L, thickness h, and width W are bonded to two rigid metal mounts. The inner mount is attached to an impedance head (a combination force and acceleration sensor), which in turn is attached to a shaker. An accelerometer is attached to the top center of the outer mount.

If f(t) is the total shear force exerted on the polymer sample then the shear stress $\sigma(t)$ is

$$\sigma(t) = \frac{f(t)}{2LW} \qquad (11)$$

and thus

$$\Sigma(\omega) = \frac{F(\omega)}{2LW} \qquad (12)$$

where $F(\omega)$ is the Fourier transform of f(t).

The force f(t) can be determined by "mass cancellation" as diagrammed in Figure 3. The equation of motion for the impedance head-inner mount assembly can be expressed:

$$f(t) = f_I(t) - (m_I + m_i)\ddot{x}_i(t) \qquad (13)$$

where $f_I(t)$ is the force measured by the impedance head, m_I the effective mass of the impedance head, m_i the mass of the inner mount, and $\ddot{x}_i$ the acceleration of the inner mount measured by the impedance head. A simple electrical circuit can be devised to multiply the $\ddot{x}_i(t)$ signal by $(m_I + m_i)$ and subtract the result from the $f_I(t)$ signal, thereby producing f(t) as its output.

The shear strain $\gamma(t)$ is

$$\gamma(t) = \frac{x_i(t) - x_0(t)}{h} \qquad (14)$$

where $x_i(t) - x_0(t)$ is found by twice integrating the acceleration difference $\ddot{x}_i(t) - \ddot{x}_0(t)$. This calculation is easily performed by a spectrum analyzer since integration in the frequency domain is equivalent to dividing the Fourier transform of the original signal by $j\omega$. Thus

$$\Gamma(\omega) = \frac{X_i(\omega) - X_0(\omega)}{h} = \frac{\ddot{X}_i(\omega) - \ddot{X}_0(\omega)}{\omega^2 h} \qquad (15)$$

where the upper case letters represent the Fourier transforms of their lower case equivalents.

Wave Effects. The foregoing treatment of stress and strain presume no standing or traveling waves in the sample, i.e. the inertia of the sample is negligible compared to viscoelastic forces. This condition is met when the length of the shear wave propagating through the sample is much greater than the critical dimension of the sample. In shear samples thickness, h, is the critical dimension.

The shear wavelength λ is given by (Fitzgerald, E. R. The Johns Hopkins University, personal communication, 1989)

$$\lambda = \frac{2\Pi C}{\omega} = \frac{2\Pi}{\omega} \left[\frac{G}{\rho} \right]^{0.5} \left[\frac{2(1 + \tan^2 \delta)}{[(1 + \tan^2 \delta)^{1/2} + 1]} \right] \qquad (16)$$

where ω is the frequency (rad/sec), ρ the mass density (kg/m^3), and C is the speed of propagation of sound through the material. This predicts that sample inertial effects will become significant at lower frequencies as the thickness increases. Also that the inertial effects will show up at lower frequencies as the modulus declines (when temperature is increased). Thus, direct measurement of stress and strain with forced vibration tests is always limited to relatively low frequencies.

FTMA provides a direct method for determining the frequency at which inertial effects become noticeable. For this purpose the outer mount is detached from the "rigid" foundation and the accelerometer is mounted on it. The free body diagram of the outer mount-accelerometer assembly is shown in Figure 4. The relevant equation of motion is

$$f'(t) = (m_a + m_0) \ddot{x}_0(t) \qquad (17)$$

where $f'(t)$ is the force exerted on the outer mount by the specimen, m_a the mass of the accelerometer, m_0 the mass of the outer mount, and $\ddot{x}_0(t)$ the acceleration of the outer mount measured by the accelerometer.

As long as the inertia of the polymer sample is negligible, then $f(t)$ should equal $f'(t)$. Then according to Equations 13 and 17

$$f(t) = (m_a + m_0) \ddot{x}_0(t) \qquad (18)$$

If the Fourier transforms of both sides of this equation are evaluated, it follows that

$$\frac{F(\omega)}{X_0(\omega)} = m_a + m_0 = \text{Constant} \qquad (19)$$

Thus, the ratio of the force measured by impedance head to the acceleration of outer mount in end-free scheme must remain constant up to the frequency at which wave effect become significant.

Bending Effects. The foregoing treatment of strain also presumes sample deformation is only simple shear, i.e. there is no bending. An approximate assessment of the effect of bending can be obtained

SHEAR SPECIMEN

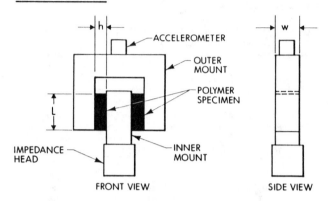

FRONT VIEW   SIDE VIEW

Figure 2. Schematic diagram of shear test samples.

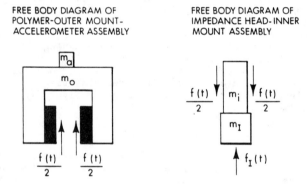

Figure 3. Analysis of forces using "mass cancellation".

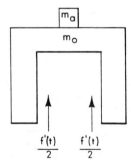

Figure 4. Analysis of forces in outer mount to investigate wave effects.

by considering the deflection of a cantilever beam shown in Figure 5.

The total deflection Δ_T is sum of deflection due to bending Δ_b and shear Δ_s and is given as (5.23)

$$\Delta_T = \Delta_s + \Delta_b \tag{20}$$

$$= 6Ph/5AG + Ph^3/3EI \tag{21}$$

where E and G are material Young's and shear moduli, A and I area and moment of inertia of the beam. For elastomers $E \approx 3G$, and for a rectangular beam $A = bL$ and $I = \frac{1}{12} bL^3$. Thus Equation 21 can be expressed

$$\Delta_T = \frac{6Ph}{5AG} \left[1 + \frac{10}{9} (\frac{h}{L})^2 \right] \tag{22}$$

where b,L, and h are sample dimensions. Equation (22) shows bending in the simple shear deformation is minimized when the sample length/thickness ratio is kept sufficiently large. If L/h is not large enough the strain and consequently stress distribution will not be uniform. Futhermore, Equation 15, used for shear strain calculation, will overestimate the shear strain resulting in lower modulus. Percent error in shear strain calculated from Equation 22 is summarized in Table I.

Table I. Percentage of Error Calculated from Equation 22

Length/Thickness	Error in Shear Strain
2	25 %
4	7 %
8	1.7%
16	.4%

This shows that sample length/thickness must be above 10 to avoid significant error in shear strain.

Experimental

A schematic diagram of the experimental apparatus is shown in Figure 6. Test specimens were made by compression molding two elastomer compounds, Neoprene rubber and NBR (nitrile rubber), between sample mounts made of aluminum. Both materials were aged for about 8 years. Old samples were used to prove the validity of the new apparatus. Details of sample preparation are given elsewhere (13,14). The exact dimensions of sample mounts varied depending on the size of the polymer specimen tested. To insure that the mounts behaved as rigid bodies, the dimensions were chosen to place the resonance frequencies of their natural modes of vibration well above the region of interest. (The lowest frequency natural mode was found to be the "tuning fork" mode of the outer mount.) (14)

A signal generator feeds band limited white noise into a power amplifier which drives an electro-mechanical shaker. A piezoelectric impedance head is mounted between the shaker and the

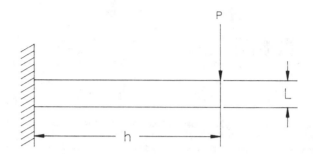

Figure 5. Deflection analysis of a cantilever beam to estimate
bending effect.

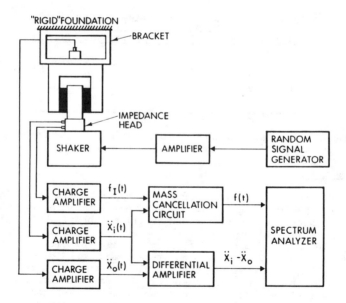

Figure 6. Schematic diagram of FTMA apparatus.

inner mount. A piezoelectric accelerometer is attached to the bracket to detect the small motion of the (supposedly rigid) foundation.

Signals from both transducers are fed through charge amplifiers. The acceleration signals $\ddot{x}_i(t)$ and $\ddot{x}_0(t)$ are then fed into a differential amplifier to obtain $\ddot{x}_i(t)-\ddot{x}_0(t)$. The force and acceleration signals $f_I(t)$ and $\ddot{x}_i(t)$ are fed into a mass cancellation circuit which produces f(t) as an output (see Equation 13). The force and acceleration difference signals f(t) and $\ddot{x}_i(t)-\ddot{x}_0(t)$ are then fed into a spectrum analyzer where they are Fourier transformed, and multiplied by the appropriate constants according to Equations 12 and 15. G* (jω) is then computed internally.

For comparison, tests were also run with the same equipment using a single frequency sinusoidal input, repeated at different frequencies over the range of interest. All comparative tests were run at 25°C. To further check the final results, the properties of the elastomer compounds were also measured using a Rheovibron.

To study effects of temperature, the entire excitation and detection assembly was placed in a chamber with controlled temperature and humidity. The test was done at 20 different temperatures from 4°C to 75°C. Each test took 4-5 sec. so there could be no significant temperature rise due to internal friction in the test sample.

Data Reduction

The raw data was strings of storage modulus, G', and frequency, ω, at various temperature levels. Data reduction was carried out with linear temperature, T, Y = log G', and X = log (ω/100). The reduced scale for X was used so that intercepts at X=0 would correspond to 100 Hz frequency and fall within the experimental range. There was noise in the modulus data attributable to mechanical and electrical perturbations. These small perturbations were removed by smoothing with simple linear and second order polynomial functions.

At high temperatures the isotherms were linear and almost level. Multiple linear regression was used to determine the separate effects of temperature and frequency:

$$Y = B_0 + B_1 T + B_2 X \tag{23}$$

At lower temperatures the fixed frequency the response of Y as a function of T was sigmoidal. Good fits over the entire range of T were obtained by regression to the function:

$$Y = B + B_1 T + H[1 - \tanh (T-T_g)/S] \tag{24}$$

where B, B_1, H, S and T_1 are parameters selected to fit the data. At elevated values of T the last term of this equation extrapolates to 0. Thus, B_1 is the same as in equation (23) and B = B_0+B_3X. Regressions at a series of X levels gave corresponding series of H, S and T_g.

The regression parameters varied with X. In the cases of S and T_g there were good linear correlations. (Statistical correlation coefficients, R> .99.) In the case of the H parameter the correlation with X was poor. This was because of the experiments did not cover sufficiently low temperatures.

Results

Test Data Validation. Samples of Neoprene and NBR with different thicknesses were used to study wave and bending effects. Figure 7 is a plot of the magnitude and phase of $F(\omega)/x_0(\omega)$ versus frequency for the 1.59mm Neoprene specimen (L/h=16). Both ratios are seen to be constant up to about 2500 Hz. Above that frequency the ratios begin to change because the inertia of the specimen was no longer negligible. Note that the phase angle appeared more sensitive to wave effects. Similar results were obtained for the other Neoprene and NBR specimens. The frequencies where the ratios began to deviate were directly proportional to the length/thickness (L/h) ratios of the samples.

Figure 8 shows apparent storage modulus spectra for different thickness samples of Neoprene. Sample inertial effects are obvious at the high ends of these spectra. For the thinnest sample (L/h=16), the wave effects began about 2500 Hz, which agrees with the results shown in Figure 7. As expected, the thicker samples showed the onset of wave effects in proportion to sample L/h. A detailed discussion of wave effects may be found elsewhere (24,25).

The frequencies corresponding to the onset of wave effects were used in equation 16 to determine wave length/thickness ratio. In all cases the critical ratios were 14 to 16. This was midway between previous estimates; Schrag (26) estimated 20 on theoretical considerations; Ferry (1) estimated 10 on the basis of experimental results.

Figure 8 also shows that the apparent storage modulus decreases directly with the sample length to thickness ratio, L/h. At frequencies below the onset of inertia effects the apparent storage moduli for the L/h = 8, 4, and 2 specimens were about 3%, 8%, and 18% smaller, respectively, than the thinnest sample (L/h = 16). These are approximately the same as theoretically predicted in Table 1 and those reported in the literature (27,28).

Figure 9 shows a typical result of directly comparing three methods for determining storage modulus on the same sample. The points were determined with a commercial Rheovibron. The dashed line was results with forced vibration using single frequency sinusoidal inputs. The solid line shows FTMA data from random noise inputs.

Model. The parameters for the NBR data are listed in Table II.

Table II. Parameters for NBR Data used in Equation 24

	100 Hz Reference	Frequency Correlation Slope per decade
B	1.32	0.0321
B_1	-0.00516	0
H	0.86	0
S	13.1	2.07
T_g	11.6	9.6

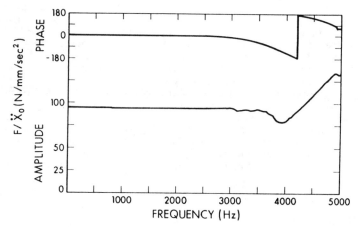

Figure 7. Point acceleration impedance of NBR, thickness 1.58 mm, showing sample inertia (wave) effects above 2500 Hz.

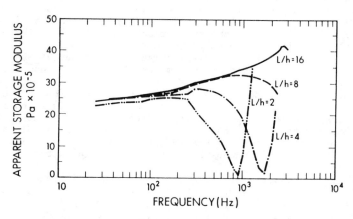

Figure 8. Real part of FTMA output (apparent storage modulus) vs. frequency for NBR samples of different thickness to show sample inertia and L/h effects.

These parameters were used in Equation 24 to determine the model contour lines in Figures 10 and 11. The points in Figure 11 are raw data at the temperatures indicated. This shows the considerable noise in the raw data. The noise was removed by smoothing each data isotherm with a second order polynomial and the data points in Figure 10 are the intercepts of the smoothed isotherms at the indicated frequencies.

Figure 12 shows damping factors for NBR (also known as tan delta, see Equation 3) at various frequencies plotted as functions of relative (or dimensionless) temperature. The dimensionless temperature is defined as $(T-T_g)/S$, where T_g and S are parameters from the model for storage modulus. Note zero on this scale corresponds to the point where storage modulus shows maximum change with temperature. The damping factor also goes through a maximum at about the same point. The points -2 to 2 on the relative scale correspond to the temperature region for 96% of the total glass transition of storage modulus.

Note that in Figure 12, both T_g and S are functions of frequency. The overlap in data for different frequencies indicates the shift in the peak and spread of damping factor as a function of frequency. Thus data for tan (δ) correlates well with the model for G'.

Discussion

FTMA is a forced vibration test method based on direct measurement of stress and strain spectra. As with all forced vibration methods, FTMA is subject to spurious wave effects at high frequencies. The lower frequency limit is determined by transducers, signal conditioners, etc. The lower limit in this research was 35 Hz as determined by the inherent properties of the piezoelectric transducers. With different transducers (for example load cell for the force and LVDT for displacement measurements) and signal conditioners, FTMA should measure material properties down to much lower frequencies.

The primary feature of FTMA is the fact that a complete isotherm is obtained in just a few seconds. This saves time and effort compared to commercial single frequency instruments. It also gives better data in the case of sensitive materials, such as natural products, which may change during long exposure in a test chamber. Changes induced by energy dissipation within the sample are minimized.

Another feature of FTMA is the fact that it readily provides direct assessment of sample inertia, bending and geometry effects. This would be very difficult with single frequency instruments.

The Fourier transform technique is perfectly general. It may be used also to study non-linear viscoelastic properties (29,30).

FTMA has great potential for applications in conjunction with a coupled theory of the effects of both temperature and strain histories on mechanical material properties. This is beyond the scope of present day theoretical capabilities (31,31) and the exact constitutive equation for a given material has to be determined experimentally.

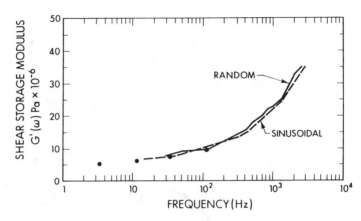

Figure 9. Comparison of test methods with NBR. Solid line from FTMA with white noise input; dotted line from sinusoidal input; points from Rheovibron.

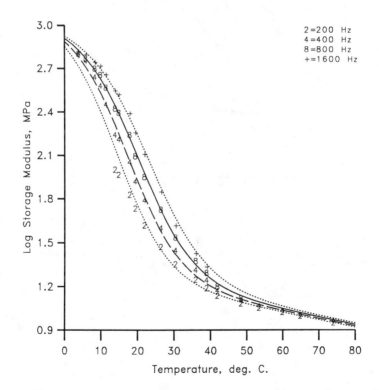

Figure 10. NBR Storage modulus contours at different frequencies. Symbols show FTMA data. Lines show model.

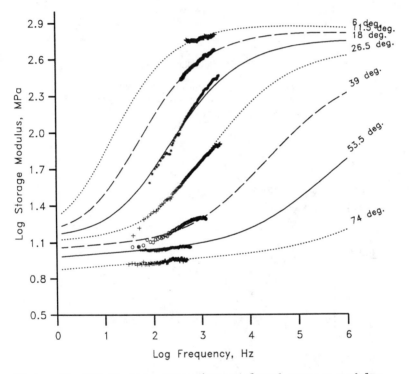

Figure 11. NBR Master curves (extrapolated storage modulus isotherms) at different temperatures. Lines show model. Symbols show raw data from FTMA.

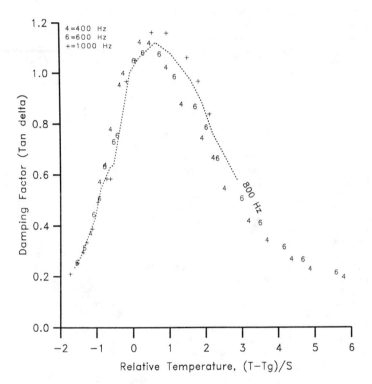

Figure 12. NBR Damping factors (tan δ) at different frequencies
plotted on scale of relative temperatures, $(T-T_g)/S$.
Dotted line connects data at 800 Hz. Symbols show
data at 400, 600 and 1000 Hz.

Phenomenological Model. The data reduction scheme developed for use with FTMA is based on a semi-empirical phenomenological model for polymeric materials with postulates corresponding to generally observed behavior. The constraints of current constitutive theory are satisfied and the model relates mechanical properties to both frequency and temperature with parameters that are material-dependent. It provides excellent interpolations of experimental results and also extrapolates to reasonable levels outside the ranges of the experimental variables.

NBR shows temperature dependent behavior typical of viscoelastic materials. This is seen in figure 10 for storage modulus, G', contours over temperature, T, for various frequencies. All materials exhibit small modulus changes as they are heated. Viscoelastic materials have an additional glass transition of two or more orders of magnitude over a specific temperature range. The model treats the transition and the other minor temperature effects as additive terms. The glass transition is characterized by an overall change, 2H, the temperature of the midpoint, T_g, and a spread factor, S. The spread is defined such that 96% of the transition (1.92H) occurs during temperature change from (T_g-2S) to (T_g+2S).

Previous workers have used several different definitions for a single point glass transition temperature (33). The most appropriate is the temperature where the rate of change of storage modulus with temperature is a maximum. The model parameter T_g conforms to this case.

Conventionally storage modulus versus frequency and temperature results are presented by extrapolated isotherms called master curves. These are plotted by shifting frequency data with a temperature dependent shift factor, α_T. Most published results for α_T are based on manual graphic methods. Computer work first aimed at systemizing and automating determinations of shift factor was initiated by a desire to efficiently process the hundreds of data points from each FTMA run.

If the model is viewed merely as good mechanism for smoothing a lot of data, it still provides a good algorithm for computing shift factors. At a reference point for T and X (temperature and log frequency), a reference value of Y (log G') is determined. Then with X fixed, T is varied to compute $(\Delta Y/\Delta T)$. Similarly, with fixed T, $(\Delta Y/\Delta X)$ is determined. The shift factor for the two temperatures is then given by:

$$\text{Log } \alpha_T = \frac{(\Delta Y/\Delta T)}{(\Delta Y/\Delta X)} \qquad (25)$$

After the model was fully developed it was seen to have utility for extrapolations. This reduced interest in shift factors because they were no longer required to get master curves.

Fig. 11 shows master curves extrapolated with the model. These have the same general features as master curves plotted by shifting data isotherms. The two differ slightly because different assumptions are involved. The conventional method of shifting data makes all isotherms congruent; the same viscoelastic processes are assumed at all temperatures. The phenomenological model is not limited to this assumption. When the thermal spread varies with frequency, the model extrapolates to isotherms like fig. 11. Note that the transition region extends over broader frequencies at the temperature rises.

A necessary condition for a good model is internal consistency. The phenomenological model meets this condition very well in the way it correlates the trends and features of loss factor with model parameters. This is shown in Figure 12 where the relative temperature scale is based on the model for storage modulus and the loss factors (tan δ) at different frequencies overlap. Thus the dependence of T_g and S on frequency predicts both storage and loss modulus variation.

Conclusions

1. Fourier transform mechanical analysis (FTMA) measures isothermal viscoelastic properties of polymers over a wider range of frequencies in a shorter period of time than is possible with other techniques.

2. A simple and flexible modeling scheme to quantitatively characterize complete viscoelastic properties of polymers is readily useable with FTMA data. It contains the essential features of the frequency-temperature equivalence of the viscoelastic behaviors of polymers and is easily adapted to modern computer systems.

3. The model provides a complete constitutive equation for NBR and Neoprene and is internally consistent with damping factor data.

Acknowledgments: The helpful support and encouragement of Philip Morris management and the individual assistance of J. L. Banyasz, J. C. Crump, III and J. F. Whidby are gratefully acknowledged.

Literature Cited

1. Ferry, J. D. Viscoelastic Properties of Polymers; 3rd. ed., John-Wiley & Sons Inc.: New York, 1980; Chs. 6, 7.
2. Gehman, S. D. Rub. Chem. Tech. 1957, 30, 1202-1250.
3. Abolafia, O. R. Survey of methods used to determine the dynamic mechanical properties of polym., U. S. Dept. of Commerce, office of Tech. Serv., Tech. rep. 2060, Sept. 1954.
4. Ganeriwala, S. N. Lecture notes on Viscoelastic properties of Biomaterials; Philip Morris Research Center, Richmond, VA., May-July, 1985.
5. Ward, I. M. Mechanical Properties of Solid Polymers, Wiley-Interscience, Inc.: New York, 1983; Chs. 5-7.
6. Boyer, R. F. Polymer Characterization, Ed. Craver, C. D.; Advances in Chemistry Series 203; ACS: Washington, D. C., 1983; pp. 3-25.
7. Medalia, A. I. Rub. Chem. Tech., 1978, 51, 3, 437-523.
8. Cramer, W. S. J. Polym. Sci., 1967, 26, 57-65.
9. Buchanan, J. L. J. Acous. Soc. Am., 1986, 81, 6, 1775-1786.
10. Yaniv, G. and Ishai, O. Polym. Eng. Sci., 1987, May, 27, 10, 731-739.
11. Enderby, J. A. Trans. Faraday Soc., 1955, 51, 106.
12. Gunderson, D. E. Proc. Mech. Cellulosic Polym. Matl., Ed. Perkins, R. W., 3rd. Joint ASCE/ASME Mech. Conf., San Diego, July, 1989.

13. Ganeriwala, S. N. Ph.D. Thesis, The University of Texas at Austin, 1982.
14. Ganeriwala S. N. and Rotz C. A. Proc. PMSE, ACS national meeting, 1984, 50, April 37-42; and a paper in comm.; Polym. Eng. Sci.
15. Williams, M. L.; Landel, R. F.; and Ferry, J. D. J. Am. Chem. Soc. 1955, 77, 3701-3707.
16. Ref. 1; ch. 11.
17. Aklonis, J. J. and MacKnight, W. J. Introduction to Polymer Viscoelasticity, John-Wiley and Sons, Inc.: New York; 1983; chs. 3,4.
18. Sperling, L. H. Introduction to Physical Polymer Science, John-Wiley & Sons, Inc.: New York; 1986; Ch. 6.
19. Ferry, J. D.; Child, W. C. Jr.; Zand, R.; Stern, D. M.; Williams, M. L.; and Landel, R. F. J. Colld. Sci. 1957, 12, 53-67.
20. Plazek, D. J. J. Phys. Chem., 1965, 69, Oct. 3480- 3487.
21. Christensen, R. M. Theory of Viscoelasticity an Introduction, 2nd. ed., Academic Press Inc.: New York; 1982; Ch. 1.
22. Greenberg, M. D. Foundation of Applied Mathematics, Prentice-hall: Englewood Cliffs,N. J.; 1978; ch. 5.
23. Popov, E. P. Introduction to mechanics of solids; Prentice-Hall, Inc.: Englewood Cliffs, N. J.; 1968; pp. 484-487.
24. Harris, C. M. and Crede, C. E. Ed. Shock and Vibration Handbook, 2nd. ed., McGraw-Hill Book Co.: New York; 1976; ch. 30, pp. 52-56.
25. Snowdon, J. C. J. Sound Vib., 1965, 2, (2), 175-103.
26. Schrag, J. L. Trans. Soc. Rheo., 1977, 21, 3, 399-413.
27. Fitzgerald, E. R. Proc. PMSE, ACS national mtg; 60, April, 1989; pp. 573-578.
28. Parsons, J. S., Yates, WAllace, and Scholoss, The measurement of dynamic properties of materials using Transfer Impedance Technique, Report 2981, Naval Ship R & D center, Washington, D. C., April, 1969.
29. Ganeriwala, S. N. Proc. 10th U.S.A. National Cong. App. Mech., Austin, Texas, 1986; T5.
30. Ganeriwala, S. N. and Rotz, C. A., Polym. Eng. Sci. 1987, Jan., 27, 2, pp. 165-178.
31. Ref. 29; ch. 3.
32. Lee, E. H., Mechanics of Viscoelastic Media and Bodies; J. Hult, Ed. Symp. Gothenburg/Sweden; Springer-Verlag; Sept. 2-6, 1974; pp. 339-357.
33. Shen, M. C. and Eisenberg, A. Rub. Chem. Tech., 1970, 43, 95-155.

RECEIVED January 24, 1990

Chapter 7

Extrapolating Viscoelastic Data in the Temperature–Frequency Domain

Peter T. Weissman and Richard P. Chartoff

Center for Basic and Applied Polymer Research, University of Dayton, Dayton, OH 45469

The viscoelastic properties of polymers give them the unique ability to be used as effective vibration damping materials. In order to evaluate damping characteristics, dynamic mechanical analysis (DMA) is used to determine viscoelastic properties. This method most often covers a restricted range of frequencies. Using the time-temperature superposition procedure and the WLF equation, a master curve is constructed in order to extrapolate DMA data to frequencies not experimentally measured. The results may be displayed on a novel graphing format known as a reduced frequency nomograph. This technique presents storage modulus as well as loss modulus (or loss tangent) data simultaneously as a function of temperature and frequency. The nomograph procedure accurately superposes data gathered using either a fixed frequency multiplexing mode or a resonant mode. The data are read directly from ASCII files generated by the DuPont 9900, 2100, or other PC based thermal analysis equipment. A computer program developed for the IBM XT personal computer then allows construction of a master curve in a nomograph format.

The viscoelastic properties of polymers make them ideally suited for use as effective vibration damping materials because of their ability to dissipate mechanical energy. Specifically, the region of transition from the glassy to rubbery state (Figure 1) has the maximum potential for vibration damping (1-2). Dynamic mechanical analysis (DMA) is used for determining a polymer's viscoelastic properties within the linear viscoelastic range. However, most DMA

0097–6156/90/0424–0111$06.25/0
© 1990 American Chemical Society

data are measured only over a narrow frequency range at various
temperatures. To use DMA data for practical design purposes, a
master curve that extends data over a broad range of frequencies
(and temperatures) must be generated using the time-temperature
superposition procedure (3-6).

Master curves can be represented on a novel graphing format
known as a reduced frequency nomograph (7). This technique displays storage modulus (E′) and loss modulus (E″) data (or loss
tangent (tan δ) data) simultaneously as a function of temperature
and frequency. In order to facilitate matters, a Fortran based
program was written that reads data generated by PC based thermal
analysis systems (such as the DuPont 9900 or 2100 T. A. Systems or
the Rheometrics Solids Analyzer) and allows direct conversion to a
nomograph format. The program can be run on any IBM PC or compatible and is menu driven. The data can be either in fixed
frequency multiplex (a method by which DMA data are determined for
several fixed frequencies in succession at a given temperature
before proceeding on to the next measurement temperature) or
resonant forms.

In this report we demonstrate the utility of the procedure for
multiplexed data. The usefulness of the procedure for resonant
data was the topic of previous publications (1,8). Using data for
typical samples of amorphous polymers, poly(methyl methacrylate)
(PMMA) and polycarbonate (PC), the procedure is shown to be well
suited to the transition region. A Poly(vinyl chloride) (PVC)
acoustical damping material is used to demonstrate the ability to
change the data reduction equation (the WLF equation) parameters
and their subsequent effects on the fit of the superposed data.

Superposition Procedure

Time-temperature superposition was first suggested by H. Leaderman
who discovered that creep data can be shifted on the horizontal
time scale in order to extrapolate beyond the experimentally
measured time frame (9-10). The procedure was shown to be valid
for any of the viscoelastic functions measured within the linear
viscoelastic range of the polymer. The time-temperature superposition procedure was first explicitly applied to experimental data by

Tobolsky and co-workers who also modified it to account for propor-
tionality of modulus to absolute temperature ($\underline{3}$). This has the
effect of creating a slight vertical shift in the data. Ferry
further modified the time-temperature superposition to account for
changes in density at different temperatures which has the effect
of creating an additional vertical shift factor ($\underline{4}$). The effect of
the temperature-density ratio on modulus is frequently ignored,
however, since it is commonly nearly unity.

Additionally, it was noticed that the shift factor for superposition fit the following empirical equation known as the WLF
equation ($\underline{5,6}$)

$$\text{Log } a_T = - C_1(T - T_0)/(C_2 + T - T_0) \qquad (1)$$

where a_T = the temperature shift factor

T_0 = an arbitrary reference temperature

C_1 and C_2 are constants

When T_0 is replaced with the glass transition temperature, T_g,
the equation takes the form of the "universal" WLF equation where
C_1 is 17.44 and C_2 is 51.60. While this equation is not truly
universal, it was developed from a large data base of various
polymers including many elastomers.

When performing superposition with reduced variables (such as
a_T), several constraints are placed on the data. With respect to
DMA multiplex data these constraints can be used to determine the
consistency of the data set. The constraints are as follows:

1) The shape of adjacent curves must match exactly. That is,
while the frequency of the curve will shift with temperature the
shape must be frequency independent.

2) The same shift factor, a_T, must superpose all of the
viscoelastic functions. One must first perform the time-
temperature shift on one of the viscoelastic functions and
determine the values of the WLF constants. The same constants must
then be applied to the other viscoelastic functions to determine
their consistency in shifting the data. This process may need to
be repeated several times in order to determine the best set of

average constants for the WLF equation that satisfies all of the
viscoelastic functions.

3) Finally, the values of the shift factor, a_T, must have a
reasonable functional form consistent with experience, i.e., the
WLF or similar equation should apply. When T_o is replaced with T_g
reasonable values for C_1 and C_2 are those quoted as the "universal"
constants.

If the above criteria are not met the principal of reduced
variables for time-temperature superposition is not valid and
should not be used.

Master Curve Representation

Viscoelastic data are commonly represented in the form of a master
curve which allows the extrapolation of the data over broad temperature and frequency ranges. Master curves have, historically,
been presented as either storage modulus and loss modulus (or loss
tangent) vs. reduced frequency. This representation requires a
table of conversions to obtain meaningful frequency or temperature
data.

Jones first suggested representing the viscoelastic master
curve on a novel graphing format known as the reduced frequency
nomograph (7). This nomograph displays modulus, E' and E" (or loss
tangent) as a function of reduced frequency but also includes axes
to read simultaneously frequency and temperature. Figure 2 represents simulated data as a typical master curve displayed in the
nomograph format.

The nomograph is created by plotting the viscoelastic function
vs. reduced frequency. Reduced frequency is defined as $f_j a_{Ti}$ where
f_j is the frequency and a_{Ti} is the value of a_T at temperature Ti.
An auxiliary frequency scale is then constructed as the ordinate on
the right side of the graph. The values of $f_j a_{Ti}$ and f, form a set
of corresponding oblique lines representing temperature.

To illustrate the use of the nomograph, assume that we wish to
find the value of E' and tan δ at T_{-1} and some frequency f (point
C) of Figure 2. The intersection of the horizontal line f = constant, line CX, with $f_j a_{Ti}$ (point X) defines a value of fa_T at
point D, of about 4×10^2. From this value of fa_T it follows from

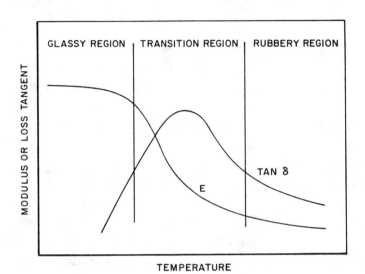

Figure 1. Typical Dynamic Mechanical Modulus and Loss Tangent as a Function of Temperature.

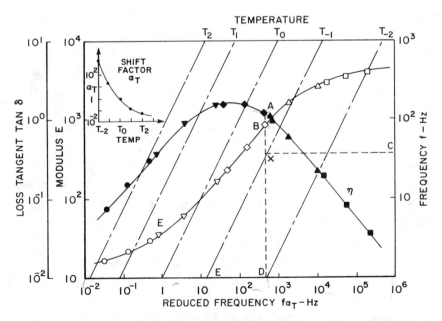

Figure 2. Illustrative Viscoelastic Master Curves Represented on Reduced Frequency Nomograph, Using Simulated Data.

the plots of E′ and tan δ that E′ = 10^{-3} N/M^2, point B, and tan δ = 1.2, point A.

Software Description and Data Reduction Scheme

A Fortran based program was developed to generate nomograph plots directly from data generated by fixed frequency multiplex DMA measurements. The program can be executed on any IBM XT or compatible and requires a Hewlett Packard plotter to generate the nomographs.

The data are first recorded via DMA to an ASCII data file which can be read directly by the nomograph program. After the DMA data are compiled either the "universal" WLF equation is chosen or the data are fit using an external software package resulting in an empirical WLF equation. The WLF equation of choice is then used to determine the appropriate values of a_T, the temperature shift factor, necessary to perform time-temperature superposition. Each individual isothermal frequency curve is then multiplied by the shift factor a_T in order to generate the master curve. The graphics sections of the program is then employed where the user has his/her choice of creating a nomograph, a fixed temperature graph or a fixed frequency graph based on the master curve previously calculated.

The reference temperature (T_0) is chosen differently depending on whether an empirical data fit or a "universal" WLF fit are carried out. If the data are fit empirically, T_0, should be taken as the value that gives the best fit of the data. The values of constants C_1 and C_2 are then calculated after superposition. This is done by shifting the curves and must be performed prior to using the nomograph program. The empirically determined constants may then be substituted in the WLF equation and a nomograph of the desired viscoelastic functions plotted.

If the "universal" values of C_1 and C_2 are used as a reasonable first approximation, the reference temperature is chosen as T_g. The glass transition temperature may be taken as the T value corresponding to the peak of E″ at an appropriate low value of frequency (0.001 to 0.1 Hz). T_g also may be determined independently (e.g., by Differential Scanning Calorimetry) and entered

in the WLF equation while still using the "universal" constants.
As before, once the form of the WLF equation has been determined a
master curve is generated and plotted in the nomograph format.

It should also be noted that while the nomograph program
reported here uses the WLF equation to calculate the shift factor,
the data reduction scheme is not limited to the WLF equation. That
is, any curve fitting equation that results in the calculation of a
temperature shift factor can easily be added to the program and
used for the generation of the master curve in the nomograph
format. The data reduction scheme is based on the reduced variable
concept and not the form of the equation.

The program entails no on screen graphics but is menu driven
and user friendly. Questions concerning the availability of this
software should be directed to the authors of this paper.

Experimental

All experimental viscoelastic data were gathered with either
DuPont's DMA 983 or a Rheometrics RSA II. The DMA 983 tests
materials by clamping a rectangular sample at each end and inducing
a flexural bending motion on one end while measuring the sample's
response at the other end. The RSA II was used in its dual cantilever bending configuration. In this mode, the sample is clamped
on each end and in the center. The ends are excited while its
response is measured via the center clamp. As previously discussed, the technique used to gather these data was a fixed
frequency multiplex method. Multiplexing is a method by which DMA
data are determined for several fixed frequencies in succession at
a given temperature before proceeding on to the next measurement
temperature.

A sample of (Poly methyl methacrylate)(PMMA), known as
Acrylite GP (a product of Cyro Ind.), was characterized using the
DMA 983. The data were collected at fixed frequencies of 0.01,
0.1, 0.5 and 1.0 Hz over a range of temperatures from RT to 140°C.
The sample size was 9.90 x 12.60 x 0.93 mm.

A standard sample of General Electric's Lexan was characterized using the RSA II. Frequencies tested were 0.1, 0.5, 1.0,

5.0 and 10.0 Hz at temperatures ranging from RT to 165°C using a
sample 45.0 x 5.12 x 2.1 mm.

Finally, a proprietary poly(vinyl chloride), (PVC) based
acoustical damping material was characterized using the DMA 983.
Frequencies tested were 0.033, 0.10, 0.320, and 1.0 Hz at temperatures ranging from -80 to 80°C. The sample dimensions were 17.22 x
14.13 x 3.17 mm.

Results

The data obtained for the PMMA sample were reduced and are displayed in the nomograph form of Figure 3. The form of the WLF
equation used is the "universal" WLF equation with T_0 replaced by
T_g and defined as the temperature at the peak value of loss modulus
for the 0.01 Hz curve. The constants C_1 and C_2 were assigned the
values of 17.4 and 51.6 respectively.

As can be seen by inspection, the time-temperature superposition for these WLF values appears quite good. The same WLF
equation was used to shift the viscoelastic functions E' and E".
The flat appearance of the E" curve is due to the compressed nature
of this particular nomograph scale. Both functions appear to fit
equally well and therefore satisfy the criteria of curve shape and
shift factor consistency for using the reduced variable time-
temperature superposition. Additionally, the criterion of
reasonable values for a_T is satisfied by virtue of using the
"universal" WLF equation.

Another method for checking the consistency of the data and
subsequent superposition is to plot a fixed frequency graph
(corresponding to one or more frequency that is represented in the
experimental data set) generated from the calculated master curve
and compare it with the actual experimental data. Figures 4 and 5
show E' and E" (respectively) vs. temperature for a fixed frequency
of 0.1 Hz. The solid line represents the actual experimental data
determined by DMA at 0.1 Hz. From inspection one can see there is
excellent agreement between the calculated data and the experimentally measured data.

Figure 6 displays master curves of E' and tan δ in the
nomograph format created from data obtained for the polycarbonate

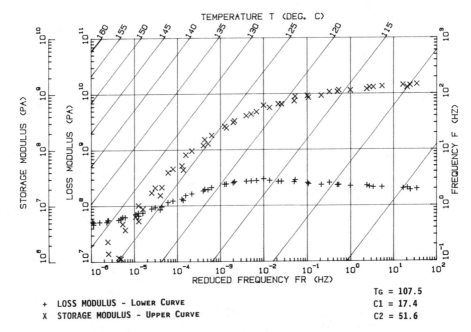

Figure 3.   Master Curves of PMMA Multiplex DMA Data Represented on
Reduced Frequency Nomograph.   Master Curves Calculated
Using "Universal" WLF Equation.

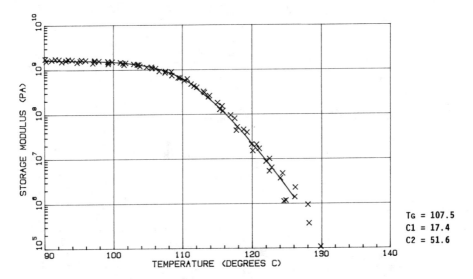

Figure 4.   Fixed Frequency (0.1 Hz) Graph of E' for PMMA Generated
from Master Curve of Figure 3.   Solid Line Represents
Experimental 0.1 Hz Data.

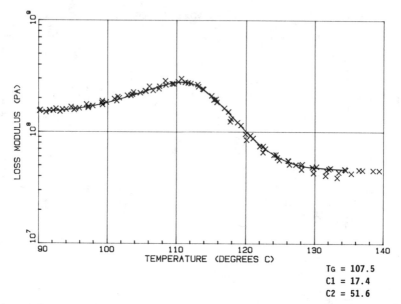

TG = 107.5
C1 = 17.4
C2 = 51.6

Figure 5.   Fixed Frequency (0.1 Hz) Graph of E" for PMMA Generated
from Master of Figure 3.   Solid Line Represents
Experimental 0.1 Hz Data.

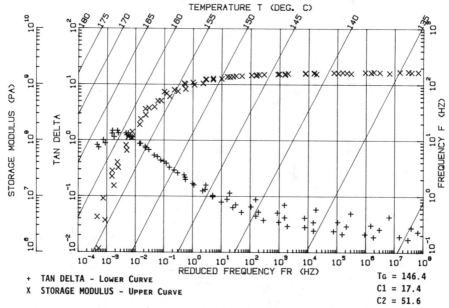

+ TAN DELTA - LOWER CURVE
X STORAGE MODULUS - UPPER CURVE

TG = 146.4
C1 = 17.4
C2 = 51.6

Figure 6.   Master Curves of Polycarbonate Multiplex DMA Data
Represented on Reduced Frequency Nomograph.   Master
Curves Calculated Using "Universal" WLF Equation.

sample. The storage modulus fits well from the glassy state
through the transition region. However, while tan δ fits well in
the transition region, the fit is not good in the glassy state.

It should be noted that polycarbonate has a strong beta transition near -100°C. The polymers least mobile state is, therefore,
below the beta transition. Further experimentation, not included
here, indicates that below the beta transition the magnitude of E"
and tan δ are far less frequency dependent than above the
transition. The poor fit seen in the "glassy" region of Figure 6
appears to be due to the presence of the beta transition. The
difference in activation energies for the α and β transitions
result in tan δ having a more complex frequency dependence in the
region between T_g and T_β.

The utility of empirically determined WLF equations was investigated using DMA data obtained on the PVC acoustical damping
material. Using a separate software package (available from DuPont
Intruments), E′, E" and tan δ were empirically fit using the time-
temperature superposition procedure. A reference temperature is
first determined by the computer software. The data are then
shifted manually and the WLF equation is fit to the resulting
temperature shift factors. Values for C_1 and C_2 are then calculated based on the WLF equation temperature shift factor curve fit.
Using this procedure, each of the viscoelastic functions is fit
separately resulting in distinct values of C_1 and C_2 being determined for each viscoelastic function. Therefore, three different
empirically determined WLF equations were obtained for the PVC,
acoustic damping material.

Using the nomograph program, the cited empirical WLF equations
and the "universal" WLF equation were used to generate master
curves of E′ and tan δ. Figure 7 shows a complete nomograph representation of E′ and tan δ master curves generated using the
"universal" form of the WLF equation. Figure 8 is an expanded view
of the transition region in Figure 7 (area A). Figures 9-11 cover
the same region as Figure 8 but each is generated by substituting a
different empirical WLF equation as discussed above. Note that
while the fit of the tan δ function is similar for all forms of the

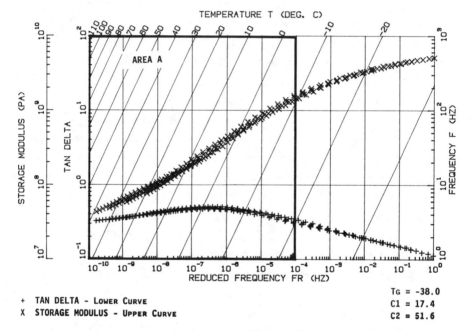

+ **TAN DELTA - Lower Curve**
X **STORAGE MODULUS - Upper Curve**

$T_G = -38.0$
$C1 = 17.4$
$C2 = 51.6$

Figure 7.   Master Curves of PVC Acoustical Damping Material
            Multiplex DMA Data Represented on Reduced Frequency
            Nomograph. Master Curves Calculated Using "Universal"
            WLF Equation.

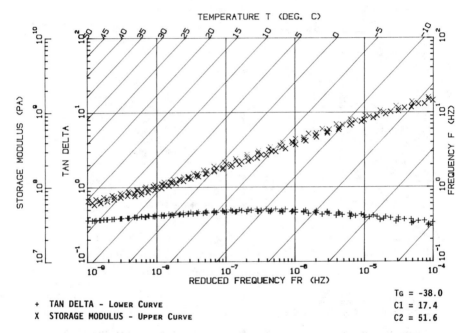

Figure 8.   Master Curves of PVC Acoustical Material Multiplex DMA Data Represented on Reduced Frequency Nomograph. Enlargement of Area A in Figure 7.   "Universal" WLF Equation.

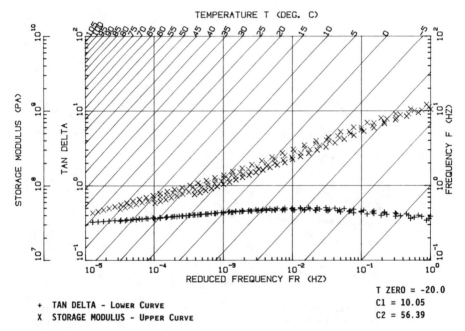

+ TAN DELTA - Lower Curve
X STORAGE MODULUS - Upper Curve

T ZERO = -20.0
C1 = 10.05
C2 = 56.39

Figure 9. Master Curves of PVC Acoustical Material Represented
on Reduced Frequency Nomograph. Enlargement of Area
A in Figure 7. Empirical WLF Equation Obtained from
Shift of Tan δ Curves.

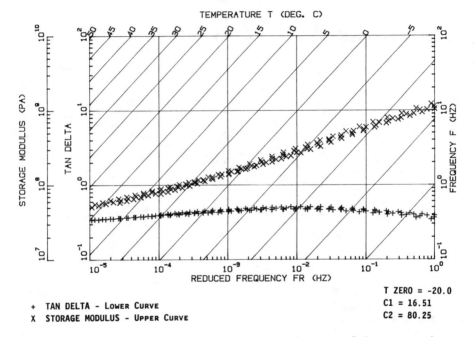

Figure 10. Master Curves of PVC Acoustical Material Represented on Reduced Frequency Nomograph. Enlargement of Area A in Figure 7. Empirical WLF Equation Obtained from Shift of E" Curves.

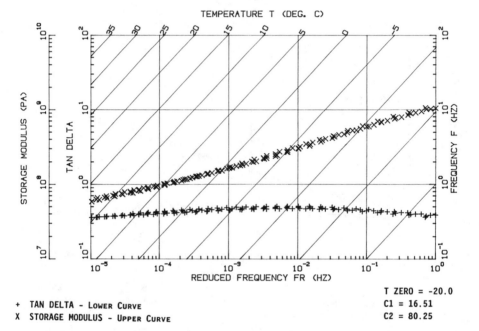

T ZERO = -20.0
C1 = 16.51
C2 = 80.25

+   TAN DELTA - Lower Curve
X   STORAGE MODULUS - Upper Curve

Figure 11. Master Curves of PVC Acoustical Material Represented
on Reduced Frequency Nomograph.   Enlargement of Area
A in Figure 7.   Empirical WLF Equation Obtained from
Shift of E′ Curves.

WLF equation, the fit of the storage modulus curve varies significantly. As can be seen from Figure 9, the empirical constants determined by shifting tan δ give a poor fit of E'. The "universal" WLF equation and the empirical WLF equation determined from shifting E" (Figures 8 and 10, respectively) give somewhat better superposition of both viscoelastic functions. Figure 11 shows master curves of E' and tan δ calculated by using the WLF equation determined from shifting E'. This appears to have the best overall fit. Similar results are found when plotting E" but are not shown here.

Using a nomograph to read storage and loss modulus values for specified temperatures and frequencies outside the range of measurement is not a new procedure. The technique has been demonstrated for numerous damping polymers by comparing experimental values with the values extrapolated from nomographs constructed from DMA data ($\underline{1},\underline{8}$). However, one must have a method for determining whether a master curve generated for a specific set of experimental data is accurate.

One method is simple visual inspection of the generated master curve. If the fit of the individual data sets is poor, subsequent extrapolation is open to significant error. Another method is to determine an activation energy plot (of the shift of T_g with frequency) for the data set and compare temperatures or frequencies calculated from the resulting Arrhenius equation to those read from the nomograph.

The following form of the Arrhenius equation can be used to determine the activation energy for shifting of the glass transition temperature as well as for defining a straight line equation characterizing the shift as a function of frequency.

$$\log f = \log b - Q/RT_{max} \qquad (2)$$

where f = the frequency

     Q = the activation energy

     T_{max} = the temperature (in deg. K) at the peak value of E" or tan δ at frequency f

     R = the universal gas constant

By plotting log f vs $1/T_{max}$ the slope of the resulting straight line can be found and the activation energy calculated. Using this technique, the values of T_{max} corresponding to the peak value of E", and T_{max} corresponding to the peak value of tan δ, were calculated for each of the three polymers discussed in this report at a frequency of 1000 Hz. Each set of DMA data was then extrapolated to 1000 Hz from the appropriate master curves and the data sets for each measured frequency. An average value of T_{max} was then determined for each polymer and each viscoelastic function. This was done by averaging the T_{max} values obtained from each individually measured frequency curve extrapolated to 1000 Hz using the "universal" WLF equation. Table 1 relates the peak temperatures determined from the Arrhenius equation with those from the reduced variable concept for 1000 Hz. Good agreement is seen for all three of the polymers with the exception of the superposition tan δ values for the PVC acoustical material. This discrepancy will be addressed later. In each case the "universal" form of the WLF equation was used to reduce the DMA data.

TABLE 1

T_{max} for the Peak Values of Viscoelastic
Functions E" and Tan δ Extrapolated to 1000 Hz

(All Superposition Data Calculated Using
the "Universal" WLF Equation)

Polymer	T_{max} Arrhenius		T_{max} Superposition	
	for E"	for Tan δ	for E"	for Tan δ
PMMA	127°C	----	129°C	----
Polycarbonate	162	170°C	162	170°C
PVC Acoustical	-19	15	-20	23

Table 2 shows the effect that changing the form of the WLF equation has on the calculated peak temperature values for 1000 Hz. In addition, it compares those values calculated from the different WLF constants with the corresponding values calculated from the Arrhenius equation. All data are for the PVC acoustical material.

The WLF constants compared are the same four sets discussed
earlier.

TABLE 2

T_{max} Obtained Using Different WLF Equations
vs. T_{max} Obtained from Arrhenius Equation

(All data for PVC Acoustic Material
Extrapolated to 1000 Hz)

WLF Equation Parameters			T_{max} from Time-Temperature Superposition		Origin of WLF Equation Parameter
T_0	C_1	C_2	for E"	for Tan δ	
-38°C	17.4	51.6	-20°C	23°C	"universal" WLF
-20°C	10.05	56.39	-21°C	33°C	from tan δ fit
-20°C	14.95	80.15	-19°C	20°C	from E" fit
-20°C	16.51	80.25	-21°C	16°C	from E' fit

For E", the form of the WLF equation has only a slight effect
on the temperature calculated from the extrapolated DMA data.
However, the value of T_{max} for tan δ varies widely with the different WLF equations. The WLF equation that gave the poor fit for
the E' curve (i.e., the WLF equation determined from shifting the
tan δ curve, Figure 9) gives a peak tan δ temperature approximately
twice the peak tan δ temperature (in °C) predicted by the Arrhenius
relation. The "universal" WLF equation and empirical WLF equation
determined for E" give T_{max} values significantly closer to those
predicted by the Arrhenius equation. However, the WLF equation
determined from superposition of the E' curve gives the best overall agreement with the Arrhenius values. In conclusion, it appears
that the most accurate form of the WLF equation for this material
is that empirically determined from superposition of E'.
Additionally, the method of comparing peak temperatures predicted
from the Arrhenius equation with those predicted from the reduced
variable master curves appears to be an effective means of checking
the form of the WLF equation.

Conclusions

A Fortran based program has been developed for the IBM XT (or compatible) that uses the principles of time-temperature superposition and the WLF equation to generate master curves of DMA data. The master curve is then represented on a reduced frequency nomograph which allows direct reading of modulus (E' or E") or loss tangent as functions of either frequency or temperature.

The form of the WLF equation used can be either the "universal" WLF equation (where $T_o = T_g$, $C_1 = 17.44$ and $C_2 = 51.6$) or an empirical relation. It has been shown that the "universal" WLF equation is well suited to the transition region and provides a good first approximation. However, by empirically shifting the DMA data, a master curve with superior precision can be calculated and plotted using the nomograph format. The Arrhenius relation has proved to be a valuable tool for validating the results of the generated master curve.

The procedure by which the nomograph is generated is not limited to the WLF equation. Since it is based on the reduced variable concept, any superposition equation that results in the calculation of a temperature shift factor may be used to calculate the needed data to create the master curve and subsequent nomograph. The software can easily be modified to calculate and display a master curve on some other superposition equation.

Literature Cited

1. Chartoff, R. P.; Graham, J. L. Advances in Chemistry Series, No. 197, on "Computer Applications in Applied Polymer Science", ed. T. Provder, Amer. Chem. Soc.: Washington, D.C., 1982, pp 367-375.

2. Analysyn Laboratories Polymers for Anechoic Coatings - The Effects of Molecular Structure, Analysyn Laboratories, A Task for Naval Sea Command, Contract #N00024-81-C-5313. Issued 1982.

3. Tobolsky, A. V. Properties and Structure of Polymers, John Wiley and Sons, Inc., 1960, pp 144-159.

4. Ferry, J. D. Journal American Chemical Society, 1950, 72, 3746.

5. Williams, M. L. Journal of Physical Chemistry, 1955, 59, 95.

6. Williams, M. L.; Landel, R. F.; and Ferry, J. D. <u>Journal American Chemical Society</u>, 1955, <u>77</u>, 3701.

7. Jones, D. I. G. <u>Shock and Vibration Bull</u>. 1978, <u>48</u> (2), 13.

8. Chartoff, R. P.; Drake, M. L.; Salyer, I. O. <u>Proceedings of the Symposium on Dynamic Mechanical Properties of Elastomers</u>, Naval Underwater Systems Center, July 1984, New London, CT.

9. Leaderman, H. <u>Textile Research J</u>. 1941, <u>11</u>, 171,

10. Leaderman, H. <u>Elastic and Creep of Filamentous Materials</u>, The Textile Foundation, Washington, D.C. 1943, pp 16,30,76 and 100.

RECEIVED January 24, 1990

Chapter 8

Selecting Damping Materials from Differential Scanning Calorimetry Glass-Transition Data

A. K. Sircar and Michael L. Drake

University of Dayton Research Institute, Dayton, OH 45469

An empirical equation was used to predict the peak
loss factor temperature of dynamic viscoelastic
measurements from DSC T_g data. The frequency for
DSC T_g was given an arbituary value of 10^{-4} Hz.
The equation offers a rough prediction of peak loss
factor temperature at dynamic frequency, thus
enabling the material scientist to select candidate
viscoelastic damping materials from DSC T_g data.
The predicted loss factor temperature is closer to
the actual temperature determined from the
nomograph plot of the viscoelastic data if the
experimental E_a is used in the calculation and the
same heating rate is used for both the dynamic and
DSC test methods.

Viscoelastic materials are frequently used for vibration and
noise damping. The properties of such materials do vary widely
as a function of temperature and frequency. Generally, a viscoelastic material is selected, such that its peak loss factor
temperature is in the region of the use temperature at a specific
resonant frequency. Since glass transition temperature (T_g) by
dynamic methods is often defined as the peak loss factor temperature and increases with increasing frequency, an obvious question
for material scientists and engineers is how does the glass
transition temperature of a polymer, as determined by static
methods (DSC, DTA, dilatometry) correlate with those by the
dynamic methods (dynamic modulus, resonant beam) that use variable frequencies? An answer to this question would enable the
material scientists and engineers to select suitable candidate
materials for high frequency applications from DSC (differential
scanning calorimetry) data, which are more easily determined and
are often available in the literature (1-3).

0097–6156/90/0424–0132$06.00/0

Experimental

Material - The polymers used in this study are given in Table I.
A general chemical identification of six of these polymers is
shown in Figure 1. Dyad 609 is a proprietary material trade
marketed by The Soundcoat Company, Inc. Hypalon, manufactured by
DuPont and Vinac B100, manufactured by Air Products, were used as
compounded. The exact formulations for the compounded polymers
are proprietary. The other polymers shown in Table I and Figure
1 were used without compounding. All materials were used in
uncured condition. The polymers and compounds in Table I were
selected for study, since the data were already available from
previous investigations. Data for both filled and unfilled
polymers will be the subject of a follow-up future investigation.

Measurement of Dynamic Modulus - Three different instruments were
used to measure dynamic modulus and calculate the viscoelastic
parameters of loss modulus, elastic modulus, and the ratio (loss
factor, tan δ). They were the Rheometrics Solids Analyzer (RSA
II), Dynamic Mechanical Thermal Analysis (DMTA), and the ASTM,
E756-83, Beam Test. The capabilities of each test method with
regard to the frequency range, temperature and test specimens are
shown in Table II. Frequency ranges used for this investigation
are 1 to 15 Hz 0.3 to 50 Hz and 200 to 4000 Hz for RSA, DMTA and
Beam Test respectively. DMA-983 at frequencies 0.01, 0.1, 0.5
and 1 Hz was used for the polymethyl methacrylate sample.
Comparison of the complex modulus data generated by the three
different measurement techniques in Table II have been discussed
recently by the authors (4). A stepwise heating of 2°C/minute,
with 3 minutes equilibration time at a constant temperature was
used for RSA II. DMTA experiments were carried out at 1°C/minute
in ramp mode. Step heating with 30 min. equilibration at each
temperature was used for beam tests. A short description of the
beam test and the procedure for obtaining a nomograph is
described below.

Beam Test

The resonant beam test technique forms the basis of the ASTM
Standard E756-83 for measuring the viscoelastic properties of
damping materials. Fundamentally, the beam test requires that
the resonant frequencies of a metal-beam, mounted in cantilever
fashion, be determined as a function of temperature and
frequency; the beam is then coated with a polymer and the
resonant frequencies and corresponding modal damping of the
composite beam are determined as a function of temperature and
frequency. From these two data sets, the vibration damping
properties of the polymer can be evaluated. The ASTM Standard
provides the necessary equations to obtain the complex modulus
data from the collected test data and also provides guidelines
for the proper choice of the specimens (1,2). The principal
difference between the beam test and the other methods used here
is that the beam test calculates the material properties from the
test results on the metal beam and the composite beam whereas the

TABLE I. PREDICTION OF DYNAMIC GLASS TRANSITION TEMPERATURE FROM DSC DATA

Sample Description	DSC T_g 10°C/Min	ΔE (Exp) from RSA II, kJ/M	Predicted T(C°) at 1800 Hz ΔE=400 kJ/Mole	Predicted T(°C) at 1800 Hz with Exp. ΔE	Peak Loss Factor Temp. °C, Obtained from Nomograph RSA II Data	Peak Loss Factor Temp. °C, Obtained from Nomograph Beam Data	Comments
Hypalon 48 + filler	-9.7	288	14	22.6	23	28	
Poly(vinylacetate) + filler	35	351	67	71	68	74	
Plexiglass GP Poly(methyl methacrylate)	111	725	160	138	132	-	Measurements were by DMA multiplex at 2°C step heating
Lexan 141 Bisphenol A Poly(carbonate)	143.7	899	202	169.6	165	-	
Ultem 1000 Poly(etherimide)	212	957	291	245	239		
Victrex 5200 P Poly(sulfone)	219.5	1177	301	247	250		
DYAD 609 Proprietory Material, Soundcoat	-8	307	16	33	54	77	

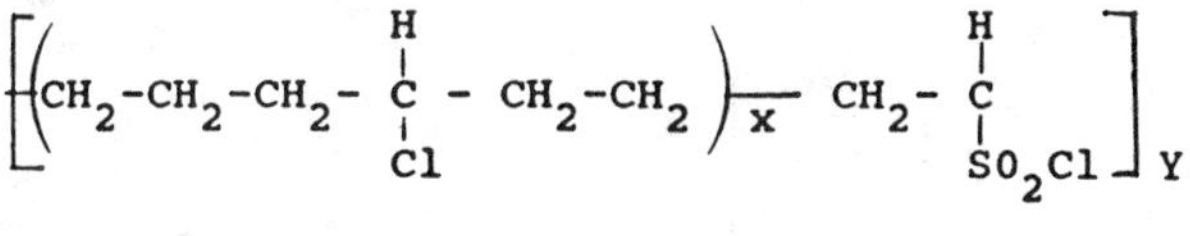

HYPALON, Chloro-sulfonated poly(ethylene)

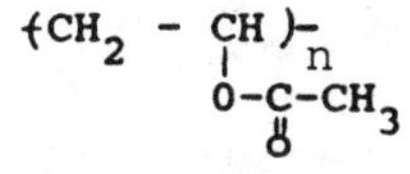

Vinac B-100, Poly(vinylacetate)

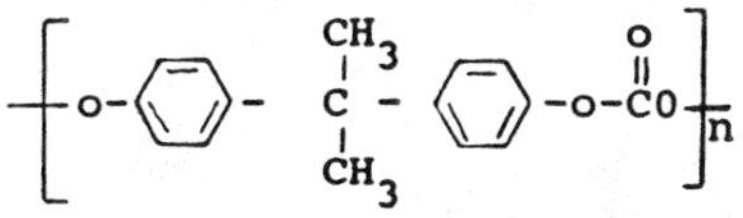

LEXAN 141, Poly(bisphenol A carbonate)

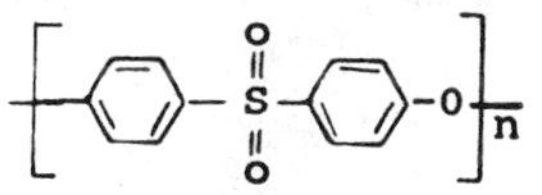

Victrex, Poly(sulfone)

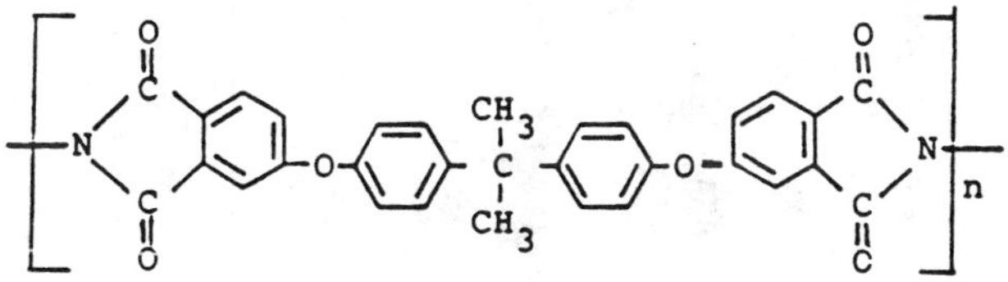

Ultem, Poly(etherimide)

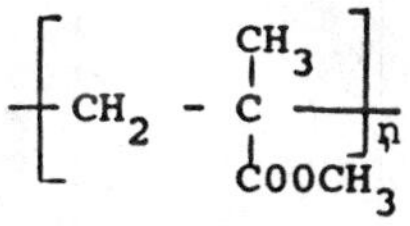

Plexiglas GP, Poly(methyl methacrylate)

Figure 1. Chemical Structure of the Polymers Under Study.

TABLE II

METHODS FOR DYNAMIC MODULUS MEASUREMENTS

TEST PARAMETERS	ASTM,E-756 BEAM TEST	POLYMER LABORATORIES DMTA	RHEOMETRICS RSA - II
Frequency Hz	100 to 12000	0.01 to 200	0.0016 to 16
Temperature (°C)	-73 to 1093°C	-150 to 800°C	-150 to 600°C
Testing Mode	Free layer or sandwich uniform	Shear & bending	Shear & bending

other methods are tests conducted directly on the material
samples.

Nomograph

The test results of a material damping test are most useful when
placed on a reduced temperature nomograph, which plots the
limited number of test results to a graph from which one can
obtain the damping properties (modulus and loss factor) at any
given combination of temperature and frequency. The WLF equation
(5) is used to obtain a nomograph for the results of each test.
A reference temperature (T_ϕ) corresponding to the best data fit
is used in the nomographic representation of the data.
    Figure 2 is a reduced temperature nomograph which
demonstrates the procedure for reading the nomograph as follows:
Select a combination of temperature and frequency, for example,
1000 Hz and 75°C. Find the point for 1000 Hz on the right-hand
frequency axis. Proceed horizontally to the temperature line for
75°C. At this intersection, draw a vertical line. Then, read
the modulus and loss factor values from the appropriate data
curve, at the point of intersection with the vertical line. In
this example, modulus G(1000 Hz, 75°C) = 8 x 10^6 N/m^2 and loss
factor (1000 Hz, 75°C) = 1.96.
    The temperature scale at the top of the reduced temperature
nomograph shows increasing temperature from right to left.
Labeling of the temperature lines is done in a uniform temperature increment. The temperature increment is identified at the
top of the nomogram. The nomograph presentation and data reduction procedure have been described in earlier publications (6,7).

Measurement of Glass Transition Temperature by DSC - Glass transition temperature by the static method (zero frequency) was
measured by the DuPont 910 Differential Scanning Calorimeter
using a DuPont 2100 Thermal Analysis System. The samples were
annealed 50°C above the T_g of the polymer and quench cooled by
pouring liquid nitrogen on the top of the cooling attachment. T_g
was determined at 5, 10, 20 and 40°C per minute rate of heating.
The multiple heating rates allowed the extrapolation of T_g to a
zero rate. The extrapolated T_g value at 0°C/min. heating rate is
comparable to that by dilatometry at a low heating rate (0.6
degree/min.) (8). Both the extrapolated T_g and the T_g measured
at 10°C/min. (determined from least square plot), were used for
the calculations, since the latter is more generally available in
the literature. The experimental conditions for the dynamic
modulus experiments, as described above, were close to zero rate
of heating, since the samples were either equilibrated at a
constant temperature or heated at a slow rate (1°C/min.). A plot
of rate of heating versus T_g for the polyvinyl acetate compound
is shown in Figure 3.

Measurement of Apparent Activation Energy - The apparent activation energy, E_a, was determined by using the Arrhenius activation
energy equation for the dependence of T_g on frequency in dynamic
mechanical measurements. In this equation, the temperature shift

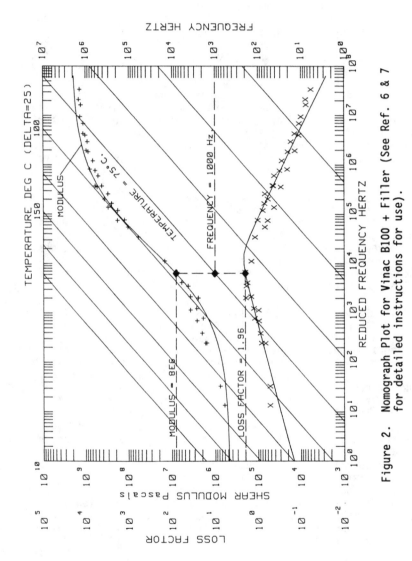

Figure 2.   Nomograph Plot for Vinac B100 + Filler (See Ref. 6 & 7
for detailed instructions for use).

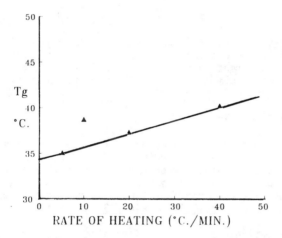

Figure 3.   Rate of Heating vs Glass Transition Temperature for Poly (vinyl acetate) Compound.

of the peak loss modulus and/or loss factor is controlled by the
activation energy of the process:

$$K = A \exp (-E_a/RT) \tag{1}$$

where:

K = reaction rate = frequency (f)
A = constant
E = activation energy
R = gas constant
T = absolute test temperature

Taking the log of the equation (1) yields:

$$\log f = \frac{-E_a}{2.303R} \frac{1}{T} + \log \tag{2}$$

From this equation it can be seen that E_a can be determined from
the slope of log f vs. 1/T plot. An example of the DMTA peak
temperature shifts at different frequencies is given in Figure 4
for the filled Vinac compound. Activation energy plots for RSA
II and DMTA exeriments were actual experimental data, while E_a
for the resonant beam method was obtained from data extrapolated
from the nomograph.

Results and Discussion

Wetton (9) has described the mechanical characteristics for
vibration damping materials in terms of the frequency and temperature dependence of the viscoelastic properties of polymeric
materials. Use of polymeric materials in free-layer and constrained layer damping configurations has been discussed in the
literature by Ungar (10-12), Kerwin (13,14), and others (15,16).

Empirical Relationship - Empirical relationships correlating
glass transition temperature of an amorphous viscoelastic
material with measurement temperature and frequency, such as the
William Landel Ferry equation (17) and the form of Arrhenius
equation as discussed, assume an affine relationship between
stress and strain, at least for small deformations. These
relationships cover finite but small strains but do not include
zero strain, as is the case for the static methods such as differential scanning calorimetry. However, an infinitely small
strain can be assumed in order to extend these relationships to
cover the glass transition temperature determined by the static
methods (DSC, DTA, dilatometry). Such a correlation which uses a
form of the Arrhenius equation was suggested by W. Sichina of
DuPont (18).

$$\ln \frac{fDMA}{fDSC} = \frac{E_a}{R} \left(\frac{1}{T_g DSC} - \frac{1}{T_g DMA} \right) \tag{3}$$

Equation (3) is nothing more than subtracting the fDMA and
fDSC frequencies in the Arrhenius equation (Equation 1). This

relationship predicts that the DMA T_g (based on the loss peak or loss factor temperature) will be higher than the DSC T_g when heated at the same rate. In an attempt to correlate DSC T_g with that measured by DMA, Sichina proposed that the magnitude of the shift of the DMA T_g versus the DSC T_g can be estimated from the following equation:

$$\Delta T_g = \frac{(T_g DSC)^2 \; 8.314}{400,000} \; \ell n \; \frac{(fDMA)}{10^{-4}} \qquad (4)$$

where:

$$\Delta T_g = T_g \, DMA - T_g \, DSC,$$

$$T_g DSC \times T_g DMA \sim (T_g DSC)^2$$

The reference frequency for DSC was assigned a low value (fDSC = 0.0001 Hz). The value does not have any physical sig-nificance and was chosen for best data fit. This equation requires the value of the activation energy for the polymers which generally lies between 200-900 kJ/mole. An intermediate value of 400 kJ/mole is suggested as an approximation, if the activation energy of the polymer is unknown. The T_g at specific dynamic frequencies can now be calculated if the DSC T_g at the same heating rate is known.

It is evident that the equation is not only empirical but is also arbitrary. Also, the activation energy for many polymers differs widely from the assumed value of 400 kJ/mole.

Test of the Empirical Equation - Table I compares the calculated T_g values at 1000 Hz using the DSC data at a heating rate of 10°C per min. T_g was calculated using both the assumed activation energy of 400 kJ/mole as well as the experimentally determined values by RSA II multiplex experiments. The predicted data were compared with the extrapolated peak loss factor temperature from the nomographs of RSA II and beam data.

It was evident that: (1) the assumed activation energy of 400 kJ/mole gives a good prediction only in the case of the polyvinyl acetate compound where the experimental activation energy was approximately the same. (2) The prediction improved when experimental E_a was substituted for the assumed value. (3) Further improvement was observed when DSC T_g at zero heating rate was substituted for 10°C/min. as will be evident from Table III. Considering the empirical nature of equation (3) and the ar-bitrary value used for DSC frequency, the correlation is quite good. Thus, the method should allow a rough prediction of the loss factor temperature at higher frequencies from DSC T_g data, provided E_a is close to 400 kJ/mole or experimentally determined E_a values are used. It may also be observed in Table I that the loss factor temperatures from the beam nomograph are slightly higher than those from the RSA II nomograph. This may be due to the different reference temperatures used to plot the nomographs. This will be discussed later.

An exception to the above was the Dyad 609. Dyad 609 is a proprietary material with unknown composition. We had difficulty

TABLE III

COMPARISON OF PREDICTED T_g WITH DSC DATA AT 10°C AND 0°C RATE OF HEATING

Sample Description	DSC T_g, °C 10°C/min	DSC T_g, °C Extrapolated to Zero Rate	ΔE (Experimental) from RSA II kJ/M	Predicted T_g (°C) at 1000 Hz			Comments
				Using Exp. ΔE value and DSC T_g at 10°C/min	Using Exp. ΔE value & DSC T_g at 0°C/min	From nomograph RSA II Data	
Hypalon 48 + Filler	-9.7	-12.5	288	22.3	19.1	23	
Polyvinyl-acetate + Filler	35	33	351	71	70.7	68	
Plexiglas GP Poly(methyl methacrylate)	111	107.8	725	138	134.6	132	Measurements of Dynamic Modulus by DMA Multiplex 2°C step heating
Lexan 141 Bisphenyl A Poly(carbonate)	143.7	141	899	169.6	166.6	165	
Ultem 1000 Poly(etherimide)	212	210.4	957	245	243	239	
Victrex 5200 P Poly(sulfone)	219.5	218.3	1177	247	246	250	

measuring a DSC T_g for Dyad 609. Supplier's data (19) show a
brittleness temperature of -30°C, but a peak loss factor temperature of 82°C. In all probability Dyad 609 is a multicomponent
system with different glass transitions and thus not amenable to
the above calculation procedure.

Table III also shows that E_a increases with increasing DSC
T_g. This would be expected from restricted segmental mobility of
the high T_g samples. Lewis (1) found that Arrhenius plots of log
frequency versus reciprocal dynamic glass transition temperature
for restricted and nonrestricted polymers converges to a different point in the frequency/temperature scale. From this
finding, equations were derived to predict static T_g from the
dynamic T_g value and vice versa.

Activation Energy Using Loss Modulus or Loss Factor Curve -
Activation energy can be determined either from peak loss modulus
or peak loss factor temperature. Table IV compares the data
obtained by either methods along with the correlation coefficient
of the linear regression plots for determining activation energy.
It is evident that either method is equally satisfactory, as
would be expected.

Activation Energy by Different Measurement Techniques - Table V
compares activation energy data obtained by three different
experimental methods for three different samples. Corresponding
Arrhenius plots are shown in Figures 5 and 6. In both figures
the log frequency vs. reciprocal temperature plot for RSA II and
DMTA are parallel to each other, giving very close E_a values.
The same reference temperature gives a good nomograph for the
complex modulus data obtained from these two methods. However,
the beam data has a good nomograph fit at a lower reference
temperature and an acceptable nomograph if the RSAII-DMTA
reference temperature is used. If the E_a is calculated from the
beam data nomograph generated with the T_g used for the other two
test methods, then the E_a values determined from all these tests
are approximately equal. However, if the nomograph generated
with the T_ϕ chosen from only the beam data is used, the E_a value
derived is about one-third of the value from the other two
methods, as is evident from Figures 5 and 6 and Table V. The
reason for this apparent anomaly is not known.

Activation Energy from DSC Heating Rate

The variation of apparent T_g with DSC heating rate has generally
been attributed to the thermal lag of the sample, which increases
in step with the heating rate (20-22). This thermal lag is
composed of a machine path error and a sample error which are
dependent on the characteristics of the instrument and the
sample, respectively (21). However, T_g itself is a rate dependent parameter, and a dependence on scan rate involving a
relaxation activation energy is to be anticipated over and above
any thermal lag errors.

The determination of activation energy from the rate expressions involving heating (8) and cooling rate (5,23) in DSC has
been attempted earlier. Unfortunately, such determinations

TABLE IV

COMPARISON OF ACTIVATION ENERGY DATA BY LOSS MODULUS
AND LOSS FACTOR PEAKS FROM RSA II DATA

Sample	Identification	Activation Energy, kJ/M		Correlation Coefficient	
		Loss Modulus	Loss Factor	Loss Modulus	Loss Factor
	Hypalon 48 Formulation + Filler	334	287	0.99	0.97
	Hypalon 48 Formulation + Filler	282	288	0.93	0.98
	Vinac B100 Poly(vinylacetate) Formulation + Filler	-	351	-	0.99
Dyad 609	Acrylate Polymer by Sound Coat	252	307	0.96	0.99
Plexiglas GP	Poly(methyl methacrylate)	725		0.99	
Lexan 141	Poly(carbonate) of Bisphenol A	933	899	0.99	0.98
Ultem 1000	Poly(etherimide)	860	957	0.98	0.99
Victrex 5200 P	Poly(sulfone)	1077	1177	0.99	0.99

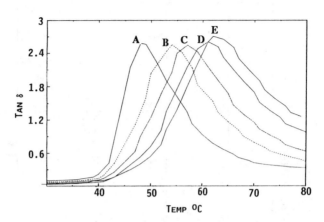

Figure 4.   Frequency Dependence of Loss Factor Peak Temperature
for Poly(vinyl acetate) + Filler.
A = 0.3 Hz; B = 3 Hz; C = 10 Hz; D = 30 Hz; E = 50 Hz
1°C/min, Dual Cant. Clamps L/C.

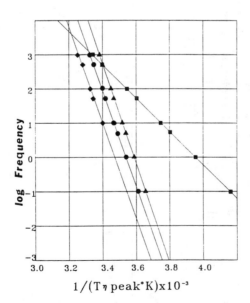

$$1/(T\eta \, peak^\circ K) \times 10^{-3}$$

Figure 5.   Arrhenius Plot for Hypalon 48 Compound.
● DMTA (T_0 = 165°C); ■ Beam Test (T_0 = 60°C); ◆ Beam
Test ($T_0 \cong$ 165°C); ▲ Rheometrics (T_0 = 165°C)

TABLE V

COMPARISON OF ACTIVATION ENERGY BY THREE DIFFERENT METHODS

Sample Description	DSC T_g, 10°C/min	Activation Energy, kJ/Mole Determined by			
		Rheometrics Data	DMTA Data	Beam Curve Same T_ϕ	Beam Curve T_ϕ, best fit for nomograph
Poly(vinylacetate), + Filler	35	351	348	325, $T_o = 200$°C	115, $T_o = 121$°C
Hypalon 48 + Filler	-9.7	287	294	262, $T_o = 165$°C	91, $T_o = 60$°C
Dyad 609	-8	307	290	249, $T_o = 180$°C	115, $T_o = 115.6$°C

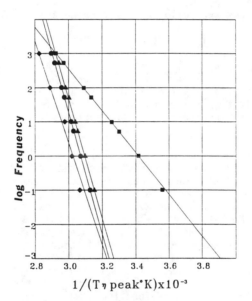

Figure 6.   Arrhenius Plot for Poly(vinyl acetate) Compound.
● DMTA (T_0 = 200°C); ■ Beam Test (T_0 = 93°C);◆ Beam
Test ($T_0 \cong$ 200°C) ▲ Rheometrics (T_0= 200°C)

suffer from some practical and theoretical limitations. Apart
from the machine and sample errorrs discussed above, T_g determination by DSC are much more dependent on the thermal history of
the sample than are the dynamic methods (24). Also, the effect
of dynamic frequency changes on T_g is more thoroughly studied and
much better understood than the effect of heating and cooling
rates, revealing a shortcoming in modern polymer theory. If it
were successful, however, one could determine activation energy
from the data similar to those in Figure 3, thus, providing the
"E_a" value in equation 3 from DSC determinations of glass transition temperature. Such expressions would be heating rate
analogues of the Arrhenius activation energy equation for the
dependence of T_g on frequency in dynamic mechanical measurements.
     Limited attempts were made during this investigation to
determine E_a from the following equation derived by Barton (8):

$$\ln (\Phi/T^2) = C - E_a/RT \qquad (4)$$

where T is the apparent T_g corresponding to the heating rate Φ.
However, the values determined do not correspond to those by the
dynamic methods (Table 4 and 5) and were not therefore included
for this publication. The reason for the deviation of the values
in E_a may be ascribed to the factors mentioned above.

Conclusions

The empirical equation (3), derived from the Arrhenius equation,
allows a rough prediction of peak loss factor temperature at
dynamic frequencies from DSC glass transition temperature data.
As is well known, DSC T_gs are very easy to measure or are often
available in the literature. Thus, an initial screening of the
potential polymeric candidates is possible from DSC T_g data.
Although the accuracy of the above prediction varies according to
the nature of T_g data available, value of energy of activation as
well as the approximations in the equation and calculation, such
derivation is of great practical importance to select polymeric
material for high frequency vibration damping applications.
     The accuracy of the estimation improves if the actual value
of energy of activation (E_a) is available from experiment or from
the literature. Also, T_g values by DSC and dynamic methods are
affected the same way by the rate of heating. Therefore, the
estimated loss factor temperatures should be for the same rate of
heating as the DSC data.
     The method does not work for nonmiscible blends such as Dyad
609.

Acknowledgments

The authors would like to thank Peter Weissman, Mary Galaska, and
Tim Montavon for help in obtaining the experimental data. They
also wish to thank I. O. Salyer for reviewing the paper and some
helpful comments.

Literature Cited

1. Lewis, A. F., J. Polymer Sci., 1963, B1, 649.

2. Nielson, L. E., Mechanical Properties of Polymers and Composites, Marcel Dekker, New York, 1974.

3. Van Krevelen, D. W., Properties of Polymers: Correlations With Chemical Structures, Elsevier, 1972.

4. Drake, M. L.; Sircar, A. K., Damping '89, West Palm Beach, FL, February 8-10, 1989, Proceedings to be published as an AFWAL Technical Report.

5. Ferry, J. D., Viscoelastic Properties of Polymers, John Wiley and Sons, Inc., New York, 1961.

6. Jones, D. I. G., Shock and Vibration Bull., 1978, 48(2), 13.

7. Chartoff, R. P.; Graham, J. L., In Advances in Chemistry Series, No. 197, on Computer Applications in Applied Polymer Science, Provder, T. ed., American Chemical Society, Washington, D.C., 1982, pp. 367-375.

8. Barton, J. M., Polymer, 1969, 10(2), 151.

9. Wetton, R. E., Applied Acoust., 1978, 11, 77.

10. Ungar, E. E., In Noise and Vibration Control, Branck, L. L., ed., McGraw Hill, New York, 1971, Ch. 14.

11. Ungar, E. E., J. Acoust. Soc. Am., 1962, 34, 1082.

12. Ungar, E. E., Machine Design, 1962, February 14, pp. 162-168.

13. Kerwin, E. M., Jr., J. Acoust. Soc. Am., 1959, 31, 952.

14. Ungar, E. E.; Kerwin, E. M., J. Acoust. Soc. Am., 1968, 36, 388.

15. Ball, G. L.; Salyer, I. O., J. Acoust. Soc. Am., 1966, 39, 663.

16. Oberst, H., Acustica, 1952, 2, AB-181.

17. Williams, M. L.; Landel, R. F.; Ferry, J. D., J. Am. Chem., Soc., 1955, 77, 3701.

18. Sichina, W. TA Hotline, Fall, 1986, p. 3.

19. Soundcoat Product Data Sheet, Bulletin 701 and 810.

20. Lambert, A., Polymer, 1969, 10, 319.

21. Strella, S.; Erhardt, P. F., J. Appl. Polym. Sci., 1969, 13, 1373.

22. Illers, K. H., Eur. Polym. J.,, 1974, 10, 911.

23. Kovacs, A. J., Adv. Polym. Sci., 1963, 3, 394.

24. Shalaby, S. W., In Thermal Characerization of Polymeric Materials, Turi, E., ed.: Academic Press, New York, 1981, p. 257.

RECEIVED January 24, 1990

Chapter 9

Emission of Acoustic Waves from Polymers under Stress

Rheo-Photoacoustic Fourier Transform–IR Spectroscopy

Marek W. Urban and William F. McDonald

Department of Polymers and Coatings, North Dakota State University, Fargo, ND 58105

A new rheo-photoacoustic Fourier transform infrared cell has been developed to perform stress-strain studies on polymeric materials. The rheo-photoacoustic measurements lead to the enhancement of the photoacoustic signal and allow one to monitor the effect of elongational forces on the molecular structure of polymers. Propagating acoustic waves are detected as a result of infrared reabsorption and the deformational and thermal property changes upon the applied stress.

Several years ago, rheo-optical Fourier transform infrared spectroscopy emerged as a method for studying the deformation phenomena during mechanical stretching of polymeric materials.[1] This approach has allowed the determination of structural changes within the polymer network. In essence, two spectroscopic observations were reported. Experiments performed by Siesler[1] revealed that the infrared band intensities change upon application of mechanical stress to polymeric materials. Wool et al.,[2] on the other hand, detected shifts of infrared bands and further supported these observations with the theoretical approach based on conformational energy minimization methods. Both harmonic and anharmonic (Morse) potential energy functions were applied to the C-C stretching modes in the valence force field. In spite of the fact that the detected wavenumber shifts were very small, the energy minimization calculations yielded good agreement with the frequency shifting coefficients obtained experimentally.

While the wavenumber shifts were primarily observed

0097–6156/90/0424–0151$06.00/0

in uncrosslinked thermoplastic systems[3], the intensity changes observed in the spectra of crosslinked natural rubber as well as poly(vinylidene fluoride)[3] were attributed to the stress induced phase changes and stress induced crystallization. The stress induced phase changes were supported by wide angle x-ray diffraction experiments.[1,4] Although the experiments of Siesler and Wool seem to provide controversial observations, it should be remembered that the above studies have been conducted on different systems. For example, the[5] mechanical deformation of polyethylene and polypropylene is quite different from that of crosslinked natural rubbers.[6]

In spite of the fact that both studies have reported important spectral features associated with the structure-property relationship, an Achilles heel of both approaches results from the necessity of using optically transparent films to allow infrared light to pass through the sample. New materials, such as fibers and composites, cannot be studied by transmission FT-IR techniques because they are often optically opaque. Thus, in order to monitor structural changes induced by external forces, it is necessary to utilize a method permitting the detection of infrared spectra on any material, regardless of its optical properties, shape or thickness.

Although the applications of photoacoustic FT-IR spectroscopy have shown several promising features in various sampling situations,[7,8,9,10] the rheo-photoacoustic measurements have been reported only recently.[11] In this work, a novel approach utilizing photoacoustic FT-IR spectroscopy to monitor the elongation processes in fibers is described. In an effort to monitor the events on a molecular level occurring upon static load on virtually all materials, rheo-photoacoustic FT-IR cell was designed.

Although the applications and theory of photoacoustic FT-IR spectroscopy have been recently reviewed,[8,9,11] here, we will only briefly mention that the detection of the photoacoustic signal is a two stage process. This is depicted in Figure 1. First, infrared light is absorbed by the sample and the reabsorption process produces heat generating acoustic waves which are detected by a sensitive microphone. The theory that governs photoacoustic detection was described by Rosencwaig and Gersho.[12] However, the second stage, that is the production of heat and subsequently acoustic waves, can be induced by external forces leading to deformational and conformational changes within the polymer. A simple example of similar phenomenon is the cracking of ice on a pond producing sounds audible to the human ear. Thus, is it our hypothesis that the molecular level movements within polymer will also affect the intensity of acoustic waves generated as a result of the light absorption-reabsorption process.[13] If stress is induced in a polymeric material and photoacoustic FT-IR measurements are performed, in addition to a "normal" photoacoustic infrared spectrum obtained as a result of the reabsorption process, an acoustic signal due to deformations within the polymer will occur. With this in mind, we will monitor the deformations of poly(p-phenylene terephthalamide) (PPTA) fibers and polyethylene films, and analyze the spectral changes occurring in infrared when the external forces are applied.

EXPERIMENTAL

The poly(p-phenylene terephthalamide) (PPTA) fibers were used as received from Du Pont de Nemours. The mechanical properties of the fibers were analyzed on an Instron Model TM mechanical tester following a common procedure to determine their elongation to break. Each tested sample consisted of 30 strands of fiber. The same procedure was employed to monitor fiber and film elongations, followed by simultaneous collection of infrared spectra. The cell design was reported elsewhere[14] and allows measurements of various samples such as fibers, composites, and films. The procedure of loading the sample consists of removing the cell top, clamping the sample in the clamping blocks, turning the lead screw the desired amount to elongate the sample, purging the cell with helium for several minutes and pressurization. In a typical experiment, each infrared spectrum consists of an average of 400 scans recorded at a 4 cm^{-1} resolution with a mirror velocity of 0.3 cm/sec. All PA FT-IR spectra were recorded on a Digilab FTS-10M FT-IR spectrometer. The coadded sample scans were ratioed against a carbon black reference. All spectra were transferred to an AT compatible computer for further spectral manipulations utilizing Spectra Calc. software (Galactic Industries).

RESULTS AND DISCUSSION

Before we begin the analysis of the photoacoustic FT-IR spectra of PPTA fibers and films recorded as a function of the sample elongation, it is first necessary to set the stage and define our approach. In monitoring the photoacoustic effect, infrared light strikes the sample surface and, due to reabsorption processes, heat is released generating acoustic waves on the surface. When the fibers are elongated, conformational changes as well as the molecular deformations occur. Although these deformational processes also produce acoustic waves which may contribute to the intensity of the photoacoustic spectrum, these are two independent acoustic processes. Thus, insights into the molecular changes that occur as a result of elongational or shear forces in polymeric materials can be gained.

With this in mind let us examine the spectral changes

detected in PPTA fibers when the fibers are subjected to external forces. Figure 2 illustrates the rheo-photoacoustic FT-IR spectra of the PPTA fibers in the 4000-2500 cm^{-1} region. Traces A through F represent the spectra of the PPTA fibers elongated at 0.0 %, 0.42%, 0.83%, 1.25%, 1.67%, and 2.08%, respectively. As seen, steadily increasing intensities of the bands in this spectral region, including the N-H stretching band at 3327 cm^{-1}, are observed.[15,16] Upon 2.08% elongation (trace F), however, the 3327 cm^{-1} band becomes weaker. The intensity of this band is of particular interest because of extensive N-H   O=C associations between neighboring chains. Thus, one would also expect that the C=O stretching band at 1656 cm^{-1} will respond similarly under the applied shear forces. Indeed, this band shows the same trend in the intensity changes as that at 3327 cm^{-1}. This is illustrated in Figure 3, A, with the Y-scale from 0 to 250 to the far left.

In an effort to further relate the molecular deformations of PPTA fibers upon their elongation, the integrated intensities of both bands were plotted as a function of percent fiber elongation. As seen in Figure 3, B, (the corresponding Y-scale from 0 to 7000 is on the right), the intensities of both bands increase as the fiber is stretched. The intensity decreases, however, when the sample breaks, indicating that this band is sensitive to the shear forces involved when the fiber is elongated. It is also interesting to note that the shapes of both curves depicted in Figure 2, resemble a load versus percent elongation curve obtained from an Instron mechanical analyzer (Figure 3, C, with the Y-scale to the inside left).

The intensity changes of the N-H and C=O bands are mostly responsible for the intermolecular forces between the polymer chains. Let us now determine how the polymer backbone is affected by the external stretching forces. In order to do so, the band due to the C-C aromatic stretching at 1408 cm^{-1} and the C-N in-plane bending at 1261 cm^{-1} modes will be examined. Figure 4, traces A through F, depicts the rheo-photoacoustic spectra in the 1450-1200 cm^{-1} region.

Similar to the N-H and C=O stretching modes, the intensities of both bands increase as the fiber is elongated. However, the maximum intensity is reached at 1.25% elongation (trace D). This is illustrated in Figure 5, traces A and B, which presents plots of the integrated intensities of both bands as a function of elongation. The behavior depicted in Figure 5 indicates that the aromatic and C-N groups of the polymer backbone are affected in the elongation region from 0.83% to 1.25 %. In contrast, the N-H and C=O groups are sensitive virtually throughout the entire elongation process. Although the breakage region is the same for all bands, the molecular processes leading up to breakage are detected later for the aromatic C-C and the C-N bands than for the N-H and C=O bands.

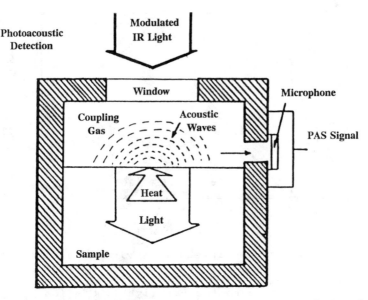

Figure 1. Schematic diagram of photoacoustic detection.

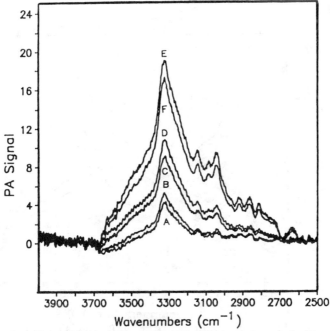

Figure 2. Rheo-photoacoustic FT-IR spectra of the PPTA fibers in the 4000-2500 cm^{-1} region at various stages of elongation: A - 0.0%; B - 0.42%; C -0.83%; D - 1.25%; E - 1.67%; F - 2.08%.

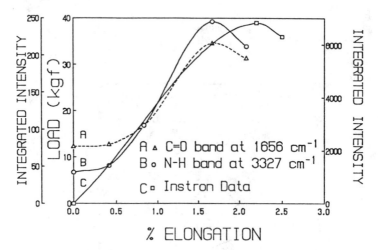

Figure 3. Integrated intensities plotted as a function of percent PPTA fiber elongation: A - carbonyl band at 1656 cm^{-1} (Y-scale to the far left); B - N-H band at 3327 cm^{-1} (Y-scale to the far right); C - load plotted as a function of percent PPTA fiber elongation from the stress - strain mechanical tester (Y-scale to the inside left).

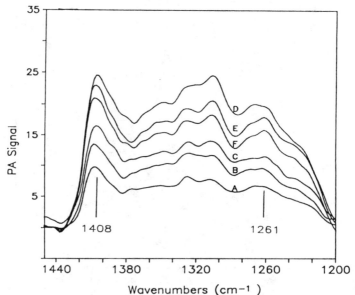

Figure 4. Rheo-photoacoustic FT-IR spectra of the PPTA fibers in the 1450-1200 cm^{-1} region at various stages of elongation: A - 0.0%; B - 0.42%; C -0.83%; D - 1.25%; E - 1.67%; and F - 2.08%.

A comparison of the N-H and C=O bands with the aromatic C-C and the C-N bands with respect to detecting the onset of fiber breakage is important because of the different contributions of each group to the chemical structure of the PPTA fiber. Figure 6 shows the structure of the PPTA fiber and indicates that the N-H and C=O groups are side groups and, therefore, may participate in hydrogen bonding with the amide carbonyl and amide N-H groups of neighboring chains. On the other hand, the C-C aromatic stretching band arises from the aromatic ring of the polyimide backbone. The aromatic ring is not capable of participating in the intermolecular interactions such as hydrogen bonding, although it does interact with the neighboring chains through Π-Π interactions. Like the aromatic C-C bonds, the C-N bond is an integral part of the backbone and should not be appreciably affected by the hydrogen bonding.

In view of the above considerations, the analysis of the integrated intensities of the N-H and C-C normal vibrations indicates that while the intensities of the N-H band (Figure 3, B) continuously change with the applied load, the C-C aromatic band (Figure 5, A) is only sensitive when the elongation reaches 0.83%. Above 1.25% elongation, the intensity of this band remains virtually unchanged. This behavior may suggest the separation of Π orbitals of the two neighboring rings or conformational changes due to stresses imposed on the fiber. Although one could propose other related phenomena, such as a movement of crystallites with respect to each other, or deformations within the crystalline regions, at this point there is no evidence as to which part of the fiber contributes to the observed phenomenon.

As was reported in earlier studies,[4] extensive hydrogen bonding between N-H and C=O groups of neighboring chains contributes significantly to the mechanical integrity of the polyaramid (Kevlar$_R$) fiber. Thus, during the stretching process, an equilibrium between H-bonded and non-H-bonded groups will shift in the direction of the non-H-bonded species. The dissociation of hydrogen bonds will enhance acoustic waves generated as a result of infrared reabsorption at the energy levels required to disrupt the bonding. As a result, the intensity of the N-H and C=O bands will increase. As illustrated in Figure 3, the intensity changes in the elongation profile reach a maximum and further elongation results in the fiber breakage. Upon breakage, however, the intensities of the N-H and C=O bands slightly decrease indicating that a new H-bonded-non-H-bonded equilibrium has been established and upon load release, a fraction of the original H-bonded species has been reformed. Such behavior is not observed for the C-C aromatic and the C-N bands in the elongational region from 1.25% to 2.08%; their intensities remain virtually unchanged.

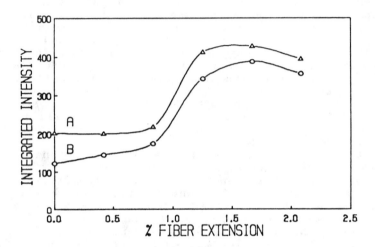

Figure 5. Integrated intensities plotted as a function
of percent PPTA fiber elongation: A - C-C aromatic
band at 1408 cm^{-1}; B - the C-N-C amide band at 1261
cm^{-1}.

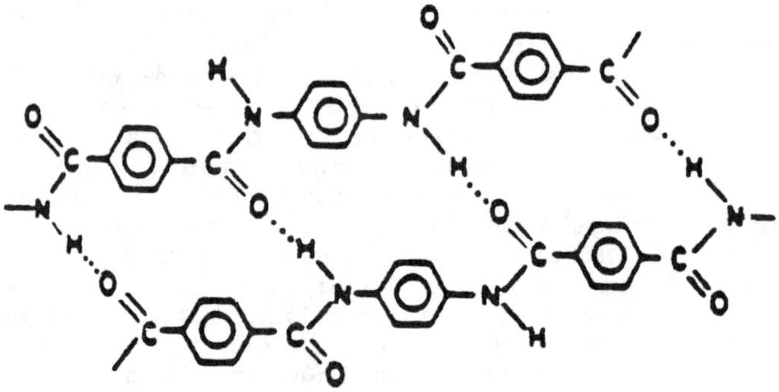

Figure 6. Structure of the PPTA fiber with hydrogen
bonding between amide groups of neighboring PPTA
chains.

In order to further correlate the response of various functional groups to the applied stress, the bond dissociation energies of the fiber functional groups were compared. One would expect the weakest bonds to break first, followed by the next weakest and so on, until the fiber sample is stressed to the breakpoint. The hydrogen bonds are the weakest bonds present in the molecule and thus, dissociate first. The dissociation energy, ΔH, of the hydrogen bonds range from 17 - 21 kJ/mole.[17,18] Once a fraction of the hydrogen-bonded amide groups partially dissociate, the N-H and C=O groups will not be appreciably affected by the elongational forces. However, the $C_{(r)}$-N and $C_{(r)}$-$C_{(o)}$(where "r" indicates the aromatic ring carbon and "o" the carbonyl carbon) bonds of the polymer backbone exhibit dissociation energies of 300 and 275 kJ/mole, respectively. The N-$C_{(o)}$ bond of the amide groups is relatively strong with a ΔH = 305 kJ/mole. The C-C bonds of the aromatic rings are stronger still, with a dissociation energy of 720 kJ/mol. Thus, a comparison of the dissociation energies further supports the analysis of the rheo-photoacoustic data indicating that the rate of dissociation of hydrogen bonds at the initial stages of stretching is high, followed by the deformation of the backbone components.

At this point it is necessary to raise the question as to why the intensity increases as the fiber is elongated followed by the decrease when the fiber breaks. If upon fiber deformation only selected bands changed intensities, this effect would have been attributed only to the structural changes within the sample. However, in the case of rheo-photoacoustic measurements, the majority of the infrared bands are being enhanced, but the changes appear at different rates. This behavior indicates that each vibrational band has different sensitivity to the deformation process. In an effort to understand further what physical properties are responsible for the observed changes, we will isolate two processes that occur during the photoacoustic detection and simultaneous elongation of fibers. First, we will consider "an ordinary" PA FT-IR spectrum resulting from absorption of light and heat release generating acoustic waves, and the second, caused by the stretching process, which results in the bulk modulus and possibly thermal property changes caused by the elongation process.

Let us represent the unstretched fiber spectrum as P', and designate P" to describe the contribution to the overall spectrum, P, arising from the stretching process. Therefore,

$$P = P' + P'' \tag{1}$$

While P' depends only on the optical and thermal properties of the material, P" will be affected by emission of acoustic waves due to bond breakage which, in

turn, will affect the thermal property changes as a
result of elongation. Since all measurements were
performed statically, the emission of acoustic waves is
relatively fast and is not as readily detectable.
However, as a result of elongation, thermal properties
of the polymer change. Therefore, we will take advantage
of this pheonomenon and describe the amount of heat
transferred to the surface by:[12]

$$H(x) = \frac{\beta I_o(1 - \eta)}{\exp[(\beta + a_s)x]} \qquad (2)$$

where $H(x)$ is the amount of heat transferred to the
surface, a_s is the thermal diffusion coefficient, β is
the absorption coefficient, η is the refractive index,
and I_o is the intensity of the modulated infrared
radiation. The heat transfer to the surface is also
dependent upon the[12] thermal diffusion coefficient, a_s,
which is given by:

$$a_s = \left(\frac{\omega}{2\alpha}\right)^{1/2} \qquad (3)$$

where ω is the modulation frequency and α is the thermal
diffusivity, which is given by:[12]

$$\alpha_s = x(\rho C_p)^{-1} \qquad (4)$$

where x is the thermal conductivity, ρ is the density,
and C is the heat capacity at constant pressure.
        As demonstrated in Figure 3 for the N-H and C=O
bands, at the initial stages of the elongational process,
an equilibrium exists between breakage and formation of
hydrogen bonds and the increase in the integrated
intensities is observed. As the polymer is elongated
further, the integrated intensity increases at a higher
rate. This behavior is attributed to the changes of the
thermal diffusion coefficient, namely, as the polymer is
elongated, the thermal diffusion coefficient decreases.
Because the thermal diffusion coefficient and thermal
diffusivity are related through Eq. (3), it follows that
the thermal conductivity increases during the
elongational process. This observation is in agreement
with the previous studies of Kline and Hansen,[19] who
provided evidence for the increase of thermal
conductivity as a result of elongation. The increase of
thermal conductivity results from the enhanced
orientation of polymer molecules during the elongational
process. Therefore, as the polymer is elongated, the heat
transfer to the sample surface is enhanced and, as
illustrated by Eq.(2), the photoacoustic bands become
stronger.

Another effect is the friction between crystallites or molecules which may also contribute to the overall enhancement of the photoacoustic signal. As the crystallites move against each other, the heat which is given off, will also contribute to the intensity of the photoacoustic signal.

In summary, rheo-photoacoustic FT-IR measurements allow us to determine the response of each individual band, as the thermal properties of the material change upon elongation. Although this work opens new avenues for in-situ spectroscopic measurements of deformations, it is obvious that in order to advance understanding of the processes governing the enhancement of acoustic waves emitted during mechanical stretching, it is necessary to improve the experimental procedures; for example, employing a high precision stress-strain gage (work in progress). Moreover, it is also necessary to develop further understanding of the nature of anisotropic changes upon elongation and correlate it with the spectroscopic data. The similarities of results obtained from stress - strain gage and photoacoustic FT-IR measurements indicate that it will be possible to determine the Young's modulus using the normalized integrated intensity of selected bands plotted as a function of percent elongation. As illustrated in Figures 3 and 5, the intensity changes are not a linear function of elongation because each band provides a different response to the elongation process. Thus, one may consider rheo-photoacoustics as a "chemical Instron" device, measuring the contribution of each chemical bond to the overall mechanical strength of a polymer. On the other hand, typical Instron stress - strain measurements provide an average value of all bonds of the bulk material and the isolation of each response is impossible.

CONCLUSIONS

In this work, we have demonstrated the utility of a newly developed rheo-photoacoustic FT-IR spectroscopy. This method allows us to study essentially any material, regardless of color, optical properties, or shape, and monitor its behavior as it is exposed to external shear forces. Although, it is clear that more work is needed to explore fully the discussed phenomena, this technique may be considered as an alternative tool to study deformational processes on the molecular level. Moreover, the processesthat govern the enhancement of acoustic waves emitted from polymeric materials have a significant effect on the ability of detecting materials under stress using infrared detectors. When a polymeric material is under stress and modulated infrared light impinges on its surface, the absorption process occurs, followed by reabsorption, generating acoustic periodic waves. The amplitude of these waves is related to the applied stress and to the wavelength of the light. Hence, the enhancement of the signal is observed due to periodic temperature increases at the sample surface. Although the effect of enhanced thermal emissivity as a result of increased temperature has been known for many years, the applied stress also changes the thermal properties which, in turn, leads to a stronger emission of acoustic waves from the surface.

ACKNOWLEDGMENTS
The authors are thankful to 3M Company for a partial support of this work.

REFERENCES

1. Siesler, H.W., Adv.Polymer Sci., 1984, v.65, 1-72.

2. Bretzlaff, R.S. and Wool,R.P., Macromolecules, 1983, 16, 1907.

3. Siesler, H.W., J.Polym.Sci.:Polym.Phys.Ed., 1985, 23, 2413-2422.

4. Northholt, M.G. and Van Aartsen, J.J., J.Polym. Sci.: Polym. Symp., 1977, 58, 283-296.

5. Lee, Y.L., Bretzlaff, R.S., and Wool, R.P., J. Polym. Sci.: Polym.Phys. Ed., 1984, 22, 681-698.

6. Siesler, H.W., Colloid Polym. Sci., 1984, 262, 223.

7. Urban, M.W., J.Coatings Techn., 1987, 59, 745, 29-34.

8. Urban, M.W., Prog.Org.Coatings, 1989, 16, 321-353.

9. Graham, J.A., Grim, W.M., and Fateley, W.G., Fourier Transform Infrared Spectroscopy, eds. J.R.Ferraro and L.J.Basile, v.4, ch.9, 345-392, Academic Press Inc., Orlando, FL, 32887,1985.

10. Urban, M.W., Gaboury, S.G., McDonald, W., Tiefenthaler, A.M, Adv.Chem.Series, Eds. Craver, C. T.Provder, ACS, Washington, DC, 1990, in press.

11. W.F.McDonald and M.W.Urban, Proc.ACS, PMSE Div.,1989, 60, 739.

12. Rosencwaig, A., Photoacoustics and Photoacoustic Spectroscopy, John Wiley and Sons, New York, 1980.

13. Middleman, S., The Flow of High Polymers: Continuum and Molecular Rheology, Interscience Publishers, New York, 1968.

14. McDonald, W.F, Goettler H. and Urban, M.W. Appl. Spectrosc., 1989, Nov-Dec. issue.

15. Chatzi, E.G., Ishida, H., and Koenig, J.L., Composite Interfaces, ed.H.Ishida, J.L. Koenig, North Holland, New York, 1986.

16. Bellamy, L.J., Th e Infrared Spectra of Complex Molecules, 3rd ed., Chapman and Hall, London, 1975.

17. Atkins, P.W., Physical Chemistry, Freeman and Co., San Francisco, 1978.

18. Douglas, B.E., McDaniel, D.H., and Alexander, J.J., Concepts and Models of Inorganic Chemistry, 2nd ed., John Wiley and Sons, New York, 1983.

19. Kline, D.E. and Hansen, D., Thermal Conductivity of Polymers, John Wiley and Sons, New York, 1984.

RECEIVED January 24, 1990

ACOUSTIC ATTENUATION

ACOUSTIC ATTENUATION

This chapter considers acoustic wave propagation in materials and especially polymers. (Such waves, when detected by the ear, are called sound waves.) In particular, we are considering the interaction between plane-waves traveling in unbounded media with the bulk properties of the material.

Acoustic waves can be launched in a material by applying a sinusoidally varying force to the material surface. If the force is applied perpendicular to the surface, a longitudinal wave is launched. As this longitudinal wave propagates into material, the particles (molecules or other structural elements) in the region are first forced in the direction of wave propagation and thereafter (for the duration of the wave) they are forced back and forth by the oscillations of the wave. This gives rise to local pressure and density fluctuations. Since the stress is longitudinal, the wave properties (speed, attenuation) are characterized in terms of corresponding aspects of the Young's modulus of the material. Alternatively, if the driving force at the surface is applied in a shearing fashion, such that the particle motion is parallel to the surface of the material, a wave can still be launched which travels perpendicular to the material surface. The particle motion associated with such a wave is perpendicular to the direction of wave propagation, and the wave properties are then characterized in terms of the corresponding aspects of the shear modulus of the material. For polymeric materials, shear waves typically travel with very low speeds and are rapidly attenuated; hence the transformation of longitudinal waves into shear waves is greatly desired.

In this section, **Jarzynski** first provides an introductory tutorial on the basic nature of sound propagation, the equations governing sound interaction with material properties, and insight into the nature of attenuation mechanisms. **Corsaro** et. al. continues this discussion, illustrating the use of this formulation in the development of several sound absorbers of classical design for underwater applications. **Madigosky and Scharnhorst** then describes a vital feature of all highly absorbing material systems - the essential presence of voids. **Harrison** provides an introduction to the two principle techniques used to evaluate the relevant acoustical properties of these materials, namely the Panel-Test and Impedance-Tube techniques. Finally, **Wendlant** provides an example of the complexity involved in numerically simulating one particular sound absorbing coating type: the resonant cavity coating.

Chapter 10

Mechanisms of Sound Attenuation in Materials

Jacek Jarzynski

George W. Woodruff School of Mechanical Engineering, Georgia Institute
of Technology, Atlanta, GA 30332-0405

This review begins with the fundamentals of
sound propagation in a lossy material. The
complex wavenumber and sound speed are
defined and are related to the complex
elastic moduli and the attenuation
coefficient of the lossy medium. The
mechanisms of sound attenuation are (1)
scattering by inhomogeneities, (2) mode
conversion at boundaries, (3) redirection,
and (4) intrinsic absorption by conversion to
heat in a viscoelastic material. Many diverse
designs, both materials and structures, have
been developed to attenuate air-borne and
water-borne sound. Viscoelastic polymers have
unique mechanical properties which make these
materials widely used in sound attenuating
coatings. The attenuation and dipersion of
sound in a lossy material are related by the
Kramers-Kronig equations, which establish
very general constraints on the maximum
absorption of sound by viscoelastic polymers.

Sound attenuation is defined as the decrease in intensity
of the sound signal as it propagates from the source to
the receiver. This definition does not, however, include
the change in intensity due to geometrical spreading of
the sound wave (such as, for example, the spherical
spreading of the sound in the far field of a finite

0097-6156/90/0424-0167$11.25/0
© 1990 American Chemical Society

source). The objective of this article is to review the
many diverse designs (materials and structures) which
have been developed to attenuate air-borne and
water-borne sound. Also reviewed are the unique
mechanical properties of viscoelastic polymers, which
make these materials widely used in sound attenuating
coatings. This review emphasizes basic concepts, and
omits practical design details and discussion of methods
for measuring the performance of coatings. The
references given here are representative of the various
areas of research, rather than complete, with some
emphasis given to more recent publications which can be
consulted for references to earlier work.

The mechanisms of sound attenuation include
(1) scattering by inhomogeneities, (2) mode conversion at
boundaries, (3) redirection, and (4) intrinsic absorption
by conversion to heat in a viscoelastic material. In the
present review the term absorption is used specifically
for the conversion of sound energy to heat by various
relaxation processes in the material (molecular and
structural relaxation), while the term attenuation is
used for any process, other than geometrical spreading,
which leads to a decrease in the sound amplitude. The
attenuation of sound has become an increasingly important
issue in many fields. In the case of air-borne sound,
attenuation is important in industrial applications such
as insulation for residential and industrial buildings,
and insulation and noise reduction for automotive and
aircraft transportation. In underwater acoustics
attenuation of sound is important for sonar and flow
noise test facilities, and insulation and baffles in
sonar systems, and for reduction of sound radiation and
echoes from ships and submarines. In all of the above
applications the medium through which the sound
propagates is a fluid (air or water) which supports only
one type of wave, namely a longitudinal wave. The

majority of the above applications involve either
reduction of the reflection (echo), or reduction of
transmission, for an air or water-borne sound wave
incident on a solid structure (such as a wall, or a duct,
in a building). Another, related, field of application
for sound attenuating materials is noise reduction, that
is, reduction of radiation of sound from solid structures
such as machinery and appliances. Here the sound
attenuating material is often applied as a damping layer
to reduce the level of vibration of selected components
of the structure.

The basic concepts and definitions relating to sound
propagation in a lossy material are reviewed. The
material may be a viscoelastic polymer which converts the
sound energy to heat by molecular relaxation, or the
material may be a composite where sound is scattered by
inhomogeneities (inclusions) in a host matrix material.
In each case the attenuation of sound can be formally
represented by defining a complex wavenumber, where the
sound attenuation coefficient is the imaginary part of
the wavenumber. The complex wavenumber also leads to the
definition of a complex sound speed and a complex dynamic
elastic modulus.

The various mechanisms of sound attenuation in
materials are described. At present, the materials most
frequently used to attenuate sound are viscoelastic
polymers where the mechanism of sound attenuation is
conversion to heat by molecular relaxation processes.
The stress-strain relations for viscoelastic materials
are reviewed. The simplest case of intrinsic absorption
in polymers is a molecular relaxation mechanism with a
single relaxation time. However, the relaxation
mechanisms which lead to absorption of sound are usually
more complicated, and are characterized by a distribution
of relaxation times. Under causal linear response
conditions the attenuation and dispersion of sound in a

lossy material or, equivalently, the real and imaginary
components of the dynamic modulus, are related by the
Kramers-Kronig equations. It is pointed out that these
relations establish very general constraints on the
maximum absorption of sound by viscoleastic polymers.

THE FUNDAMENTALS OF SOUND PROPAGATION IN A LOSSY MATERIAL

The simplest and at the same time a very basic type of
sound signal is a plane harmonic wave. Fig.1a shows a
plane harmonic wave travelling in the x direction in a
lossless medium. The acoustic pressure, p, as a function
of time and position is [1,2],

$$p = p_0 \cos(kx - \omega t + \phi_0) \tag{1}$$

where p_0 is the amplitude, and ϕ_0 is the initial phase,
of the wave. The total phase ϕ at position x and time t
is $\phi = kx - \omega t + \phi_0$, and the wave is plane because the
wavefronts (surfaces of constant phase) are planes
perpendicular to the x axis. The wavenumber, k, and the
angular frequency, ω, are related respectively to the
wavelength, λ, and frequency, f, of the wave as follows:

$$k = \frac{2\pi}{\lambda} \quad , \quad \omega = 2\pi f. \tag{2}$$

The wave is harmonic because it has the single frequency
f associated with it. The plane harmonic wave is an
idealization. Actual sound waves are transient signals
with a finite time of duration. However, the harmonic
wave is a basic signal since any transient signal can be
represented as a superposition of harmonic waves with
different frequencies.

It is often convenient to use a complex number
representation for sound waves [2]. The harmonic wave is
represented by the complex exponential,

$$p = p_0 e^{i(kx - \omega t + \phi_0)} \tag{3}$$

where the physical wave (Eq.1) is the real part of the complex quantity representing the sound wave.

In a lossless medium sound waves propagate with a constant speed, characteristic of the medium. For a harmonic wave the speed of sound is simply related to the wavelength and the frequency as follows [1,2],

$$c = \lambda f = \omega/k \; . \tag{4}$$

From Eq.4 it follows that the sound wavelength is inversely proportional to the frequency. For example, the speed of sound in air is 346 m/sec (at 25^{0}C and atmospheric pressure) and so the wavelength λ = 3.46 meters at a frequency of 100 cycles/sec (Hertz,Hz) and λ = 34.6 cm at 1 kilohertz (kHz). In water at the same temperature and pressure c = 1496 m/sec so λ = 14.96 m at 100 Hz and λ = 1.469 m at 1 kHz. The increase in the wavelength of sound with decreasing frequency has important implications for the attenuation of sound. Experiment shows that when a homogeneous material is used to attenuate sound it is necessary to use a layer of the material at least $\lambda/2$ thick (where λ is the sound wavelength in the material) to obtain a significant reduction in the sound level. Therefore, in general, treatments for attenuation of sound become more bulky with decreasing sound frequency.

The discussion which follows, of sound propagation in a lossy material, is limited to the linear case. That is, it is assumed that the stress-strain relation in the lossy material is linear. In this case a plane, harmonic sound wave propagating through the lossy material decays exponentially with distance [3],

$$p = p_{0}e^{-\alpha x} \cdot e^{i(k'x-\omega t)} \tag{5}$$

as shown in Fig.1b, where p is the acoustic pressure at distance x from the reference point where the pressure is

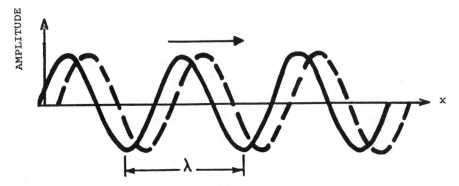

Figure 1a.   Undamped progressive harmonic wave.

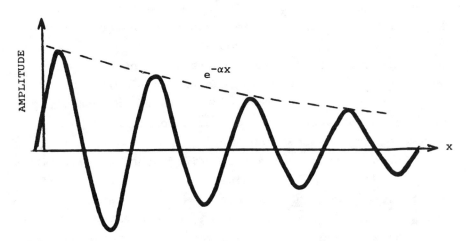

Figure 1b.   Damped progressive harmonic wave.

p_0, $k' = 2\pi/\lambda$ is the real part of the wavenumber, where λ is the wavelength, and α is the sound attenuation coefficient in nepers per meter. From Eq.5 it follows that sound attenuation can be formally included in the wavenumber by defining a complex wavenumber,

$$k^* = k' + ik'' = k' + i\alpha \qquad (6)$$

The expression for the sound wave now has the same form, $p = p_0 \exp[i(k^*x-\omega t)]$, as in a lossless medium. The complex wavenumber, substituted in Eq.4, defines a complex sound speed for the material as follows,

$$c^* = \frac{\omega}{k^*} \qquad (7)$$

However, as shown in Fig.1b, the propagation of the wave as a function of space and time is still determined by the phase velocity, c, which is related to the real part of the wavenumber, $c=\omega/k'$.

The elastic moduli of a lossy material can also be represented as complex quantities [3,4]. The complex dynamic modulus M^* is related to the corresponding complex sound speed c^* as follows [3],

$$M^* = M' - iM'' = \rho c^{*2} \qquad (8)$$

where ρ is the density. In Eq.8 the imaginary part of the modulus, M'', has a negative sign. Later in the text, in Eq.23, the sign of the imaginary component of the modulus is positive. Both signs are possible, and the choice of the sign is determined by the fundamental requirement that a harmonic sound wave propagating through a lossy material decays exponentially with distance. If the time dependence for the harmonic wave is taken to be $e^{-i\omega t}$, as in Eq.5, then the spatial dependence is e^{ik^*x} and the imaginary part of the complex wavenumber has a positive sign (Eq.6) to ensure exponential decay of the sound wave. The positive k''

then leads, via Eqs.7 and 8, to a negative sign for M''. Conversly, if the time dependence is taken to be $e^{i\omega t}$ then M'' has a positive sign.

Eq.8 has the same form as the relation between the (real) elastic modulus and sound speed in a lossless medium. From Eqs.6-8 it follows that the sound attenuation coefficient is related to the complex elastic modulus,

$$(k'^2 - \alpha^2) + i(2k'\alpha) = \frac{\rho\omega^2}{M'(1+r^2)}(1+ir) \qquad (9)$$

where r is called the loss factor, and is the ratio of the imaginary to the real part of the modulus,

$$r = \tan\delta = \frac{M''}{M'} \qquad (10)$$

When $r \ll 1$ (and $\alpha \ll k'$) the above relations simplify to

$$k'^2 = \frac{\rho\omega^2}{M'} \qquad \text{and} \qquad \alpha\lambda = \pi r \qquad (11)$$

where the wavelength $\lambda = 2\pi/k'$. Various experimental techniques are available for measuring the complex moduli of lossy materials [4,5]. The experimental data shows that both the real and the imaginary parts of a complex modulus are frequency dependent. Once data on the complex modulus M^* is available, then Eqs.9 and 11 are very useful for predicting from the data the attenuation coefficient α for the lossy material.

The loss factor r is an important quantity for sound attenuating materials. Eq.11 shows that r is directly related to the attenuation per wavelength for a harmonic wave propagating through the material. An alternate expression for r can be derived in terms of the energy dissipated per one cycle (period) of the sound wave. This expression is derived by considering a control volume (for example, unit volume) in the material through which the sound is propagating. The total acoustic

energy (kinetic and potential [2]), E_{st}, stored in the
control volume is calculated, as well as E_{diss}, the
energy removed by attenuation from the sound wave in the
control volume during one wave period. It can be shown
that [3],

$$r = \tan \delta = \frac{E_{diss}}{2\pi E_{st}} \qquad (12)$$

A sound wave transports energy and the acoustic
intensity, I, is defined as the acoustic energy
transported per unit area and time in the direction of
propagation. For a plane travelling wave in a medium
with density ρ and phase velocity c [2],

$$I = \frac{p^2}{\rho c} \qquad (13)$$

Both the total sound level, and the attenuation of sound,
are often expressed in a dimensionless unit, the decibel,
where the sound level (SL) in decibels is,

$$SL = -10 \log \left\{ \frac{p^2}{p_0^2} \right\} = -20 \log \left\{ \frac{p}{p_0} \right\} \qquad (14)$$

where p is the sound pressure at the observation point.
For the attenuation of sound (relative SL) p_0 is the
pressure at some appropriate reference point, whilst for
the total sound level p_0 is a fixed reference pressure
(usually p_0 = 20 μPa for air-borne sound, and p_0 = 1 μPa
for underwater sound). It is difficult to specify any
general criteria on what constitutes good performance for
sound attenuation. However, in many applications a
reduction in echo level (or in the level of transmitted
sound, depending on the application) by 10 dB, 15 dB, or
20 dB is considered as significant, good, or very good,
respectively. Note that a 20 dB reduction in sound level
means a reduction of the acoustic pressure p by a factor
of 10, which (according to Eq.13) requires attenuation of
99% of the incident sound energy.

From Eqs.5 and 14 it follows that the attenuation of
a sound wave propagating through a medium can be
expressed either in terms of the coefficient α (nepers
per meter), or as a reduction in sound level in decibels
per meter. The relation between these units of
attenuation is obtained by inserting x = 1 meter in Eq.5,
and substituting the result in Eq.14,

$$-20 \log \left\{ \frac{p(1m)}{p_0} \right\} \left[dB/meter \right] =$$

$$20 \log(e) \cdot \alpha \left[nepers/m \right] \qquad (15)$$

where $20 \log(e) = 8.686$.
Most of the publications on sound attenuating
materials assume linear conditions for the propagation of
sound in the material. Non-linear effects can introduce
additional attenuation mechanisms [6]. For example,
sound energy can be transferred from the original

(fundamental) frequency component to higher harmonics of
the propagating sound wave. However, there has been no
study so far to determine at what sound pressure level
nonlinear effects may become important. Comparisons
between theoretical predictions based on linear
conditions and experimental data, in the somewhat limited
number of cases where such a comparison was made, suggest
that the linear approximation is adequate for most cases
of practical interest.

The attenuation coefficient α is not the only factor
which determines the performance of an anechoic coating.
Equally important is the initial transmission of sound
from the surrounding medium (usually air or water) into
the coating. The simplest case is when the boundary
between the medium and the coating is a plane surface, as
shown in Fig.2a, and the sound is incident normally. The

sound reflection and sound transmission ratios then are
[1,2],

$$\frac{p_r}{p_i} = \frac{Z_a - Z_o}{Z_a + Z_o} \quad , \quad \frac{p_t}{p_i} = \frac{2Z_a}{Z_a + Z_o} \tag{16}$$

where p_i is the incident, p_r is the reflected and p_t is
the transmitted sound pressure, $Z_a = \rho_a c_a^*$ is the
characteristic impedance of the coating material, and
$Z_o = \rho_o c_o$ is the characteristic impedance of the
surrounding medium. The characteristic impedance is an
important quantity related to reflection and transmission
of sound at boundaries. The characteristic impedance of
a medium is defined as the ratio of the acoustic pressure
to the particle velocity in a progressive plane wave
propagating through the medium.

When the sound is incident obliquely on a plane
interface (as shown in Fig.2a), and if both media support
only one type of wave (a longitudinal wave if the medium
is air or water), then Eq.15 can still be used to
calculate the reflection and transmission coefficient.
However, the expressions for Z_a and Z_o are modified as
follows [2,7],

$$Z_o = \frac{\rho_o c_o}{\cos\theta} \quad , \quad \text{and} \quad Z_a = \frac{\rho_a c_a^*}{\cos\phi} \tag{17}$$

where θ is the angle of incidence, and ϕ is the angle of
refraction. Note that $\sin\theta / \sin\phi = c_o / c_a^*$ and $\sin\phi$ is
complex if the coating is lossy. The wave transmitted
into a lossy coating is an inhomogeneous plane wave.
If the coating material is solid, such as a viscoelastic
polymer, then an incident longitudinal wave may generate
both longitudinal and shear waves in the coating. The
expressions for the reflection and transmission
coefficients are then more complicated than Eqs.16 and 17
[8].

Eq.16 is valid when both the surrounding medium and

the lossy material are of semi-infinite extent. In
practice the acoustical coating is usually a layer of
finite thickness, or a series of layers, applied over a
solid structure. In a layer of finite thickness multiple
reflections of sound occur, as shown in Fig.2b, and a
standing wave pattern develops in the layer [2,3,9]. At
normal incidence an equation of the same form as Eq.16
still applies, but Z_a is now the input impedance at the
interface between the surrounding medium and the
acoustical coating. In this case Z_a is the ratio of the
acoustic pressure to the particle velocity at the front
surface of the coating, and is a function of (1) the
complex moduli and the density of the coating material,
(2) the thickness of the coating, and (3) the input
impedance of the backing structure over which the coating
is applied [10]. When the sound is incident at some
angle θ to the normal at the surface of the coating the
expressions for p_r/p_i and p_t/p_i are more complicated than
Eq.16. The complete expressions for the reflection and
transmission coefficients of a multilayer coating, at
arbitrary angle of incidence, and including generation of
both shear and longitudinal waves in the coating, are
given by Folds [10].(The reflection coefficient for a
finite thickness coating layer is discussed in more
detail in the chapter by Corsaro, Covey and Spryn in this
book).

From Eq.16 it follows that complete transmission
(zero reflection) occurs when the impedances of the two
materials match, $Z_a = Z_o$. However, when the surrounding
medium has a low sound attenuation coefficient, like air
or water, the sound speed c_o is a real quantity, whilst
the c_a^* in the coating has an imaginary part. A high
attenuation in the coating material leads to a large
imaginary component for c_a^*, therefore a large impedance
mismatch and a strong reflection from the front surface.
Various coating designs have been developed to minimize

the above problem. One successful design is a multilayer, gradual transition coating [11,12]. In this design the outer layer is low-loss material with a good impedance match to the surrounding medium. Successive layers are then designed to have a gradually increasing attenuation. Another successful design is a wedge structure consisting of an array of sound absorbing wedges [11,13]. The wedge coating can be approximately modeled as a gradual transition layer [14]. The outer layer of the wedge coating is mainly the surrounding medium with a small fraction of the lossy wedge material. Therefore the outer layer is a good impedance match to the surrounding medium. Successive layers of the coating are more attenuating as the fraction of the wedge material increases. Wedge coatings are widely used in the design of anechoic chambers for air acoustics measurements. The performance of a typical wedge coating in air is shown in Fig.3. It can be seen that this type of coating is fairly broadband, with peak performance in the frequency range where the wedge length is a least $\lambda/2$, where λ is the sound wavelength in the surrounding medium.

Most attenuating coating designs use materials with a high intrinsic sound absorption to finally convert the sound energy to heat. This material may be configured in different ways, such as multiple layers, or absorbing wedges. The absorbing materials used in many coating designs are viscoelastic polymers. These polymers are usually isotropic solids characterized by two elastic moduli, for example the bulk modulus K, and the shear modulus G. The two types of acoustic waves which can propagate in the unbounded material [8] are the longitudinal (dilatational) wave with sound speed

$$c_L^2 = \frac{K + \frac{4}{3} G}{\rho} \qquad (18)$$

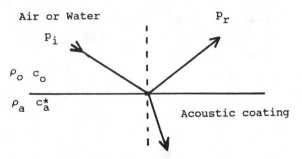

Figure 2a.   Reflection and transmission of sound at the fluid-acoustic coating interface.

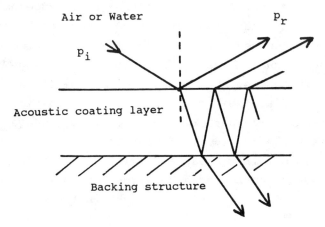

Figure 2b.   Reflection of sound from an acoustic coating layer on a backing structure.

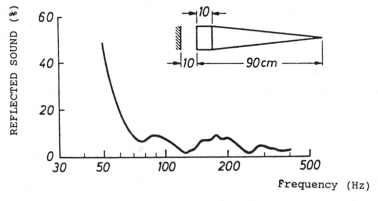

Figure 3.   Wedge absorber in air.   Glass wool wedge.

(where ρ is the density and, at sonic frequencies, K is
the adiabatic bulk modulus), and the shear wave with
sound speed

$$c_s{}^2 = \frac{G}{\rho} \qquad (19)$$

For a viscoelastic material both K and G are complex
quantities. When the material sample has finite
dimensions other modes of wave propagation may occur as a
result of multiple scattering from the material
boundaries. Mode conversion from longitudinal wave to
shear wave, and vice versa, occurs on reflection at a
solid boundary. For material samples in the form of thin
rods or plates the modes of wave propagation are
extenional waves (with speed determined by the Young's
modulus, or the plate modulus), and flexural (bending)
waves [8].

In a longitudinal wave the particle motion is
parallel to the direction of wave propagation, whilst in
a shear wave the particle motion is transverse to the
direction of propagation. The shear modulus G = 0 for
fluids. Therefore only longitudinal waves propagate in
fluids such as air or water. It follows from Eq.18 that
the particle motion produced by a longitudinal wave can
be expressed as the sum of compressional and shear
strains. The sound speed is related to the adiabatic
bulk modulus of the fluid by Eq.18. The shear strains in
the longitudinal wave in a fluid lead to viscous forces
which convert some of the acoustic energy to heat. The
attenuation coefficient for viscous absorption of sound
in a fluid is proportional to the viscosity of the fluid
and to the acoustic frequency squared [15]. In fluids
with low viscosity, such as air or water, the viscous
absorption is negligible at frequencies below the
megahertz range [15]. Therefore, in all calculations

related to sound attenuating coatings the speed of sound, c_o, in air or water is taken to be a real quantity.

In conclusion, it follows from the above discussion that the design of anechoic coatings requires both development of materials with a high intrinsic absorption of sound, and the development of composite structures (such as the wedge structure) to ensure a good impedance match between the coating and the surrounding medium, to optimize the attenuation of sound. A discussion of various composite structures for optimizing the attenuation of sound is presented in the next section. The most widely used materials for absorption of sound are viscoelastic polymers, and these material are discussed in a separate section.

MECHANISMS OF ACOUSTICAL ATTENUATION

One of the mechanisms for acoustical attenuation listed in the Introduction, is redirection of sound. Direct reflections from a surface can be reduced by treating the surface with acoustical tiles with input impedance Z_a which is alternately higher and lower than the impedance Z_o of the surrounding medium. When $Z_a < Z_o$ it follows from Eq.16 that the phase of p_r is reversed on reflection, and the reflected sound interferes destructively (in the backscatter direction) with sound reflected from an adjacent high impedance tile. If the acoustic tiles are arranged in a random pattern, with a range of values of Z_a, the incident sound is converted to a diffuse reflected sound field. This arrangement has application in architectural acoustics [16,17,18].

A second mechnism for sound attenuation is mode conversion. The sound waves in air or water are longitudinal waves, where the particle motion is parallel to the direction of sound propagation. With appropriate boundary conditions the longitudinal deformation can be converted to shear deformation or to viscous flow. Conversion to viscous flow is most readily achieved at

the boundaries in porous or fibrous materials [11,19,20].
The viscous flow is then converted to heat by molecular
collisions. The end result is attenuation of the sound
wave as it propagates through the porous composite.
Significant mode conversion is achieved when the motion
of the medium relative to the porous framework is
maximum, that is, when the porous framework is rigid
relative to the medium. This condition is easier to
achieve in air, where there is a large impedance mismatch
between the air and any solid porous framework. Also, it
is easier to achieve an impedance match between air and a
porous, open-cell, material (into which air can flow),
than between air and a closed-cell or solid material.
Therefore, porous viscous absorbers are frequently used
to attenuate air-borne sound. The performance of a
viscous absorber can be optimized by placing the porous
layer in a region of maximum particle motion. Relative
to the characteristic impedance of air, the surface of
any solid structure is approximately a solid boundary.
Therefore, in air the porous layer should be placed at
approximately $\lambda/4$ from the surface of the backing
wall. Alternately the coating can be designed as a
honeycomb panel with small holes in the front surface.
The honeycomb cavities act like Helmholtz resonators.
Near the resonance frequency there is large particle
motion at the holes in the front surface which leads to
strong viscous attenuation of the incident sound [7].

An interesting type of viscous absorber, developed
by Meyer et.al. [21] for underwater operation, is shown
in Fig.4. It is clear that in this case the longitudinal
particle motion is converted to a transverse viscous
flow. The peak performance (and frequency range) of this
absorber are functions of the mass of the cover plate,
and the viscosity of the liquid between the cover plate
and the backing structure. Tests in a water-filled
impedance tube have shown that a relatively broadband

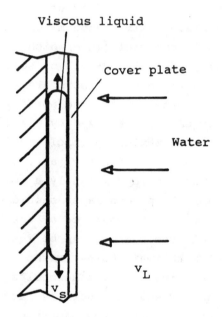

Figure 4.   Thin viscous absorber for water-borne sound.

performance (in the range of 250 Hz to several kilohertz)
with a peak echo reduction of 20 dB can be achieved.
This design is particularly interesting because a high
echo reduction is achieved with a very thin coating. The
thickness of the coating is typically $\approx \lambda/100$, where λ is
the wavelength, in water, of the incident sound. A major
disadvantage of the absorber in Fig.4 is that it is very
pressure sensitive, and is unable to support any increase
in ambient pressure. This absorber is also sensitive to
changes in temperature, because the viscosity of the
liquid used in the coating changes with temperature.

Another example of mode conversion is the conversion
of longitudinal deformation to shear deformation at the
boundaries of a soft (air-filled) cavity in a solid
[13,22,23]. This conversion is particularly pronounced
in materials such as viscoelastic polymers, whose shear
modulus is at least an order of magnitude lower than the
bulk modulus. If the host material is viscoelastic the
shear deformation energy is converted to heat by
molecular relaxation, and the sound wave is attenuated.
The cavities may be microscopic (dimension much smaller
than the wavelength of sound), or macroscopic with
dimensions comparable to at least the shortest sound
wavelength (usually the shear wavelength) in the host
material, in the frequency range of interest. The effect
of microscopic cavities is discussed in the next section.
With macroscopic cavities maximum sound absorption is
obtained at frequencies in the vicinity of the resonance
frequencies of the cavity [22,23]. The advantage of a
coating with macroscopic cavities is that large sound
absorption can be obtained with a relatively thin coating
layer. However, a disadvantage of this type of coating
is the relatively narrow frequency range over which the
absorption occurs. Also, the absorption can be strongly
temperature dependent if the loss tangent of the
viscoelastic coating material is strongly temperature

dependent. An example of a coating with macroscopic
cavities, and its performance as a function of frequency,
are shown in Fig.5. The frequency dependence shown in
Fig.5 is for the case where the cavity has a single
resonance frequency. Performance over a broader range of
frequencies can be achieved either by designing the
cavity to have several resonances, or by introducing
several cavities of different size, and in each case
staggering the resonance frequencies.(Viscoelastic
coatings with macroscopic cavities are also discussed in
the chapter by Corsaro, Covey and Spryn in this book).

The above case of longitudinal to shear conversion
can be considered as a particular example of the
scattering of sound by inhomogeneities in a host medium.
The complete treatment of the scattering of sound in a
composite material involves both the calculation of the
scattering cross-section of a single scatterer in the
host medium, and the multiple scattering effects due to
surrounding scatterers. There are many theoretical
publications on this topic [24,25,26,27]. In most of the
studies so far the inclusions are assumed to have a
simple geometrical shape, often spherical. The objective
of the theory is to derive an expression for the complex
wavenumber, κ, of the sound wave propagating through the
composite medium, in terms of the wavenumber k in the
host medium and the far-field amplitude, $f(\theta)$, for the
scattering of sound by a single scatterer in the host
medium. For example, the expression derived by Waterman
and Truell [24] is,

$$\left(\frac{\kappa}{k}\right)^2 = \left[1 + \frac{2\pi n_o f(0)}{k^2}\right]^2 - \left[\frac{2\pi n_o f(\pi)}{k^2}\right]^2 \qquad (20)$$

where $f(0)$ and $f(\pi)$ are the (complex) scattering
amplitudes in the forward and the back directions
respectively, and n_o is the number of scatterers per unit

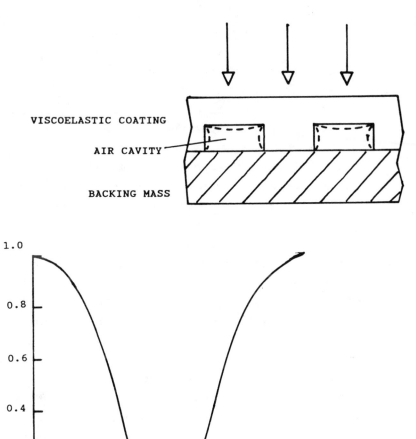

Figure 5.   Thin rubber coating with resonant
macroscopic cavities.   The dashed lines show motion
at the cavity walls.

volume. The description of sound propagation in the
composite in terms of multiple scattering of sound from
the inclusion is the most fundamental description
available, and should lead to a complete model for sound
propagation in composite media. At present, however, the
development of useful approximations for multiple
scattering effects, particularly at high densities of
inclusions, is still under research. The attenuation of
the sound wave is described by the imaginary part of κ.
Examination of Eq.20 shows that the scatterers can
attenuate sound by any one of the following mechanisms
[24]: (1) intrinsic dissipation of sound within the
scatterer itself, (2) mode conversion at the boundary of
the scatterer, and (3) scattering of sound to a back
propagating wave. However, there has been no
quantitative study so far to determine the optimum
parameters, and the maximum sound attenuation which may
be achieved with each of the above mechanisms. A
composite material with high sound attenuation is
described by Madigosky et.al. [28]. In this absorber
both the host material and the inclusions have an
acoustic impedance approximately matched to water, but
there is a large difference in the speed of sound in the
two materials. The dimensions of the inclusions are
comparable to the wavelength of the incoming sound. It
is probable that all three of the sound attenuation
mechanisms described above contribute to the performance
of the absorber in Ref.28.

It is important to note that, in general, scattering
of sound from an inclusion in a medium does not become
significant until the dimensions of the inclusion are
comparable to the wavelength of sound in the medium. For
example, Fig.6 shows the normalized scattering
cross-section σ_s for a rigid sphere immersed in water
[29]. It can be seen that σ_s reaches its maximum value
when ka $\approx$ 1, where k is the wavenumber of the incident

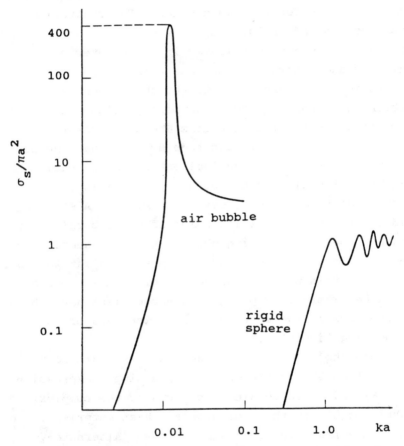

Figure 6.   Normalized scattering cross section for an air bubble and a rigid sphere.

sound in water and a is the radius of the sphere. When
the inclusion is an elastic solid the frequency
dependence of σ_s is more complicated, with maxima and
minima at frequencies corresponding to internal
resonances of the sound in the inclusion [30,31].
However, the average trend as a function of frequency is
similar to that for the rigid sphere.(The scattering
cross-sections of inclusions and their relation to sound
attenuation are also considered in the chapter by
Madigosky and Scharnhorst in this book).

An important exception to the above results is the
scattering of sound by very compliant inclusions (usually
gas-filled cavities) in a medium where the shear modulus
is at least an order of magnitude smaller than the bulk
modulus. In this case the incident sound can excite
large radial motion at the boundary of the inclusion,
which leads to a large σ_s even when ka<<1. A striking
example of this type of inclusion is a gas bubble in a
liquid. The corresponding scattering cross-section is
shown in Fig.6. There is a large peak in σ_s at ka≈0.01.
Physically this peak is the result of a "mass-spring"
resonance, where the spring is the compliant gas bubble
and the mass is the entrained radiation mass of the
surrounding liquid [7,29].

When the dimensions of the scatterers are much
smaller than the wavelength of sound simple expressions
for $f(\theta)$ and κ are obtained in terms of the complex
elastic moduli of the inclusion and host materials, and
the volume fraction of the inclusions. Alternately,
static self-consistent mean field models can be used to
derive expressions for the complex effective moduli of
the composite material in terms of the complex elastic
moduli and volume fractions of the component materials
[32,33,34,35]. The propagation wavenumber κ can then be
expressed in terms of the effective complex moduli of the
composite using Eqs.9 and 10. Particularly interesting,

for sound attenuation, are the composite materials obtained when the host material is a lossy viscoelastic polymer and the inclusions are soft microbubbles, usually air. The elastic properties of the gas in the

microcavity are not important, and the main effect of the cavity is to provide pressure release conditions at the cavity boundary. Scattering of sound from a small cavity in a viscoelastic polymer is similar to the scattering of sound from an air microbubble in a liquid. However, the compliant "spring" element in the polymer is the extensional (or shear) deformation of the polymer around the wall of the cavity.

The air microbubbles convert longitudinal strain to shear strain. The need for this type of conversion is as follows. It is shown in the next section that in a viscoelastic material the loss tangent for compressional strain is relatively small (tan $\delta \leq 0.1$), while the loss tangents for both extensional and shear strains can be large (tan $\delta \simeq 1$). A longitudinal sound wave propagating through a viscoelastic material produces both compressional and shear strains. However, since the bulk modulus is one or two orders of magnitude greater than the shear modulus, the fraction of acoustic energy stored in the shear strain is small. Therefore, to achieve high attenuation of the longitudinal wave it is necessary first to convert the compressional strain energy to either shear or extensional strain energy. Note that an isotropic material has only two independent elastic moduli. Therefore a given conversion of energy can be regarded as being from compressional to either shear or extensional strain, depending on whether the shear or extensional modulus is chosen as the second independent modulus.

It is instructive to consider the predictions of the mean field model for the above type of composite. A frequently used mean field model is that of Kerner [32], who develops expressions for the effective elastic moduli of the composite in terms of the elastic moduli of both the host material and the material in the inclusions. When the inclusion is an air bubble in a viscoelastic solid the shear modulus of the air is zero, and also it is reasonable to neglect the bulk modulus of the air since it is several orders of magnitude less than the bulk modulus of the solid. The Kerner mean field model then predicts the following expression for the effective bulk modulus of the air-polymer composite,

$$K^* = K_0^*(1-\phi)/(1 + 3K_0^*\phi/4G_0^*) \qquad\qquad (21)$$

where K_0^* and G_0^* are the bulk and shear moduli, respectively, of the viscoelastic polymer and ϕ is the volume fraction of air. For viscoelastic polymers K_0^* is typically one to two orders of magnitude larger than G_0^*. Also, as discussed in the next section, shear waves are strongly absorbed in viscoleastic polymers, and therefore G_0^* has a large imaginary component. Therefore, both the magnitude and the imaginary part of the term $3K_0^*\phi/4G_0^*$ are large. Physically this term corresponds to conversion of dilatational to shear (or extensional) deformation at the boundaries of the air microbubbles, and subsequent dissipation of the shear deformation as heat in the viscoelastic polymer. According to scattering theory the conversion of energy occurs as follows. When the longitudinal wavelength is much greater than the diameter of the cavity, the scattering of sound by the cavity is due to a combination of two motions at the boundary, namely a radial motion and a translational motion, as shown in Fig.7. Both motions contribute to conversion of compressional strain energy. The radial motion is analyzed in detail in Ref.22. In the far field region of the cavity the scattered wave due to the radial motion is again a longitudinal wave. However, in the near field region there is strong conversion to extensional strain.

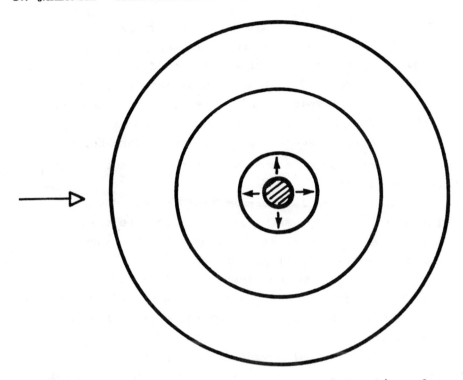

Figure 7a.   Scattering of sound by radial motion of an air microbubble.

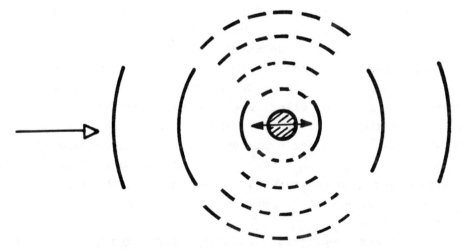

Figure 7b.   Scattering of sound by transverse motion of an air microbubble. Both longitudinal (solid line) and shear (dashed line) waves are generated.

From Eq.21 it follows that the imaginary party of the effective bulk modulus K^* is much larger than the imaginary part of K_o^*. Therefore, sound waves are strongly attenuated in this type of material. The viscoelastic polymer-air microbubbles composites are particularly useful in the design of broadband, transition type anechoic coatings for underwater applications, such as the wedge type structures described earlier [11,13].

The self-consistent mean field models developed so far have been very useful in describing the trends in the mechanical properties of the composite as a function of the volume fraction of inclusions, and in giving some physical insight on how given inclusions modify the mechanical response. However, the mean field models have not been extensively tested against experimental data, particularly at high volume fraction of inclusions.

Various inclusion-viscoelastic polymer matrix combinations have been investigated for use as sound absorbing materials. Among the inclusions used were sawdust [11], metal oxide microparticles [37], and phenolic microballoons [12]. In some cases [12] two different inclusions were used, for example phenolic balloons and iron oxide particles. By using two inclusions it is possible to adjust both the speed of sound and the density of the composite to desired values, and so achieve a better impedance match to the surrounding water. It is probable that in each of the above cases some air microbubbles were also introduced during mixing of the inclusions with the polymer matrix. Therefore, the observed high absorption of sound in the above materials may be partly due to the presence of air microbubbles.

In many applications acoustical coating are used over a range of temperatures, and it is important to estimate the temperature dependence of the attenuation in

the sound absorbing material used in the coating. For
materials with microscopic inclusions this type of
estimate can be made using an expression (such as Eq.21)
for the effective bulk modulus, and inserting the
temperature dependence of K_o^* and G_o^*. The temperature
dependence of the bulk and shear moduli of viscoelastic
polymers is discussed in the next section. For some
applications it is also important to determine the
pressure dependence of the effective bulk modulus. The
mean field models for composites, described above, can be
extended to predict the change of the effective bulk
modulus with hydrostatic pressure as shown, for example,
in Ref.38.

ABSORPTION OF SOUND IN VISCOELASTIC MATERIALS

Many materials, particularly polymers, exhibit both the
capacity to store energy (typical of an elastic material)
and the capacity to dissipate energy (typical of a
viscous material). When a sudden stress is applied, the
response of these materials is an instantaneous elastic
deformation followed by a delayed deformation. The
delayed deformation is due to various molecular
relaxation processes (particularly structural
relaxation), which take a finite time to come to
equilibrium. Very general stress-strain relations for
viscoelastic response were proposed by Boltzmann, who
assumed that at low strain amplitudes the effects of
prior strains can be superposed linearly. Therefore, the
stress at time t at a given point in the material depends
both on the strain at time t, and on the previous strain
history at that point. The stress-strain relations
proposed by Boltzmann are [4,39]:

$$\sigma(t) = M_1 e(t) - \int_0^\infty e(t-\Delta t) \, M_r(\Delta t) \, d(\Delta t) \qquad (22)$$

where $\sigma(t)$ is the stress and $e(t)$ is the strain at time

t, and e(t-Δt) is the strain at a time t-Δt earlier. M_1
is the elastic modulus (bulk or shear) which describes
the instantaneous response of the material, and M_r(Δt) is
a memory function (generally different for bulk and shear
strains) which determines the contribution of the
previous strain history to σ(t). The relation proposed
by Boltzmann is called the linear viscoelastic model.
Sometimes it is also called isothermal linear
viscoelasticity. Eq.22 is not exact [40] since it fails
to take proper account of thermal effects associated with
the dissipation of mechanical energy during the

relaxation process. However, experiment shows that Eq.22
is a very good approximation for the mechanical response
of the viscoelastic polymers used in acoustical
coatings.

When a sinusoidal (harmonic) sound wave propagates
through a viscoelastic material, the stresses and strains
in the material vay sinusoidally. Eq.22 predicts, in
this case, a phase lage between the stress and the
strain, which leads to conversion of acoustic energy to
heat. From the Fourier transform of Eq.22 it follows
that the sinusoidal stress and strain are related by
complex, frequency-dependent elastic moduli as follows,

$$K^*(\omega) = K'(\omega) + iK''(\omega)$$

and
$$G^*(\omega) = G'(\omega) + iG''(\omega) \tag{23}$$

Any two complex, frequency dependent, elastic moduli are
sufficient to describe the mechanical response of a
viscoelastic polymer. The two moduli which are most
frequently measured are the bulk modulus [41], and the
Young's modulus $E^*(\omega) = E'(\omega) + iE''(\omega)$ [42]. Poisson's
ratio is close to 0.5 for polymers both in the rubber
state, and the rubber-glass transition region.
Therefore, in these regions $E^* \simeq 3G^*$, and the frequency
and temperature dependence of the two moduli is very

similar. In the discussion which follows all statements
relating to G^* also apply to E^*.

Fig.8a shows the variation of the real and imaginary
part of the shear modulus, $G'(\omega)$ and $G''(\omega)$, with
frequency for a polymer which shows no flow. It can be
seen that at low frequencies the polymer is rubber-like
and has a low frequency (relaxed) modulus G_r of
~10^6dyne/cm^2, and at high frequencies the polymer is
glassy with an unrelaxed modulus G_u of ~10^{10}dyne/cm^2. In
the rubber-glass transition region the polymer exhibits
viscoelastic behavior, with a peak in the imaginary part
of the modulus G'' and a peak in the loss factor tan δ.
The fractional change in the modulus from the relaxed to
the unrelaxed value, $(G_u-G_r)/G_r$, is called the relaxation
strength of the viscoelastic transition. Figs. 8a and 8b
show that the frequency dependence of the bulk modulus
$K^*(\omega)$ is similar to the frequency dependence of $G^*(\omega)$.
However, the relaxation strength and the peak value of
the loss factor are much smaller in the case of the bulk
modulus. For example, values of tan δ ≃ 2 are possible
for $G^*(\omega)$, whilst tan δ ≤ 0.1 for $K^*(\omega)$. This means that
elastic shear waves are rapidly attenuated in
viscoelastic materials, whilst longitudinal sound waves
are weakly attenuated. Therefore, to attenuate
longitudinal waves it is necessary to insert inclusions
which convert the longitudinal wave to shear waves, which
are then dissipated in the viscelastic polymer. As
discussed in the previous section, the inclusions which
are most effective for conversion from longitudinal to
shear deformation are air cavities, either in the form of
small microbubbles or as macroscopic cavities.

The frequency dependence of G' and G'' in the
viscoelastic transition region has been the object of
many experimental and theoretical studies [4,5]. The
simplest model for the transition region is the standard
linear solid, which is characterized by a single

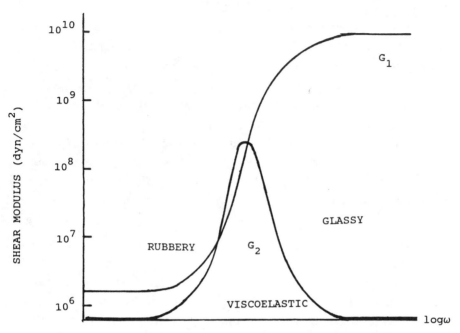

Figure 8a.   Frequency dependence of the complex
shear modulus of a viscoelastic polymer.

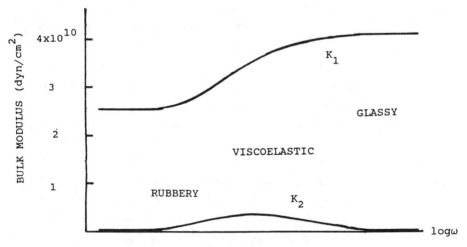

Figure 8b.   Frequency dependence of the complex
bulk modulus of a viscoelastic polymer.

relaxation time [5]. This model predicts that the rubber
to glass transition will occur over a frequency range of
approximately one decade. However, measurements show
that in polymers this transition occurs over a broader
frequency range. One approach to describe the observed
broad frequency dependence is to introduce a spectrum of
relaxation times [4,5]. So far, however, it has not been
possible to find any simple physical interpretation for
the proposed relaxation time spectra. Other models for
viscoelastic behavior include the use of fractional
derivatives [43], and a four-parameter model developed by
Hartmann [44]. Models for the frequency dependence of G^*
are important in the design of acoustical coatings since
these models provide the input data for calculations of
the performance of the coating as a function of
frequency. Alternately, one can use as input data the
measured frequency dependence of G^*. In addition to
frequency dependence, it is also usually necessary to
estimate how the performance of the coating changes with
temperature. The input data for this estimate is the G^*
as a function of temperature. The assembly of this data
is greatly facilitated by the time-temperature
superposition principle [4,5,44]. This principle in its
simplest form implies that the viscoelastic behavior at
one temperature can be related to that at another
temperature by a change in the time scale only. A more
accurate application of the superposition principle
includes taking into account changes with temperature in
the relaxed (G_r) and unrelaxed (G_u) moduli respectively.

In many practical applications of anechoic materials
the specifications demand not only high attenuation of
sound, but also good performance over a range of
temperatures and pressures. The extent to which these
demands can be simultaneously realized is limited by some
general relations which govern the behavior of linear
damped systems. These relations, known as the

Kramers-Kronig relations, are derived from the basic
causality condition that the output strain cannot precede
the input stress in any physical material. The
Kramers-Kronig relations are [8],

$$G'(\omega) - G_u = \frac{1}{\pi} \, PV \int_{-\infty}^{\infty} d\omega' \, \frac{G''(\omega')}{(\omega' - \omega)}$$

$$G''(\omega) = -\frac{1}{\pi} \, PV \int_{-\infty}^{\infty} d\omega' \, \frac{G'(\omega') - G_u}{(\omega' - \omega)} \qquad (24)$$

where PV is the principal value of the integral. For the
purpose of the present discussion it is more convenient
to use an approximate local version of the Kramers-Kronig
relations, developed by O'Donnel et.al.[45]. This
approximation relates the attenuation coefficient α to

the frequency derivative of the phase velocity, $dc/d\omega$, as
follows,

$$\alpha(\omega) = \frac{\pi\omega^2}{2c^2} \frac{dc}{d\omega} \qquad (25)$$

where α and c are related to G' and G'' by Eqs.7-10. It
follows from Eq.25 that a high value of α requires a
large $dc/d\omega$ in the viscoelastic region, which points to a
material with a large relaxation strength, $(G_u-G_r)/G_r$,
and a narrow range of relaxation times. However,
experience shows that in materials where the viscoelastic
region occurs over a narrow range of frequencies the
relaxation times, and therefore α, are strongly
temperature dependent. The unrelaxed modulus, G_u, has
approximately the same value for polymers, $\sim 10^{10}$ dyne/cm^2.
Therefore, to increase the relaxation strength it is
necessary to choose materials with a low relaxation
modulus G_r. However, for composite materials with air
cavities a low G_r of the host material leads to strong
pressure dependence for the effective moduli of the

composite. Therefore, it follows from the above
discussion that some trade-offs have to be made in
choosing a viscoelastic material for any specific sound
attenuation or damping application.

STRUCTURAL DAMPING

The discussion so far has centered on anechoic coatings,
that is coatings designed to reduce the reflection of
sound from elastic structures, such as the walls of an
anechoic tank in air or underwater. Another important
area of application of viscoelastic polymers is the
damping of flexural waves in structures such as plates
and beams. The particle displacement in a flexural wave
is normal to the surface of the plate. Strong radiation
of sound can occur from flexural waves in plates and
beams which are incorporated in the structures of
vehicles, machines, and buildings. Damping of the
flexural waves is necessary for noise reduction, hearing
protection, and the inhibition of structural fatigue.

A most useful measure of the degree of damping is
the loss factor, tan δ, as defined in Eq.12 in terms of
the energy dissipated per cycle, E_{diss}, and the total
stored energy of the vibration, E_{st} (E_{diss} and E_{st} may
apply to a complete vibrating system, or to a typical
unit area or unit length).

Two widely applied damping configurations that use
viscoelastic materials are the free viscoelastic layer
and the constrained viscoelastic layer, as shown in
Fig.9a and 9b. The deformation of the viscoelastic layer
is extensional in the first case and shear in the second
case. Both these deformations are highly damped by
intrinsic absorption in the viscoelastic polymer. In the
case of the free viscoelastic layer (Fig.9a) it flexes
with the plate participating in the bending stiffness as
part of a two-layer beam. The viscoelastic layer must be
tightly bonded to the plate and must be continuous over a

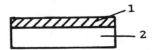

Figure 9a.   Free viscoelastic damping layer.

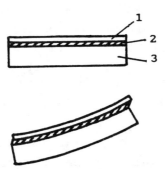

Figure 9b.   Constrained viscoelastic damping layer.

distance of at least 1/2 flexural wavelength. The
performance is described approximately as follows [46].

$$(\tan \delta)_{composite} = \text{const.} \frac{Y_2}{Y_1} \left(\frac{h_2}{h_1}\right)^2 (\tan \delta)_2 \qquad (26)$$

where Y and h are the Young's modulus and thickness
respectively, and subscripts 1 and 2 refer to the plate
and the viscoelastic layer, respectively. The composite
loss factor increases with the modulus and loss factor of
the viscoelastic layer, and with the square of the layer
thickness. This was recognized by Oberst who designed a
series of filled elastomer damping materials with high
modulus - loss factor products.

In contrast to the free layer, the operation of the
constrained viscoelastic layer (Fig.9b) involves shear
deformation of the layer [47]. Significant dampling can
be achieved in the frequency range where there is a
balance between the shear stiffness of the viscoelastic
layer (2) and the extensional stiffness of the
constraining layer (3). In this region the composite
loss factor varies approximately as follows,

$$(\tan \delta)_{composite} = \text{const.} \frac{Y_3 h_3}{Y_1 h_1} \cdot (\tan \delta)_2 \qquad (27)$$

where the subscripts 1, 2, and 3 refer to the plate, the
constrained viscoelastic layer, and the constraining
layer respectively. From Eq.27 it follows that good
damping requires the constraining layer to be
extensionally stiff, and the viscoelastic layer to have a
high loss factor. Although the modulus of the
viscoelastic layer does not appear explicitly in Eq.27,
it determines the layer thickness required to place the
maximum damping in the desired frequency-temperature
region.

Composite loss factors in the range 0.05 - 0.2 can

be realized with the above damping treatments. Good damping can also be achieved with thick viscoelastic layers at frequencies for which the layer is resonant in its thickness direction [48].

CONCLUSIONS

It is obvious from the above review that a wide variety of composite materials and composite structures have been developed for attenuation of sound. Nevertheless, there is a continuing need and opportunity for new concepts for attenuation of air-borne and water-borne sound waves, and to reduce structural vibrations. In particular there is a continuing need for coatings which operate over a wide range of frequencies, over a range of temperatures, and in some cases over a range of pressures.

On the theoretical side there is a need to develop better models for sound propagation in composite materials, including such structures as wedges. The models should include both microscopic and macroscopic inclusions, particularly at high densities. Both the long wavelength region, and the region where the wavelength is comparable to the dimensions of the inclusion should be studied. In the macroscopic case inclusions of different shapes should be evaluated. The theoretical models should identify the various possible mechanisms which can attenuate sound in a composite, and establish the optimum parameters and the maximum attenuation which can be achieved with each mechanism.

In the specific area of development of viscoelastic polymers for sound attenuation and vibration damping, the need is to identify materials with high intrinsic absorption which are also relatively temperature and pressure independent. These conditions can be optimized within the constraints of the general causal relations between the real and imaginary parts of the moduli.

REFERENCES

1. L.E. Kinsler, A.R. Frey, A.B. Coppens and J.V
    Sanders, "Fundamentals of Acoustics", J. Wiley
    and Sons, New York, 1982.

2. A.D. Pierce, "Acoustics", McGraw-Hill, Inc., 1981.

3. L. Cremer and M. Heckl, "Structure-Borne Sound",
    translated by E.E. Ungar, 2nd Ed. Springer-Verlag,
    1987.

4. J.D. Ferry, "Viscoelastic Properties of Polymers,
    3rd Ed., John Wiley and Sons, New York, 1980.

5. I.M. Ward, "Mechanical Properties of Solid
    Polymers", J. Wiley and Sons, New York, 1971.

6. R. Beyer, "Nonlinear Acoustics". (G.P.O.,
    Washington, DC, 1974) Naval Sea Systems Command,
    Stock No.: 0-596-215.

7. A.P. Dowling and J.E. Ffowcs Williams, "Sound and
    Sources of Sound", Ellis Horwood Limited, 1983.

8. A.I. Beltzer, "Acoustics of Solids",
    Springer-Verlag, 1988.

9. C. Zwikker and C.W. Kosten, "Sound Absorbing
    Materials", Elsevier Publishing Co., Inc., New York,
    1949.

10. D.L. Folds and C.D. Loggins, J.Acoust.Soc.Am., $\underline{62}$,
    1102 (1977).

11. K.Tamm, Ch.6 in "Technical Aspects of Sound", E.G.
    Richardson Ed., Elsevier Publishing Co., New York,
    1957.

12. R.D. Corsaro, J. Klunder and J. Jarzynski,
    J.Acoustic.Soc.Am., $\underline{68}$, 655 (1980).

13. E. Meyer, Applied Acoustics, $\underline{3}$, 169 (1970).

14. N.B. Miller, J.Acoustic.Soc.Am., $\underline{30}$, 967 (1958).

15. K.F. Herzfeld and T.A. Litovitz, "Absorption and
    Dispersion of Ultrasonic Waves", Academic Press, New
    York, 1959.

16. K.Schroeder, Physics Today, 30, 24, (Oct. 1979).

17. H.W. Strube, J. Acoust. Soc.Am., 70, 633 (1981).

18. M.C. Junger, J. Acoust. Soc.Am., 56, 1347 (1974).

19. J.A. Moore and R.H. Lyon, J. Acoust. Soc.Am., 72, 1989 (1982).

20. R.F. Lambert, J. Acoust. Soc.Am., 73, 1131 and 1139 (1983).

21. E. Meyer, K. Brendel, and J. Richter, Acustica 19, 8 (1967-68), and 21, 260 (1969).

22. E. Meyer, K. Brendel, and K. Tamm, J. Acoust. Soc.Am., 30, 1116 (1958).

23. H. Oberst, Ch. 7 in "Technical Aspects of Sound," E. G. Richardson Ed., Elsevier Publishing Co., New York, 1957.

24. P.C. Waterman and R. Truell, J.Math. Phys.,2, 512 (1961).

25. V. Twersky, J. Acoust. Soc.Am., 77, 29 (1985).

26. V.K. Varadan, Y. Ma, and V.V. Varadan, J. Acoust. Soc.Am., 77, 375 (1985).

27. L.W. Anson and R.C. Chivers, J. Acoust. Soc.Am.,85, 535 (1989).

28. W.M. Madigosky, R.W. Harrison and K.P. Scharnhorst, p. 615 in "Multiple Scattering of Waves in Random Media and Random Rough Surfaces," V.V. Varadan and V.K. Varadan, Eds., The Pennsylvania State University, 1985.

29. C.S. Clay and H. Medwin, "Acoustical Oceanography," J. Wiley and Sons, 1977.

30. C.F. Ying and R. Truell, J. Appl. Phys.,27, 1086 (1956).

31. R. Hickling, J. Acoust. Soc.Am., 36, 1124 (1964).

32. E.H. Kerner, Proc. Phys. Soc. (London), 69B, 808 (1956).

33. Z. Hashin and S. Shtikman, J. Mech. Phys. Solids, 11, 127 (1963).

34. R.M. Christensen, "Mechanics of Composite Materials," J. Wiley and Sons, 1979.

35.   E.P. Honing, P.E. Wierenga, and J.H.M. van der
      Linden, J. Appl. Phys. 62, 1610 and 1613 (1987).

36.   G.C. Gaunaurd and W. Wertman, J. Acoust. Soc.Am.,
      85, 541 (1989).

37.   Manufactured by B.F. Goodrich Co. under the trade
      name SOAB.

38.   G.C. Gaunaurd, E. Callen, and J. Barlow, J. Acoust.
      Soc.Am., 76, 173 (1984).

39.   R.M. Christensen, "Theory of Viscoelasticity," 2nd
      ED.,Academic Press, 1982.

40.   R.N. Thurston, Ch. 2 in "Mechanics of Solids,"
      Vol 4, C. Truesdell Ed., Springer-Verlag, 1974.

41.   J.C. Piquette, J. Acoust. Soc.Am., 77, 1665 (1985).

42.   W.M. Madigosky and G.F. Lee. J. Acoust. Soc.Am., 66,
      345 (1979), and 73, 1374 (1983).

43.   L. Rogers, J. Rhoel., 27, 351 (1983).

44.   B. Hartmann, article in the present book.

45.   M. O'Donnel, E.T. Jaynes, J.G. Miller, J. Acoust.
      Soc.Am., 81, 696 (1981).

46.   D. Ross, E.E. Ungar, and E.M. Kerwin, Jr., in
      "Structural Damping," J.E. Ruzicka,Ed., Am. Soc.
      Mech. Engrs., New York (1959).

47.   E.M. Kerwin, Jr., J. Acoust. Soc.Am., 31, 952
      (1959).

48.   E.E. Ungar and E.M. Kerwin, Jr., J. Acoust. Soc.Am.,
      36, 386 (1964).

RECEIVED January 24, 1990

Chapter 11

Acoustic Coatings for Water-Filled Tanks

Robert D. Corsaro[1], Joel F. Covey[1], Rose M. Young[2], and Gregory Spryn[2]

[1]Naval Research Laboratory, Washington, DC 20375–5000
[2]SFA, Inc., Lanham, MD 20785

Applications requiring underwater sound absorption usually make use of the high damping capability of polymeric materials. This paper describes the trade-offs necessary in selecting materials for sound absorbing coatings for use in reducing extraneous wall echoes in water filled pools and tanks. Coatings for this purpose are called anechoic, literally meaning no echoes, or non-reflective. Satisfactory anechoic coatings have been developed using a variety of different approaches including: simple and multi-layer absorbers; wedge-shaped designs; and resonant cavity approaches. This paper reviews the advantages of each and the material requirements involved. Data is presented on various commercially available underwater sound absorbing coatings, including three new coatings which have been only recently developed. Two examples of typical coating applications are presented: a tank for use in medical studies at ultrasonic frequencies (0.5 to 10 MHz) and one designed for calibration use at lower frequencies (6 to 100 kHz).

Underwater acoustics is routinely used in laboratory-scale test facilities for flaw detection, transducer calibration, material property evaluations, and acoustic visualization. In a typical underwater acoustic study, an object of interest is submerged in a water filled tank and acoustically illuminated (insonified). The acoustic signals scattered by the object are then measured and analyzed. If the tank used is not sufficiently large, these measured acoustic signals will include spurious echo components due to extraneous wall reflections. Since the effect of these contaminating echoes usually cannot be removed from the resulting data set by post analysis, they must be prevented from occurring at their source. One cost effective approach toward reducing these extraneous echoes is to apply sound absorptive treatments to the tank walls.

Several anechoic coatings are commercially obtainable for this purpose. Because their availability is not widely recognized, their characteristics are reviewed in this paper. We also include consideration of coatings which are not immediately commercially available but whose construction and performance is sufficiently documented in the literature to be reproduced on demand.

Some general design principals for each coating class are also included here to provide physical insight as to the operating principles of each coatings class. These can also be used for feasibility estimates of coating dimensions and properties. However it must be emphasized that in the final stages of a coating design, such model calculations are no substitute for direct measurements on candidate materials and structures.

Reflection at a Boundary

In this section we very briefly review the principles of sound reflection. Additional introductory material is presented in the preceeding papers in this publication, and a more complete analysis appears in standard acoustics textbooks such as that of Kinsler and Frey (1), or Pierce (2), or on a broader introductory level that of Crawford (3).

An acoustic wave is a traveling periodic pressure disturbance. This wave travels at a speed c dependent on the properties of the medium and the type of motion associated with the wave. The periodic nature of the acoustic wave is (for present purposes) taken to be a sinusoidal oscillation occurring at a frequency f. At any location x and instant in time t, the pressure associated with this traveling wave can be expressed as a cosine wave, or in a mathematically equivalent form as the real part of a complex exponential:

$$P = P_0 \cos(kx - \omega t) = P_0 \, Re\{\exp j \, (kx - \omega t)\} \tag{1}$$

The terminology used here is conventional:

k is the wavenumber ($k = 2\pi/\lambda = \omega/c$)

ω is the angular frequency in radians ($\omega = 2\pi f$)

λ is the wavelength, ($\lambda = c/f$)

and where P_0 is the maximum sound pressure amplitude, which is considered to occur at location x=0 and time t=0.

For plane waves propigating in an isotropic homogeneous medium, three acoustic properties are important: the speed of sound, the attenuation coefficient (to be discussed), and the characteristic impedance of the media. This impedance z is defined as the ratio of the acoustic pressure to the particle velocity associated with the wave motion in the material. For simple free-field plane waves, this is simply the product of the sound speed and density ρ.

$$z = \rho c \tag{2}$$

When sound traveling in media 1 strikes an interface with another material, media 2, a portion is reflected while the remainder is transmitted into media 2. As shown in Figure 1, if P_i is the incident pressure, P_r is the reflected pressure, and P_t is the transmitted pressure, then the reflectivity R and the transmissibility T are defined:

$$R = P_t / P_i, \qquad T = P_t / P_i \tag{3}$$

The property of the interface which controls the relative amounts reflected and transmitted is termed the "specific acoustic input impedance" Z which is defined as the ratio of the acoustic pressure to the particle velocity associated with the wave motion at the boundary. Note that this definition is similar to that used previously in defining the specific impedance of the media z. The use of the term impedance for both "input impedance at the boundary" and "specific impedance of the media" occasionally leads to some confusion if care is not taken.

Considering perpendicular sound incidence, the specific acoustic input impedance Z_2 at the boundary is simply

$$Z_2 = z_2. \tag{4}$$

Hence in this case it is immaterial whether we refer to impedance as Z_2 or z_2. It can then be shown that the reflectivity and transmissibility are simply related to the ratio of the impedance of media 1 and the input impedance of the boundary:

$$R = (1-z_1/Z_2) / (1+z_1/Z_2), \qquad T = 2 / (1+z_1/Z_2) \tag{5}$$

For consistency with latter equations, we will also express these two in the form:

$$R = M /M', \qquad T = 2 / M' \tag{6}$$

where in this case, $M = (1-z_1/Z_2)$ and where (in general) M' is simply M but with the sign of z_1 changed. While this usage is not common, it is preferred by this author since it simplifies the subsequent analysis of more complicated cases.

From equation 5 we see that if $Z_2 > z_1$ then R is positive and the incident acoustic signal simply reflects with a reduced amplitude. If $Z_2 < z_1$ the sign of R is negative indicating that the incident signal is inverted (180° phase shift) on reflection. Finally, if $Z_2 = z_1$ there is then no reflection at the boundary and all acoustic energy striking the surface is transmitted into medium 2. This latter case is of particular interest here, since it indicates that we can eliminate reflections from a surface by properly selecting or adjusting the density and sound speed of the material of medium 2.

While reflectivity and transmissibility are simple ratios, it is more common to present these parameters in decibel units. The corresponding parameters are then echo reduction (ER) and transmission loss (TL):

$$ER = -20 \text{ Log}|R|, \qquad TL = -20 \text{ Log}|T| \tag{7}$$

where the logarithms are base 10 and where the absolute values of R and T are used.

Note that in Equation 3, both R and T are defined in terms of pressure ratios. While this is unimportant for reflectivity (because both P_i and P_r exist in the same media), it can lead to difficulties in interpreting transmissibility. Consider, for example, the case of sound in water impinging on a steel surface. Since the density and speed of sound in steel are 7.7 g/cc and 6.1×10^5 cm/s, the impedance ratio z_1/Z_2 is 0.032. Then R = 0.938, indicating most (93.8%) of the incident pressure was reflected. But T = 1.94 indicating a pressure increase upon transmission to steel. While this may at first appear unclear, it is understandable if one considers that the impedance of water is much less than that of steel; hence although only a small fraction (12%) of the acoustic energy was transferred across the water-steel boundary, that small energy fraction corresponds to a much higher acoustic pressure in steel then it would in water. For following acoustic signals across boundaries, it is safer to use acoustic intensity rather than pressure, where intensity is proportional to the pressure squared. It can be shown that the intensity transmission coefficient and its corresponding transmission loss are:

$$T_I = 4 \ (z_1/Z_2) \ / \ (1+z_1/Z_2)^2, \qquad TL = 10 \ Log \ (T_I). \qquad (8)$$

As mentioned previously, the reflectivity and transmissibility of the surface is entirely controlled by the ratio of the specific impedance of media 1 and the input impedance at the boundary with media 2. As additional complexities are included in the formulation, they simply present themselves as modifications to these impedances.

For example, the previous discussion was for sound incident "normal" or perpendicular to the media boundary. If the acoustic signal strikes the surface at some other angle, as shown in Figure 1, the effect can be included simply as an increase in impedance. Hence, for an angle of sound incidence ϕ_1 the effective impedance of media 1 can be represented as:

$$z_1(\ \phi_1) = z_1 \ / \ cos(\ \phi_1). \qquad (9)$$

The effective impedance of media 2 (or the effective input impedance at the boundary) is represented in a similar manner, but with the angle of transmission in media 2, ϕ_2, replacing ϕ_1. These two angles are related by Snell's law (which might be familiar from geometric optics):

$$sin \ \phi_1 \ / \ c_1 = \ sin \ \phi_2 \ / \ c_2 \qquad (10)$$

These effective impedances then replace their counterpart in all previous equations.

Similarly, if one or both of the media are absorptive the analysis becomes more complicated. As we will now demonstrate, the sound speed then becomes a complex quantity. Hence the impedance of the media will also become a complex quantity, and the reflection and transmission coefficients will similarly be complex.

The complex nature of the sound speed can be derived as follows: For an absorptive media, the amplitude of the acoustic wave is exponentially reduced as it propagates. Hence

$$P_o = P_o' \exp(-\alpha x) \tag{11}$$

where α is the attenuation coefficient in nepers/cm. Combining equations 11 and 1 (using the complex exponential form of equation 1), we find that attenuation can be included in equation 1 simply by making the wavenumber complex:

$$k^* = k - j\alpha. \tag{12}$$

Hence attenuation is considered as an imaginary part of the wave number. (There are also fundamental physical justifications for this, but these need not be addressed here.) This correspondingly forces sound speed to also become a complex quantity:

$$c^* = \omega/k^* \quad = c_o/(1 - j\alpha/k). \tag{13}$$

Thus, in an attenuating media, we see that sound speed is no longer a simple quantity. It will be frequency dependent, its magnitude being:

$$|c^*| = c_o / (1+\gamma^2)^{1/2}, \qquad \gamma = 0.0183 \; \alpha'\lambda. \tag{14}$$

where α' is the attenuation coefficient now expressed in dB units.

Since sound speed is a frequency dependent complex quantity, it therefore follows that the characteristic impedance of the media will also be frequency dependent and complex. If the frequency dependence of sound speed is not known, it can be estimated from the attenuation coefficient as follows. For the rubber composites of interest here, usually $\alpha'\lambda$ is essentially independent of frequency. Using Kramers-Kronig relationships (5) it can then be shown that:

$$c = c_o [1+ (\alpha'\lambda/\pi^2) \ln(f/f_o)] \tag{15}$$

where c_o is the sound speed at any reference frequency f_o.

Anechoic Treatments

Simple Absorbing Layer   The simplest anechoic wall treatment available involves covering the wall with a layer of absorptive material. Such a case is shown in Figure 2. To be effective in absorbing the incident sound without reflection, the reflected echos labeled $\mathcal{A}$ and $\mathcal{B}$ must both be small. Here echo $\mathcal{A}$ is called the specular or front face reflectivity (described in the previous section), echo $\mathcal{B}$ is the first internal reflection, and echo $\mathcal{C}$ is the first multiple internal reflection. If echo $\mathcal{B}$ is small then all multiple internal reflections (such as echo $\mathcal{C}$) will be even smaller and can be neglected.. The material used must therefore perform two functions: it must allow all incoming sound to pass into the coating, and it must absorb this acoustic energy before it can be re-emitted.

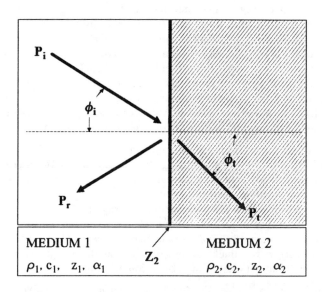

Figure 1. Reflection and transmission at a boundary between two materials.

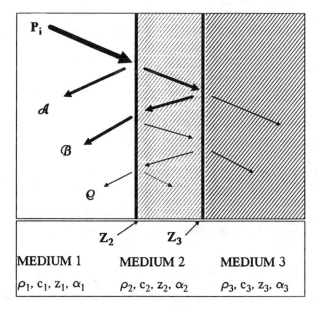

Figure 2. Major echoes reflected from a single finite layer of material.

Consider first the signal labeled $\mathcal{A}$. As discussed in the previous section, for the incident sound energy to penetrate the outer surface of the coating without reflection, the material used must have an acoustic impedance close to that of water. Materials having impedance values near that of water include many common rubbers. For example, arbitrarily selecting four commercial rubber samples from stock, we found the particular natural rubber sample had a surface reflectivity of 24 dB (6% reflective), the urethane sample measured 20 dB (10% reflective), and both the neoprene and nitrile rubbers were 12 dB (25% reflective). If lower reflectivity were required, the impedance of any of these rubbers could be altered simply by adding low density or low sound speed filler material.

Consider now the reflected signal component labeled $\mathcal{B}$ in Figure 2. For this to be small, the acoustic energy entering the sample should be largely absorbed before being re-emitted. Achieving a specific level of reduction therefore involves knowing the attenuation coefficient, and using an appropriate thickness of material. For example, at 1 MHz the natural rubber sample mentioned above has an attenuation coefficient of 1.1 dB/cm. Hence, reducing the contribution of internally reflected waves by 20 dB would require a layer of material 9 cm thick. This is admittedly rather thick for practical use at these high frequencies, and lower frequency coatings will require proportionately thicker coatings.

A simple way to increase the attenuation coefficient of a polymeric material is to add a very compliant filler such as air. The presence of microscopic air bubbles couples the traveling longitudinal wave to shearing motions at the bubble interface. Since in rubbers the damping factor ($\tan\delta$) for shear motion is usually orders of magnitude larger than that of longitudinal, the effect of the air bubble is to dissipate sound energy via the shear damping factor of the rubber. Additionally, some of the acoustic energy will be incoherently scattered by these bubbles, and will be subsequently dissipated via multiple scattering. Note that the use of air bubbles also lowers the sound speed and density of the material thus altering the acoustic impedance, however this change can be offset by the concurrent addition of a dense filler.

While this approach can increase attenuation by orders of magnitude, it does not follow that the required coating thickness can be proportionately reduced. As shown in the previous section, sound speed is a complex quantity having an imaginary part directly related to the attenuation coefficient. The impedance of an absorptive material will correspondingly also contain a significant imaginary component. Since the impedance of water is essentially real (having very little attenuation at frequencies considered here), the impedance match between the material and rubber necessarily becomes progressively worse as the material becomes more absorptive.

The design equations for predicting the performance of single layer coatings are uncomplicated . For the case of a continuous plane wave incident normal (perpendicular) to the surface, the reflectivity and transmissibility of a layer can be shown to be:

$$R = (M + jN) / (M' + jN') \tag{16}$$

where:

$$M = (1-z_1/Z_3) \cos Q_2, \qquad N = (z_2/Z_3 - z_1/z_2) \sin Q_2$$

$$Q_2 = k_2 L_2, \qquad L_2 = \text{layer thickness}$$

and M' and N' are simply M and N but with the sign of z_1 reversed (in the manner of equation 6). Sound absorption within any of the materials can be included as an imaginary part of the impedance or sound speed, exactly in the previous manner.

For coating design, it is usual to design for the worst case, namely a rigid backing structure. In this case Z_3 approaches infinity, $M = \cos Q_2$, $N = z_1/z_2 \sin Q_2$, $M' = M$, and $N' = -N$. Then equation 16 becomes:

$$R = (z_2 - j \, z_1 \tan Q_2) / (z_2 + j \, z_1 \tan Q_2) \tag{17}$$

which is seen to be simply the reflectivity of a single boundary (equation 5) but using an input impedance at the boundary of $Z_2 = z_2 /j \tan Q_2$.

This observation illustrates a general principle: the total influence of any structure behind a boundary can be combined and represented as the acoustic input impedance of the boundary. Hence the input impedance at the interface between media 1 and 2 is defined by a rearrangement of Equation 5 as:

$$Z_2 = z_1 (1+R) / (1-R) \tag{18}$$

There are many applications where this principle greatly reduces the complexity of the mathematical analysis, a few of which are mentioned in this paper.

For sound incident at other than the perpendicular direction the equations for reflectivity and transmissibility become significantly more complicated. If the deviation from perpendicular is small and the speed of sound in all three materials is similar, then the behavior can be closely approximated by correcting the specific impedances of the media using Equation 9. However when these conditions are not met, an expression such as that derived by Barnard (6) must be used.

Some general guidelines have been reported (7) which are useful in comparing the single layer coating with other coating types. A single layer 15 dB coating must be (nominally) at least 1 wavelength thick. Such a coating designed for use at 1 MHz (using a material with a sound speed near that of water) must therefore be at least 1.5 mm thick, while for 100 kHz it must be at least 1.5 cm thick. Coating thickness becomes more of a concern at lower frequencies or when greater than 15 dB ER is required. For example, continuing the above discussion, a 15 dB coating at 10 kHz is seen to require a thickness of 15 cm. If greater than 15 dB reduction is needed, the required thickness increases considerably, roughly doubling the thickness for each additional 5 dB ER. Note that these are theoretically imposed limits. Actual coatings must be somewhat thicker to allow for fabrication variations. To include some tolerance for material property variations during fabrication, typical coatings are designed for 3 to 5 dB greater performance than actually required.

While the design of such coatings is straightforward, selection of appropriate materials is not. Usually materials with the properties required for a particular application are not readily available, and some custom laboratory fabrication is necessary. This usually involves selecting a polymer composite which somewhat approximates the required physical properties. Then minor alterations to the chemical constituents or fillers are used on a trial basis and the acoustic properties (some combination of Young's Modulus and damping factor, sound speed, attenuation, density, and front-face reflectivity) of these sample formulations are measured. This continues until a suitable formulation is achieved.

One example of a commercially available absorptive material designed specifically for anechoic use is the product called "Soab" produced by B. F. Goodrich (8). It has been used in single layer coatings designed primarily for use in the 20 to 500 kHz frequency range. It consists of a butyl rubber to which aluminum powder has been added. Apparently this filler carries microscopic air pockets with it, thus introducing some sound attenuation. The material is a sufficiently close impedance match to water that echo reductions greater than 13 dB have been achieved, and the attenuation coefficient can be as high as 6 dB per wavelength, depending on the particular formulation used. It is temperature and pressure sensitive, and as with most commercially available acoustic materials, processing variations have reportedly been responsible for reduced performance on occasion.

To summarize, the advantages of the single layer coating are its simple construction and good predictability. Disadvantages include the rather tight material tolerances needed to obtain good impedance matching, particularly if greater than 15 dB echo reduction is required. These coatings also have relatively poor performance at angles of sound incidence other than that for which they were originally designed (usually normal or perpendicular incidence).

Multi-Layer Absorbers.   One approach toward reducing coating thickness is to use multiple layers of material. For illustrative purposes, consider the case where an outer layer of material is used which, because of its moderate attenuation, has an impedance 80% that of water. This impedance match contributes a front surface echo which is 20 dB reduced. The second layer of material can then have higher loss, since its impedance need only be 80% of the first material (or so this illustrative model suggests).

In actuality, the design of two layer coatings is somewhat more complicated. For the case of a plane acoustic wave incident normal to the surface, and for the worst case of a rigid backing, the reflectivity is as given by Equation 16, but where:

$$M = z_3 \cos Q_2 \cos Q_3 - z_2 \sin Q_2 \sin Q_3 \qquad (19)$$

$$N = (z_1 z_3 / z_2) \cos Q_3 \sin Q_2 + z_1 \cos Q_2 \sin Q_3,.$$

As before, $Q_3 = k_3 L_3$, M' and N' are simply M and N but with the sign of z_1 reversed, and sound absorption within any of the materials is included as an imaginary part of the impedance and sound speed.

Coatings with more than two layers can be developed using a stepwise approach, where the backing material and the layer closest to it are first treated as a single coating configuration and an equivalent acoustic input impedance is calculated at the front face of this layer. This then is used as the equivalent backing impedance for the next layer, and so on until all layers have been included. This stepwise approach is cumbersome. More efficient computational approaches use the formulation of Brekhovskikh (9).

The use of additional layers can reduce the thickness required, but this is not without limits. The simplest member of this class is the two layer coating, whose required thickness for 15 dB reduction is approximately 0.75 wavelengths, or 25% less than that of the single layer coating. The addition of more layers generates only progressively smaller thickness reductions. Usually a two layer design is a reasonable compromise, since the thickness reduction obtained with more layers rarely justifies the additional manufacturing complexity involved. When a three layer design is used, frequently only the first two layers are optimized, with the third layer being added in an ad-hoc manner to extend the frequency range or provide a sound barrier.

Examples include the "Wallgone" series of two and three layer coatings, which were based on the material system reported elsewhere (7). Although no longer commercially available, these coatings are mentioned here because their production is sufficiently documented that they have been successfully reproduced by various users, and they remain the most suitable coating available for applications at frequencies near 1 MHz. Additional information regarding their construction and use is presented later in this paper.

Another example of a commercially available three layer tile is available from Burke Rubber Co. as part number BR-8899-117. This tile is a foamed natural rubber with added fillers, and was designed to have an echo reduction of 7.5 dB at frequencies above 7.5 kHz. Its performance as measured in our laboratory is shown in Figure 3.

To summarize, the principal advantages of multiple layer coatings over single layer designs are reduced thickness (reduced materials cost) and less stringent requirements on the material formulation. The principal disadvantage is the additional manufacturing steps required.

Resonant Cavity Coatings.   The thinnest and least costly acoustic coatings are of the resonant cavity design, as illustrated in Figure 4. Such coatings are rubber

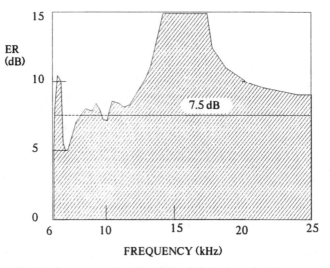

Figure 3.   Performance of a three-layer 7.5 dB (nominal) ER coating.

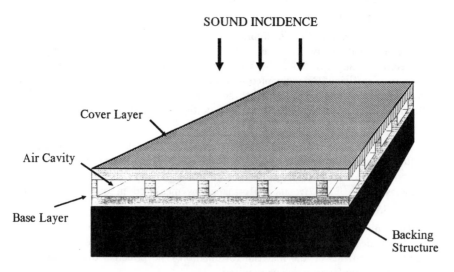

Figure 4.   Illustrative drawing of the cross-section of a typical resonant cavity coating.

layers containing large air voids. The incoming acoustic wave deforms this cavity, thus coupling some of the incoming wave to shearing motions in the rubber. These shearing motions can be rapidly attenuated due to the high shear damping factor. The utility of this coating lies in the design of the cavity, which can be made resonant at the frequency of interest. Such resonance behaviour generates enhanced deformations which very efficintly disipate the incoming acoustic wave.

Although such coatings have been used for many years, the acoustic mechanism involved is still not completely understood. The design of such coatings is difficult since their performance must be tuned to the particular operating frequency of interest, and factors such as the size and geometry of the air cavity, the modulus of the supporting rubber, and the inertia of the rubber covering the cavity all can play major roles. While attempts at understanding these factors have been made (10), the design of practical coatings of this type still involves considerable trial and error.

The design of these coatings also involves a trade-off between bandwidth and performance. A coating of this type with a maximum ER of 20 dB would be said to have a high Q (where the Q factor is a measure of the damping of the resonance) and would then correspondingly have a rather narrow bandwidth. On the other hand, one designed to have a maximum ER of only 10 dB would have a low Q and a bandwidth which might extend a few octaves (where an octave is a factor of two in frequency).

Depending on how high the Q factor, the performance of resonant cavity coatings can be highly dependent on environmental factors. Small changes in temperature or pressure will change the modulus of the rubber or the size of the air cavity and can therefore shift the frequency range of best performance. Additionally, the impedance of the backing material and the firmness of the adhesive used often plays an important role in influencing the Q and frequency of the cavity resonance.

A resonant cavity coating called Saper-T is commercially available from B. F. Goodrich (8). This coating has a relatively low Q and is designed to provide 10 dB ER to frequencies as low as 500 Hz. According to the manufacturer's literature, best performance occurs at a frequency given by f_c(KHz) = .56/ad, where a is the tank wall thickness in inches, and d is the wall density in pounds per cubic inch. Although the bandwidth is reported to range from 1/2 to 8 times f_c (1 octave below and 3 octaves above), this is probably optimistic for most applications. Because resonant cavity coatings are highly dependent on production and environmental factors, a trial evaluation of a test tile should always be performed in-situ to determine its suitability.

Wedge Coatings.   Wedge shaped coatings are familiar from their common use in air acoustics as wall coverings for sound studios and acoustic or microwave anechoic chambers. This coating geometry is also applicable underwater, where the shape of the surface reduces the importance of matching the impedance of the material to that of water. More precisely, the shape of the surface plays a dominant role in influencing the characteristic impedance of the surface. Since the specific impedance of the material then does not have to be a close match to water, materials with high acoustic loss can be used.

The operation of this coating type can be illustrated as follows. At low frequencies, where the spacing of the wedges is smaller than the wavelength of sound, the effect of the wedge is to cause the apparent acoustic input impedance to tend to be an area weighted average of the specific impedances across the wavefront. Hence near the tips of the wedges the input impedance encountered by the incoming acoustic wave is nearly that of water, while near the trough of the wedge the input impedance is essentially that of the material. This smearing of the wavefront more smoothly couples sound into the absorptive material. At high frequencies, where the spacing of the wedges is much greater than one wavelength in size, the incoming acoustic wave will undergo at least a few surface reflections (depending on wedge angle) before being returned. Since the loss contributed by each reflection is cumulative, the importance of impedance matching the material is considerably reduced.

Wedge type coatings have been constructed in many different shapes, some of which are illustrated in Figure 5. These shapes range from distributions of pyramids or cones with linear or optimally curved sides, to a simple linear run with a saw-tooth cross section or even flat slabs of material of various width extending out from the surface. A mathematical description of wedge coatings is not presented here because of the great variety of shapes and since presently such descriptions are of questionable usefulness. Some of the theoretical complications inherent in these models include the need for incorporating grazing angle of sound incidence, multiple scattering contributions, bulk attenuation in the wedge material, the influence of an absorptive backing layer (which is usually included for improved high frequency performance), and the wide frequency range typically required.

The design of such coatings therefore usually relies on using a few simple rules. For example, in our laboratory we have used the following in designing sawtooth-shaped wedge coatings. The coating is selected to be one wavelength thick (using the wavelength in water) at the lowest frequency of interest (although 0.5 wavelength thickness is theoretically usable). For good high frequency performance, a peak to peak wedge spacing of 20 to 50% of the peak height is reasonable. The material used should preferably have a sound speed somewhat less than that of water, since this will tend to bend the acoustic wave into the material, as shown in equation 10. Materials with sound speeds much higher than water should be avoided since beyond the critical angle of incidence they exhibit total reflection at the wedge surface and no sound can penetrate the absorbing material.

To obtain an adequate impedance match, the desired bulk impedance of the material used can be calculated from equation 9, knowing the speed of sound in the material and the wedge angle. For materials with sound speeds less than that of water the desired bulk impedance is seen to be somewhat less than water. In practice, however, the range of acceptable impedance values is so wide that this

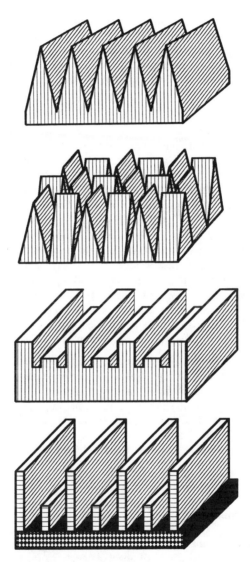

Figure 5. Some geometries which have been used in wedge-type anechoic coatings.

calculation is almost unnecessary. Consider that the total reflected signal will undergo typically at least three reflections at the surface of the sawtooth-shaped wedge. Then an 18 dB coating need only achieve 6 dB reduction for each surface reflection. To obtain 6 dB or better reflectivity, the specific impedance of the material (adjusted for the angle of incidence on the wedge surface) then need only be in the range 0.33 to 3.0 times that of water. This is a very wide range of suitable material impedances. Hence highly attenuating materials can be used with little fear of significantly upsetting the impedance matching at the boundary. (Of course, in practice one would not design a coating which approached these extreme limits.)

Because the material requirements for this class of coating are not very restrictive, wedge-shaped absorbers can be constructed using a variety of commercial available materials. For example, they have been formed of rot-resistant wood such as Cypress or Redwood (11). Wooden wedges are particularly low in cost and their fabrication is simple, however, they must be pressure treated underwater to fully saturate the wood pores and thereafter they must be kept wet at all times to avoid cracking. Additionally, these woods leach contaminating oils into the water thus requiring frequent cleanup. Wood impregnated concrete is used in the "Insulcrete" wedges developed by USRD/NRL (12). These are rather massive, but work particularly well under pressure. The ATF facility at USRD uses these wedges extensively, and typically 20 dB echo reductions are measured at frequencies from 15 kHz using wedges 18 inches long. Wedges have also been formed of simple polymeric materials, such as Soab (13-14). These are typically more compact and more easily mounted, but are more costly. Resonant cavity type coatings and voided rubbers (15) have also been used in wedge designs, offering significant cost savings at low frequencies.

In summary, wedge-shaped coatings are typically well behaved broad band echo reducers with excellent high frequency performance and excellent performance at off-normal angles of sound incidence. They can be reproducibly formed with little processing control since their material properties need not be tightly specified. However they also have a large surface area which can cause cleaning and contamination (air bubble) problems in some environments, they are relatively thick, and the process of forming the desired shape can involve special molds or operations.

Application Examples

Ultrasonic Inspection Tank   The material system described in reference 1 was previously used in the construction of anechoic coatings for a small 4 liter tank for use at 2 to 10 MHZ. In this paper, we consider a somewhat larger 40 liter tank requiring coatings with at least 20 dB ER at frequencies from 0.5 to 10 MHz. The approach used is similar to that described previously, but with the following exception. Since the silicone rubber previously used is no longer available, we selected RTV-31 (General Electric Co.) as an adequate (though less desirable) substitute. As presented previously, Bakelite Microballoons (Union Carbide Corp.) are added to the rubber base to adjust the attenuation coefficient, and iron oxide powder (United Mineral Inc.) is added to match the acoustic impedance to that of water. To reduce material costs, a three layer coating was developed. The compositions of these layers is shown in Table I, where the toluene is added to permit mixing and air removal. (The toluene evaporates from the resulting tile within the first few days after curing, with some consequent shrinkage.) The construction is straightforward, however the rubber mixture must be evacuated (as per manufacturer's product usage literature) at least two minutes to remove entrapped air bubbles, and the tiles must be formed in vertically positioned molds so that small (1 cm) vertical displacements of the microballoons during curing will not alter the properties of the resulting tile (except for a small region at the top and bottom, which can be discarded).

Table I. Formulation used for Ultrasonic Frequency Coating . (Quantities shown are sufficient for constructing 12 x 18 x 1/8 inch layer. Catalyst and toluene must be adjusted to obtain suitable working viscosity)

Component	units	Layer A (Inner)	Layer B (Middle)	Layer C (Cover)
RTV-31	gm	984	913	919
Microballoons	gm	16.3	6.9	1.4
Iron Oxide	gm	250	230	213
Catalyst for RTV-31	ml	2	1.4	2
Toluene	ml	250	210	230

The Cover and Middle layers of this coating were designed using the equations presented previously, however each layer can also be used independently. Hence, the Cover layer alone can ideally provide 25 dB ER from 2.5 to 20 MHz, and the Middle layer alone can ideally give 20 dB ER from 0.7 to 5 MHz. When these two layers are used together the calculated reduction is 25 dB from 700KHz to 20 MHz. This is 5 dB higher that that actually required in this application, and is considered an appropriate safety factor considering the likely laboratory variations during material formulation (temperature, mixing time, solvent evaporation). The Inner layer is added as an ad hoc addition to provide greater loss at frequencies to 0.5 MHz. An advantage of this type of construction is that the fabrication of a range of coatings can be accomplished with little change in setup, mold design, or raw materials.

Acoustic Pool Facility   As an example of a typical lower frequency installation, we recently applied anechoic coatings to our Acoustic Pool Facility. In this pool, acoustic studies have occasionally been hampered by reflections from the concrete pool walls. At frequencies above 25 kHz, the pool is sufficiently large (11.5 x 8.0 m and 6.1 m deep) that these wall echoes can generally be excluded from the data record by using time-domain filtering. However this filtering is not helpful for many interesting test and calibration transducer arrangements at frequencies below 25 kHz. We therefore wished to extend the usable range of this pool for general and routine use down to 6 kHz.

Because of the large size of this facility, it would have been prohibitively expensive to cover all walls with the best available coating. The most economical treatment meeting our requirements was found to require the use of five anechoic coatings, each with different characteristics and each covering selected regions of the pool. Four of the coatings used were new designs developed specially for this application, but which have since been made available by the manufacturers for other applications. The other commercial tile selected for less critical pool regions was Saper-T.

The application involved the following: 7.3 m² of rubber wedge-type tiles were mounted in the two most critical regions: the central region of the back wall (behind the test region) and the central floor region. These tiles have the best performance available in this frequency region (Figure 6), but the cost of these tiles ($1,000/sq.ft.) limits the extent of application. Surrounding these were 14.5 m² of a two-layer rubber tile. The performance of these tiles (Figure 7) was acceptable in this less critical region, and the lower tile cost ($400/sq.ft.) made their use economically attractive. Additionally 150 Burke tiles (previously mentioned), each 22 inches square, were installed on the central region of each side wall. Their cost is slightly lower ($350/sq.ft.) and their ER performance is somewhat poorer at normal incidence, however these tiles are nearly neutrally buoyant and hence easily supported, and their low acoustic impedance will result in improved ER performance at off-normal angles of sound incidence.

Finally 10 m² of the surface was covered using an absorptive polymer-epoxy composite with slight negative buoyancy formed into a 25 cm long wedge-shape and suspended from free floating rafts. The design of coatings for an air-water surface encounters one additional implementation feature not previously mentioned. Since there can be no net pressure difference across the water surface (3) (the pressure being fixed at 1 atmosphere), sound absorption mechanisms which rely on acoustic pressure dissipation will be less effective near this low pressure boundary. Hence these wedges are suspended 13 cm below the surface to locate them in a region where the efficiency of the attenuation mechanism is not significantly reduced. The performance of this coating is shown in Figure 8.

Much of the remainder of the pool was covered with the SAPER-T tiles. These were comparatively cheap ($50/sq.ft.) but contributed only 4 to 5 dB ER, as shown in figure 9. These tiles were used in this application primarily as an inexpensive low frequency reverberation reduction coating. (The reduced performance measured for these tiles in this application is a consequence of the unfavorable backing impedance presented by the thick concrete walls and method of attachment which used double stick tape instead of a rigid adhesive bond.)

An example of the performance actually measured in this pool facility after applying these coatings is shown in Figure 10.

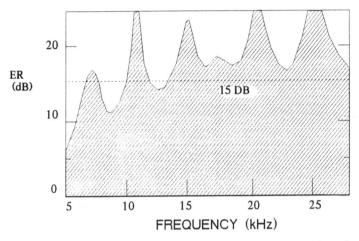

Figure 6. Performance of the 15 dB (nominal) wedge-shaped coating developed for the NRL pool facility.

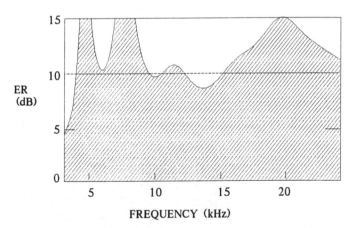

Figure 7. Performance of a two-layer 10 dB (nominal) tile, developed for the NRL pool facility.

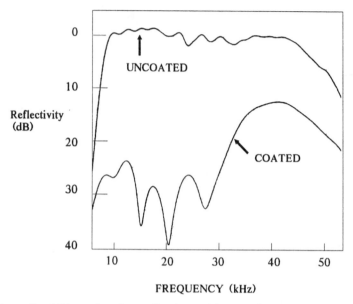

Figure 8. Measured surface reflections with and without the wedge-type surface treatment.

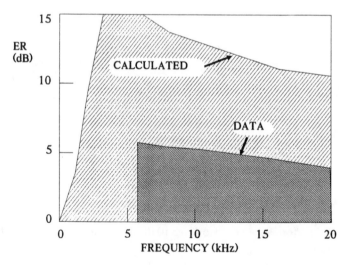

Figure 9. Performance of the Saper-T tile. "Calculated" is from B. F. Goodrich literature; "Data" is as measured in this particular application (see text).

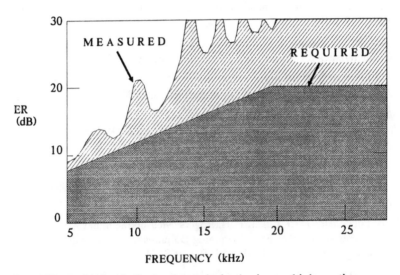

Figure 10. Reduction in floor echoes obtained using multiple coating treatment. Performance is for one particular downward-looking test geometry and includes some multiple reflections.

Literature Cited

1. L. A. Kinsler, A. R. Frey, A. B. Coppens and J. V. Sanders, Fundamentals of Acoustics, John Wiley & Sons, NY, 1982.

2. A. D. Pierce, Acoustics: An Introduction to its Physical Principles and Applications, McGraw-Hill Book Co., NY, 1981.

3. F. S. Crawford, Jr, Waves (Berkely Physics Course - Volume 3), McGraw-Hill Book Company, NY, 1968.

4. P. L. Edwards, "Ultrasonic Signal Distortion and its Effects on Velocity Measurements in Dispersive Constant-Group-Velocity Media", J. Acoust. Soc. Am. 1983, 73, 1608.

5. M. O'Donnell, E. T. Jaynes, J. G. Miller, "Kramers-Kronig Relationships Between Ultrasonic Attenuation and Phase Velocity", J. Acoust. Soc. Am. , 1981, 69, 696.

6. G. Barnard, J. Bardin, J. Whiteley, "Acoustic Reflection and Transmission Characteristics for Thin Plates", J. Acoust. Soc. Am., 1975, 57, 57.

7. R. D. Corsaro, J. D. Klunder, J. Jarzynski, "Filled rubber material system: Application to echo absorption in waterfilled tanks", J. Acoust. Soc. Am., 1980, 68, 655; and Erratum: 1982,71, 224.

8. Products-Materials for Underwater Sound Applications, B. F. Goodrich Aerospace and Defense Products Publication

9. L. M. Brekhovskikh, Waves in Layered Media, Academic Press, NY, 1960.

10. R. Lane, "Absorption Mechanisms for Waterborne Sound in Alberich Anechoic Layers", Ultrasonics, 1989, 28

11. L. B. Poche', Jr., L. D. Luker, P. H. Rogers, Measurement and Analysis of Echolocation Clicks of Free-Swimming Dolphins in a Tank with Echo-Reducing Wood Lining, 1981, NRL Report 8434

12. C. L. Darner, "An Anechoic Tank for Underwater Sound Measurements under High Hydrostatic Pressures", J. Acoust. Soc. Am., 1954, 26, 221.

13. W. J. Toulis. "Simple Anechoic Tank for Underwater Sound", J. Acoust. Soc. Am., 1955, 27, 1221.

14. W. S. Cramer, T. F. Johnston, "Underwater Sound Absorbing Structures", J. Acoust. Soc. Am., 1956, 28, 501.

15. J. Tanzosh, W. Reader, "Low-frequency Reflectivity and Transmissivity of Gradual Transition Tank Linings" (Abstract), J. Acoust. Soc. Am., 1988 84, S193

RECEIVED January 24, 1990

Chapter 12

Acoustic Wave Propagation in Materials with Inclusions or Voids

W. M. Madigosky and K. P. Scharnhorst

Naval Surface Warfare Center, Silver Spring, MD 20903–5000

The acoustic properties of homogeneous, isotropic
viscoelastic materials are modified significantly in the
presence of randomly distributed voids or viscoelastic
inclusions. Given a frequency-dependent effective
medium theory, effective wave speeds and attenuation
properties of the resulting inhomogeneous materials may
be calculated by introducing complex valued material
parameters for the constituent materials. The resulting
effective dynamic parameters are found to reduce to the
quasi-static ones originally derived by Kerner, in the
long wavelength, low volume concentration limit. The
data presented here suggest that the Kerner equations
adequately predict the properties of materials with
microscopic voids even at moderately high volume concentrations. With increasing volume concentrations and
decreasing wavelengths, theoretical results generally
diverge. However, in the resonance region of the
inclusions, considerable qualitative insight into the
dynamic response of these materials may be obtained by
studying a certain simple, closed form expression for
the effective wave vector. Spherical inclusions are
considered throughout this paper.

Many problems in ultrasonic visualization, nondestructive evaluation, materials design, geophysics, medical physics and underwater
acoustics involve wave propagation in inhomogeneous media containing
bubbles and particulate matter. A knowledge of the effect of voids
or inclusions on the attenuation and velocity of sound waves is
necessary in order to properly model the often complex, multilayered
systems.

Two factors will determine the accuracy of the modeling. The
first is the accuracy with which the dynamic mechanical properties
of the constituent materials is known, and the second is the degree
to which the effective modulus theory actually models the properties
of the inhomogeneous material.

The complex shear modulus is by far the most important dynamic
property to influence the acoustic properties of inhomogeneous materials. The accurate determination of the dynamic shear properties
is not an easy task as they are often very temperature and frequency
dependent. In addition, the dynamic properties are difficult (if
not impossible) to measure at the frequencies of interest. Consequently, shifting, using a time-temperature principle such as that
of the WLF shift (1) is employed to obtain data over a frequency
range larger than that used in the actual experiment. Recently,
several methods have emerged which greatly improve the reliability
of the dynamic data and offer a much wider actual measurement range
(2-4).

The second problem facing the researcher, and the one which is
the subject of this paper, is to determine which, if any, effective
modulus theory accurately predicts the acoustic wave velocity and
attenuation in a microscopically or macroscopically inhomogeneous
material.

The approaches are divided into those which do not take frequency into account (except for relaxation phenomena) and therefore
are "static" theories and those which take frequency into account,
generally by a scattering approach. The latter, "dynamic" theories,
when reduced to the low frequency or quasi-static limit, usually
compare favorably with the static theories. Some approaches take
multiple scattering into account and cannot be solved in closed
form. These require elaborate computer number crunching techniques
(5).

Finally, it is important to mention the basic limitation of all
of the theories with regard to the concentration of inclusions.
Except for the large number crunching approach, which requires a
knowledge of the pair correlation function, all theories are
essentially limited to low or dilute concentrations of inclusions.
The exact limit is not known, but it would appear that something on
the order of 10 percent is reasonable. Most of the theories using
different approaches agree up to that concentration. In practice,
however, the material designer uses these theories up to much higher
concentrations.

In what follows, we shall attempt to give the reader a sense of
the available theories and approaches, and the assumptions and
limitations of each. We will also compare the different theories
with some experimental data.

Spherical Microscopic Inclusions

Static Theories. The simplest effective modulus theory is that
which assumes the components behave as an ideal liquid mixture of
one substance in another (6). In this case, both the density and
the compressibility are additive properties of the corresponding
quantities of the two materials depending upon the proportional
amount of each substance in the mixture. Thus,

$$\tilde{\rho} = (1 - \phi)\, \rho + \phi \rho_1 \tag{1}$$

and

$$\tilde{\beta} = (1 - \phi)\, \beta + \phi \beta_1 \tag{2}$$

where ρ and β are the density and compressibility of the medium, ρ_1 and β_1 are the density and compressibility of the inclusions, and ϕ is the volume fraction of inclusions. The tilde indicates effective properties. Hence, the effective compressional modulus, M, is given by

$$\tilde{M} = \tilde{\rho}\tilde{c}^2 = \frac{1}{\tilde{\beta}} \tag{3}$$

This approach correctly predicts the observed effective sound velocity, $\tilde{c}$, in the case of metal inclusions in a soft viscoelastic medium (7), but cannot be applied when the medium has a large shear modulus.

For a fluid medium or a soft rubber, the effective velocity is

$$\left(\frac{c}{\tilde{c}}\right)^2 = (1 + A\phi)(1 - B\phi) \tag{4}$$

where

$$A = \frac{\rho_1 - \rho}{\rho} \text{ and } B = \frac{\beta - \beta_1}{\beta} \tag{5}$$

The velocity has a minimum (assuming A, B>0) at

$$\phi_{min} = (A - B)/(2\,AB) \tag{6}$$

For solids in a soft rubber the loss (see Equations 35-39) is

$$\bar{a}\lambda = \frac{4\pi}{3}\frac{\mu}{M}\delta_\mu \tag{7}$$

and is small. Here $\bar{a}$ is the attenuation coefficient, λ the wavelength, μ the shear modulus, and δ_μ the shear loss factor.

Major effects which greatly modify the acoustic properties of materials are due to inclusions which are in fact voids. This case will now be considered.

One of the most successful theories is that of Kerner (8). Kerner employs a three phase model consisting of an average size spherical inclusion surrounded by a shell of the host material and imbedded in the equivalent homogeneous medium. The inclusions are distributed randomly and there is no interaction between them. Under the effect of uniform hydrostatic pressure and uniform tension, the effective bulk modulus, $\tilde{K} = 1/\tilde{\beta}$, and effective shear modulus, $\tilde{\mu}$, are calculated. The result is

$$\widetilde{K} = \frac{\dfrac{(1-\phi)K}{3K+4\mu} + \displaystyle\sum_i \dfrac{\phi_i K_i}{3K_i+4\mu}}{\dfrac{(1-\phi)}{3K+4\mu} + \displaystyle\sum_i \dfrac{\phi_i}{3K_i+4\mu}} \tag{8}$$

and

$$\widetilde{\mu} = \mu \frac{\displaystyle\sum_{i \neq 1} \dfrac{\phi_i \mu_i}{(7-5\sigma)\mu+(8-10\sigma)\mu_1} + \dfrac{1-\phi}{15(1-\sigma)}}{\displaystyle\sum_{i \neq 1} \dfrac{\phi_1 \mu}{(7-5\sigma)\mu+(8-10\sigma)\mu_1} + \dfrac{1-\phi}{15(1-\sigma)}} \tag{9}$$

where K and μ are the bulk and shear moduli of the host medium, and
the subscript i refers to the ith inclusions. Also, $\sigma = (3K-2\mu)/$
$(6K+2\mu)$ is Poisson's ratio.

The effective compressional modulus is given by

$$\widetilde{M} = \widetilde{K} + 4\widetilde{\mu}/3 \tag{10}$$

When the inclusions are thin-walled, hollow spheres, such as the
glass microspheres used in a syntactic foam, then the moduli of the
inclusions are given by (9)

$$K_1 = \frac{2E_g t}{3(1-\sigma_g)R} \tag{11}$$

and

$$\mu_1 = \frac{5E_g t}{2(1+5\sigma_g)R} \tag{12}$$

Here E_g and σ_g are Young's modulus and Poisson's ratio for the shell
material, and t and R are the wall thickness and outer radius of the
sphere.

When the inclusions are voids, Equation 10 becomes

$$\widetilde{M} = \frac{K(1-\phi)}{1+\dfrac{3K\phi}{4\mu}} + \frac{\dfrac{4\mu}{3}}{1+15(1-\sigma)\phi/[(7-5\sigma)(1-\phi)]} \tag{13}$$

Other investigators (10) have obtained results similar to those of
Kerner. The results of Dewey (11) which are valid for a dilute
solution agree with the Kerner equation in the dilute solution
limit. Christensen (12) reviews and rederives the effective modulus
calculations for spherical inclusions. The three models which are

presented are the dilute suspension, non-dilute composite spheres, and three phase models. His dilute suspension model agrees with the work of Dewey and the dilute suspension limit of the Kerner equation. His composite model is identical to that of Hashin (13). Christensen's treatment of the three phase model differs somewhat from that of previous authors (8,10). His results differ only in the shear modulus from those of Kerner, although under dilute suspension conditions the two agree.

The Kerner equation, a three phase model, is applicable to more than one type of inclusion. Honig (14,15) has extended the Hashin composite spheres model to include more than one inclusion type.

Starting with a dynamic theory and going to the quasi-static limit, Chaban (16) obtains for elastic inclusions in an elastic material

$$\widetilde{M} = \frac{\lambda + 2\mu}{1 + 3\phi \dfrac{\lambda + \dfrac{2}{3}\mu - \lambda_1 - \dfrac{2}{3}\mu_1}{3\lambda_1 + 2\mu_1 + 4\mu} + \dfrac{10}{3}\phi\left(1 - \dfrac{\mu_1}{\mu}\right)E} \tag{14}$$

where, $\lambda = K - 2\mu/3$ and

$$E = \frac{6}{2\dfrac{\mu_1}{\mu}\dfrac{3\lambda + 8\mu}{\mu} + \dfrac{9\lambda + 14\mu}{\mu}} \tag{15}$$

For a voided material

$$\widetilde{M} = \frac{\lambda + 2\mu}{1 + \dfrac{3(\lambda + \dfrac{2}{3}\mu)\phi}{4\mu} + \dfrac{20\mu\phi}{9\lambda + 14\mu}} \tag{16}$$

Equations 13 and 16 yield the same results when $\phi <$ 10 percent.

Dynamic Theories. Dynamic theories take into account the scattering of acoustic waves from individual inclusions and generally include contributions from at least the monopole, dipole, and quadrupole resonance terms. The simpler theories model only spherical inclusions in a dilute solution and thus do not consider multiple scattering. To obtain useful algebraic expressions from the theories, the low concentration and the low frequency limit is usually taken. In this limit, the various theories may be readily compared.

The theory of Chaban (16) involves a self-consistent field technique for randomly distributed inclusions at low concentrations. The field at each inclusion is equal to the sum of the incident field and the fields scattered by all other inclusions minus its own self field. Thus, the self consistent approach attempts to improve the modeling of the scattering process but still does not take into account multiple scattering.

Chaban has considered the case of resonant voids in a low shear modulus material, as in the case of resonant air bubbles in water. In this case, the dipole, quadrupole, and higher order terms may be dropped compared to the monopole term. The effective modulus when $\lambda \gg \mu$ is given by

$$\widetilde{M} = \frac{\lambda}{1 + 3\lambda\phi/4\mu \left(1 - \left(\dfrac{\omega}{\omega_0}\right)^2 + i\eta\right)} \qquad (17)$$

where

$$\omega_0 = \frac{2}{a}\sqrt{\frac{\mu}{\rho}} \qquad (18)$$

is the resonance frequency (17,18) of a cavity of radius a, and η is a loss term associated with the damping of cavity vibrations. For an air cavity in water this will be a sum of viscous, thermal, and radiation terms whereas for a cavity in rubber it will be essentially the loss associated with the damping of shear waves, i.e., the shear loss factor.

Kligman and Madigosky (19) extended the work of Chaban to the case of a solid medium. Based on this, it can be shown that when only the lowest order terms in frequency of the monopole and quadrupole contributions are retained (the dipole contribution appears in the effective density), the effective modulus is

$$\widetilde{M} = (\lambda + 2\mu) / \left\{ 1 + \phi \left[\frac{(\lambda - \lambda_1) + \dfrac{2}{3}(\mu - \mu_1) - \left[\dfrac{1}{6}\left(\lambda + \dfrac{6}{5}\mu\right) - \dfrac{1}{10}\left(\lambda_1 + \dfrac{2}{3}\mu_1\right)\right](ka)^2}{\dfrac{4}{3}\mu + \lambda_1 + \dfrac{2}{3}\mu_1 - \left[\dfrac{\lambda}{3} - \dfrac{1}{2}(\lambda_1 + \dfrac{2}{3}\mu_1)\right](ka)^2} \right] \right\} \qquad (19)$$

where $k = \omega/c = \omega/\sqrt{(\lambda + 2\mu)/\rho}$ and the radiation damping term in the denominator has been omitted. This has a resonance frequency at

$$\omega_0 = \frac{c}{a}\left(\frac{\dfrac{4}{3}\mu + \lambda_1 + \dfrac{2}{3}\mu_1}{\dfrac{\lambda}{3} - \dfrac{1}{2}\lambda_1 - \dfrac{2}{3}\mu_1}\right)^{1/2} \qquad (20)$$

and for solid inclusions is generally higher in frequency than the dipole resonance.

For voids, the monopole is the lowest frequency resonance and Equation 19 reduces to

$$\widetilde{M} = (\lambda + 2\mu)/\left\{1 + 3\phi\left[\frac{\lambda + \frac{2}{3}\mu - \left(\frac{\lambda}{6} + \frac{\mu}{5}\right)(ka)^2}{4\mu - \lambda(ka)^2}\right]\right\} \tag{21}$$

This resonance is at

$$\omega_0 = \frac{2}{a}\left(\frac{\mu}{\rho}\left(1 + \frac{2\mu}{\lambda}\right)\right)^{1/2} \tag{22}$$

or

$$k\,a = 2(\mu/\lambda)^{1/2} \tag{23}$$

which is essentially the monopole result of Meyer, Bremdel and Tamm (17) when $\mu \ll \lambda$, but otherwise predicts a higher resonance frequency. Similarly, for the dipole resonance of a cavity, we find

$$k\,a = [12\mu/(\lambda + 6\mu)]^{1/2} \tag{24}$$

In the case of a solid inclusion, the dipole resonance frequency is generally lower than the monopole resonance frequency. For low shear modulus material with dense particles, Chaban finds the dipole resonance frequency

$$\omega_0 = \frac{3}{a}\left(\frac{\mu}{\rho}\right)^{1/2}\left(\frac{\rho}{2\rho_1 + \rho}\right)^{1/2} \tag{25}$$

Recently we have calculated this term when the shear rigidity of the matrix material is taken into account. We find the dipole resonance frequency

$$\omega_0 = \frac{1}{a\sqrt{2\overline{\rho}}}\left(\frac{\lambda + 2\mu}{\rho}\right)^{1/2}\{g(\beta_0, \overline{\rho}) - [g^2(\beta_0, \overline{\rho}) - 36\overline{\rho}\beta_0^2]^{1/2}\}^{1/2} \tag{26}$$

provided the inclusion material is rigid and fairly incompressible. Here

$$g(\beta_0, \overline{\rho}) = \beta_0^2(2 + \overline{\rho}) + 9\beta_0 + (1 + 2\overline{\rho}) \tag{27}$$

Where $\beta_0^2 = \mu/(\lambda + 2\mu)$ and $\overline{\rho} = \rho_1/\rho$. Equation 26 reduces to Equation 25 in the limit $\beta_0 \to 0$. To first order in β_0, Equation 26 becomes

$$\omega_0 = \frac{3}{a}\left(\frac{\mu}{\rho}\right)^{1/2}\left(\frac{\rho}{2\rho_1 + \rho}\right)^{1/2}[1 - \beta_0\left(\frac{9\rho}{2\rho_1 + \rho}\right)]^{1/2} \tag{28}$$

Finally, we note that Kuster and Toksoz (20) introduced a new method in which the displacement fields, expanded in series, for waves scattered by an "effective" composite medium and individual

inclusions are equated. The coefficients of the series expansions
of the displacement fields provide relationships between the elastic
moduli of the effective medium and those of the matrix and the
inclusions. Both spherical and spheroidal inclusions were considered. They derived their model only for the long wavelength limit
and, in that limit, their results for spherical inclusions are
identical to the static theory of Kerner. Recently, attempts (21)
have been made to bypass the low frequency limitation of the Kuster-
Toksoz method. Although well known results were obtained for the
case of air bubbles in water, generally the results are believed to
be fortuitous as the assumptions made in dropping the term containing the radius of the effective scatterer have recently been
questioned (22,23).

All of the above results for $\widetilde{M}$ and the various resonance
frequencies are contained in the long wavelength, low concentration
expression for the effective compressional wave vector derived from
the Foldy-Twersky theory for the coherent effective field (24-26),
namely

$$\widetilde{k}^* = k + (3\phi/2ka^3)f = \omega/c^* = (\omega/\widetilde{c}) + i\widetilde{\alpha} \qquad (29)$$

where f is the forward scattering amplitude of an isolated
scatterer. This equation contains contributions from all resonances
of the scatterers via f. It is perhaps most reliable at frequencies
below the lowest resonance. In special situations it is sufficient
to consider only the lowest resonance at low frequencies, either the
monopole or the dipole. The results then are equivalent to the
expressions for $\widetilde{M}$ quoted above. In general, at least several resonances should be retained in Equation 29 since their contributions
usually overlap. As ω approaches zero, the static limit, discussed
in the previous section, is reproduced at low concentrations of
inclusions. If other than spherical scatterers or distributions of
scatterer sizes and orientations are to be considered, f/a^3 should
be replaced by its average value.

The appearance of the forward scattering amplitude in Equation 29 enables us to give a conceptually appealing interpretation
of the attenuation coefficient associated with $\widetilde{k}^*$. Using the
forward scattering theorem, which is valid for any scatterer
geometry, one finds (27-30)

$$\mathrm{Imag}(f)/k = (\sigma_{pp} + \sigma_{ps} + \sigma_a)/4\pi \qquad (30)$$

σ_{pp} is the longitudinal to longitudinal or "p to p-wave" scattering
cross section of an isolated scatterer. At low frequencies it is
primarily associated with the monopole scattering mode. σ_{ps} is the
mode conversion or longitudinal to shear wave or "p to s-wave" cross
section. It is primarily associated with the dipole mode at low
frequencies. σ_a is the absorption cross section which describes
energy dissipation inside viscoelastic inclusions (embedded in an
otherwise elastic matrix material). It is active in all modes of
excitation. Combining Equation 29 and 30 yields

$$\widetilde{c} = c(1 + 3\phi\,\mathrm{Real}\,(f)/2k^2a^3)^{-1} \qquad (31)$$

and

$$\tilde{\alpha} = 3\phi\,(\sigma_{pp} + \sigma_{ps} + \sigma_a)\,/\,8\Pi a^3 \qquad (32)$$

The effective attenuation coefficient is therefore proportional to the total (extinction) scattering cross section, or equivalently, to the fraction of the energy flux density scattered in all directions ($\sigma_{pp} + \sigma_{ps}$) per unit distance travelled, plus the fraction converted into heat by the inclusions (σ_a).

Other closed form expressions for $\tilde{k}^*$ exist (31,32). These reduce the Equation 29 at low frequencies and in the limit of low concentrations of inclusions. The forward scattering amplitude plays an important role in such theories (see also Twersky (25), where multiple scattering is taken into account). More fundamentally, it is the unperturbed single particle resonance spectrum which is mirrored in $\tilde{k}^*$ via f. In the case of high concentrations of inclusions, correlated motions of scatterers may become important and dynamic loading may alter the spectra to some extent. Such effects are picked up by theories which explicitly attempt to treat near fields and multiple scattering interactions [5]. The resulting expressions for $\tilde{k}^*$ are implicit and depend on appropriately modified single particle spectra, rather than the (far field) appropriately modified single particle spectra, rather than the forward scattering amplitudes of isolated scatterers. At low concentrations of inclusions, Equation 29 should be yield accurate predictions. Expanding $\tilde{c}$ to first order in [3ϕReal(f)/$2k^2a^3$), shows that both $\tilde{\alpha}$ and $\tilde{c}$ increase linearly with ϕ when ϕ and f are small, since f does not depend on ϕ.

There are indications that correlation in scatterer responses and interactions between scatterers may not be important generally, even at moderately high volume concentrations of scatterers (33,34). Equations like Equation 29 should therefore yield useful qualitative information about the low frequency structure in $\tilde{c}$ and $\tilde{\alpha}$ over extended ranges of values of ϕ, depending in the particular system under consideration. A study of the cross sections adds further information about the types of waves scattered (longitudinal or shear) and the relative importance of energy dissipation by the inclusions.

One finds that solid inclusions tend to define the low frequency structure in $\tilde{k}^*$, the attenuation edge and the resonance in $\tilde{c}$, by means of the dipole resonance, whereas cavities tend to define it by means of the monopole resonance (35). Since the dipole resonance of a solid inclusion involves center of mass motion of the inclusion, this resonance tends to be at lower frequencies than that due to the monopole of a cavity which merely involves cavity wall motion. The heavier the inclusions in the case of solid inclusion and the softer the matrix material in the case of cavities, the lower in frequency in which significant attenuation is achieved (attenuation edge) and the larger the corresponding scattering cross sections.

In rigid matrix materials, which support shear waves because of their shear rigidity, mode conversion scattering by heavy solid inclusions predominates at low frequencies ($\sigma_{ps} \gg \sigma_{pp}$). In that case $\tilde{\alpha}$ reduces to

$$\tilde{\alpha} \simeq 3\phi\,(\sigma_{ps} + \sigma_a)\,/\,8\Pi a^3 \qquad (33)$$

Figure 1 shows an example; 30 percent lead inclusions (a = 2mm)
in epoxy, Epon 828Z. Here $\sigma_a \simeq 0$ and $\tilde{\alpha}$ is seen to be essentially
due to σ_{ps} at low frequencies. The volume concentration of inclusions in this example is, however, known to be rather high for
Equation (29) to yield exact predictions. It has been pointed out
that Equation (31) does not predict the shift in resonance frequency
with concentration that is experimentally observed for Pb in epoxy
at volume concentrations above 5 percent. Note the steep rise of $\tilde{\alpha}$
and the attenuation edge (predicted by Equation (30) at all
concentrations of inclusions), approaching the characteristic ω^4
(Rayleigh) dependence of σ_{ps} below the dipole resonance frequency;
Equation (26). At ka>1, attenuation is controlled by high frequency
resonances above the quadrupole resonance.

Soft matrix materials do not support shear waves and
consequently dipole scattering by solid inclusions and mode
conversion effects are weak in such materials. The corresponding
resonance frequency is given by Equation (19). Mode conversion is
also weak in the case of cavities in soft materials. The dipole
resonance frequency is given by Equation (28). In the case of
cavities in soft materials, monopole scattering predominates at low
frequencies. The lowest resonance frequency is given by Equation
(18) and $\tilde{\alpha}$ reduces to

$$\alpha \simeq 3\phi\,(\sigma_{pp} + \sigma_a)\,/\,8\pi a^3 \qquad (34)$$

An example is shown in Figure 2 for 30 percent voids in an
elastic rubber material ($\rho c^2 = 3.04 \times 10^9$ N/m^2, $\mu^* = \mu = 1.13 \times$
10^8 N/m^2, $\rho = 1.19 \times 10^3$ Kg/m^3, a = 2mm) Note that mode conversion
is virtually absent and the attenuation edge varies as ω^4, as in
Figure 1. In a viscoelastic material one would observe a transition
from the ω^4 dependence to a more gradual variation at low frequencies ($\tilde{\alpha} \propto \omega$ in the case of fairly flat loss factors). Overlapping of
the two regions may mask the frequency dependence of the effective
low frequency loss factor of the inhomogeneous material. A wide
distribution of inclusion sizes will further complicate this picture
by spreading out the peak in $\tilde{\alpha}$ and moving some resonance effects to
low frequencies. Similar considerations apply to Figure 1, although
in the case of solid inclusions and strong mode conversion
scattering, the matrix material is more or less elastic below the
resonances because of this rigidity and scattering is also weaker
than in the case of cavities in soft rubber.

In a rigid matrix material, the low frequency resonance structure of cavities tend to be smeared out and overlapping. Monopole
scattering is then relatively weak, but dipole and quadrupole
scattering, with accompanying mode conversion, enhance attenuation
throughout the low frequency region.

In the case of solid inclusions in rigid materials, Equation 29
not only predicts that the position of the lowest (dipole) resonance
and the magnitude of the scattering cross section will be sensitive
to the mass density of the inclusion, it also predicts that it will
be sensitive to the compressibility of the inclusion. The higher

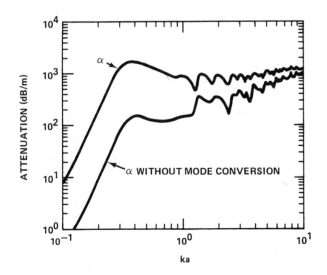

Figure 1. The Contribution of Mode Conversion Scattering to Longitudinal Stress Wave Attenuation by Spherical Pb Inclusions in Epoxy (Attenuation $\equiv 20\tilde{a} \log 10(e) = 8.69\tilde{a}$)

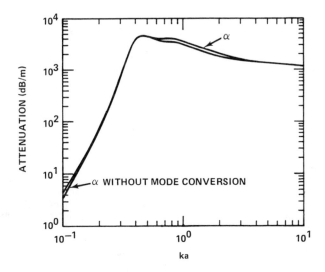

Figure 2. The Contribution of Mode Conversion Scattering to Longitudinal Stress Wave Attenuation by Spherical Cavities in Soft Rubber

the compressibility, the lower the dipole resonance frequency and the smaller the mode conversion cross section.

With respect to energy dissipation by solid inclusions in rigid elastic matrix materials, the situation according to Equation 29 is that in the dipole mode, α_a is relatively small compared to σ_{ps} regardless of the magnitude of the loss factor of the inclusion material. Hence, its contribution to attenuation is relatively weak. This is true regardless of whether or not the real parts of the characteristic impedances of the matrix and inclusion materials are matched.

Regarding the effective wave speed, $\tilde{c}$, Equation 31 predicts a gradual decrease with increasing frequency below the lowest resonance. This is followed by a rapid rise and if the scattering cross sections are small, $\tilde{c}$ will go smoothly through a maximum and then approach the speed of the matrix material at high frequencies. The latter property points toward the fundamental limitation of this equation with respect to the magnitude of ϕ; as the volume fraction of inclusions increases, the speed should approach that of the inclusion material. If, on the other hand, the scattering cross sections are large and ϕ not too small, a gap occurs in the propagation speed near the lowest resonance frequency, regardless of whether this resonance is due to the monopole or the dipole mode of oscillation. It is difficult to measure the phase of the received signal under these conditions (36); the formation of a coherent wave appears to be impossible in the gap. Another effect which has been observed experimentally (36), but cannot be explained by Equation 31 or any of the other closed form solutions for $\tilde{k}*$. (31,32), is a shift of the resonance in $\tilde{c}$ to higher frequencies with increasing concentrations of lead in epoxy.

Extension of Elastic Theories to Viscoelastic Materials

Theories of elastic materials may be extended to include viscoelastic materials by introducing complex moduli of the form

$$\mu_i^* = \mu_i(1 + i\delta_{\mu i}) \tag{35}$$

$$K_i^* = K_i(1 + i\delta_{Ki}) \tag{36}$$

$$M_i^* = M_i(1 + i\delta_{Mi}) \tag{37}$$

where $\delta_{\mu i}$, δ_{Ki}, and δ_{Mi} are the respective loss factors of the shear, bulk, and compressional moduli of the constituent materials. A useful relationship between the loss factors was introduced by Madigosky (37).

$$\delta_K = \delta_\mu \frac{(1-2\sigma)}{(1+\sigma)} \tag{38}$$

$$\delta_M = \delta_\mu \frac{(1-2\sigma)}{(1-\sigma)} \tag{39}$$

The frequency dependences of the storage and loss components of the complex moduli must be introduced (2-4). These must obey causality, as discussed by Hartmann in the present volume. Hence, a frequency dependent modulus always implies a frequency dependent loss factor and vice versa.

A particularly useful expression for the complex dynamic moduli was introduced using operators and fractional derivatives for the viscoelastic constitutive equations (38). For a rheologically simple system the complex shear modulus is

$$\mu^* = \mu_g + (\mu_r - \mu_g)/(1 + (i\omega/\omega_0)^q) \tag{40}$$

where μ_r and μ_g are the limiting rubbery and glassy state moduli, q is a real number between zero and one and is related to the slope of the real part of the modulus in the transition region and ω_0 is the frequency where the imaginary part of the modulus is a maximum. The quantity q may be related to the maximum loss factor δ_μ and ω_0 to the glass transition temperature. Note that all of the constants have a physical meaning.

In the quasi-static case, effective frequency dependent moduli and loss factors may be calculated from Equation 8. With respect to Equation 29, a lossy matrix material implies that k is now a complex number. The new expressions for $\tilde{c}$ and $\tilde{\alpha}$ differ from Equations 31 and 32, but follow straightforwardly. Equation 30 is usually cited only for elastic matrix materials, but, of course, it need not be used to interpret $\tilde{\alpha}$. The potential problem (also with viscoelastic inclusions) is that the derivation of Equation 30 is based on homo-geneous stress waves, whereas in viscoelastic materials one should, strictly speaking, consider inhomogeneous waves. The results obtained from Equation 29 are reasonable in the sense of yielding the expected superposition of scattering and dissipation effects.

Comparison With Experimental Data

The effect of mass loading in silicone rubber, Sylgard 184 (General Electric Corp.), was studied (7) using aluminum, lead, and iron oxide fillers. The experimental results are shown in Figure 3. The lowering of the sound velocity with the weight percent is clearly visible. The solid curves are those predicted using Equation 3 with additional results for tungsten and iron.

Similarly, the effects of bubbles in viscoelastic materials were studied by preparing rubber samples containing microvoids. Microvoids were used in order to avoid the effects of bubble resonance and to compare theories in their quasi-static limit.

Urethane samples were prepared using Solithane 13 (Thiokol Chemical) (4). The dynamic properties were measured at 25 kHz in the materials using a pulse-echo ultrasonic technique for the sound velocity and a resonance method (3) for the Young's modulus. The results are: $M = \rho c^2 = \lambda + 2\mu = 3.83 \times 10^9$ N/m², and $E = \mu(3\lambda + 2\mu)/(\lambda + \mu) = 1.0 \times 10^9$ N/m², and λ and μ may be calculated from these values.

Similarly, foamed urethane samples were prepared with various air contents, sound velocity measurements were made on the foamed

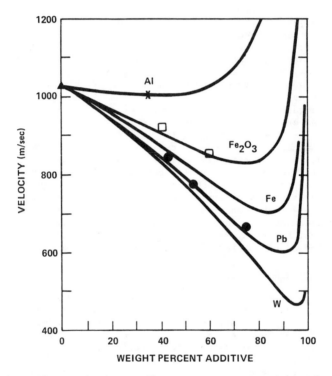

Figure 3. Effects of Mass Loading on the Wave Speed in Sylgard 184. Solid Curves Theory Equation 3.

samples and, together with the density values, this allowed the effective modulus to be calculated. The results are shown in Figure 4. Although the theories of Kerner and Chaban are strictly valid only for materials with low air content, they are presented here over the entire range where they are generally used. Note that both of the theories shown predict a rapid decrease in effective modulus with increasing air content up to 20 percent, and a much slower decrease in modulus thereafter.

Note that the experimental data agree best with the Kerner theory and that, at high void content, this theory correctly predicts an effective modulus of zero. Up to 15 percent in void fraction, the Kerner and Chaban predictions are essentially identical. Finally, the fact that the Kerner theory agrees with the experimental data up to 45 percent void content suggests that, currently, this is the best theory for predicting the effective modulus in microscopically voided materials.

Some data for the case of solid inclusions in a solid medium are shown in Figure 5. The figure shows a general lack of agreement between measured ([36]) and calculated ([39]) effective wave speeds in the dipole resonance region for 15 percent lead in epoxy. The curve was derived from

$$\tilde{k}^* = k[1 + (3\phi/k^2 a^3)f]^{1/2} \tag{41}$$

Equation 29 and other closed form solutions for $\tilde{k}^*$ ([31,32]) give very similar results; note the structure in $\tilde{a}$(30 percent Pb) in Figure 1. The dipole resonance frequency is given by Equation 20. Finally, we also note that at lower concentrations of Pb, agreement between experiment and theory is much improved ([39]).

Summary

We have reviewed various results in the theory of the effective material parameters of inhomogeneous materials with random distributions of spherical inclusions. Limitations to low volume concentrations have been stressed repeatedly although ranges of applicability are expected to vary from system to system. In the case of microscopic voids in rubbery materials, the range of applicability of Kerner's quasi-static parameters appears to be larger than commonly assumed. As the radii of the inclusions are increased and resonances encountered, the structure of the effective parameters becomes more problematic. A qualitative analysis, based on the single particle forward scattering amplitude which occurs in closed form solutions for the effective wave vector, is possible, however. Studies of the lowest resonances and their dependence on material combinations yield useful information about the attenuation edge, wave speed, and scattering cross sections as well as scattering and absorption mechanisms. At large volume concentrations numerical approaches must be used. The transition from elastic to viscoelastic materials is achieved by simply introducing complex valued, casually related material parameters. Arbitrary inclusion shapes, sizes, and orientations may in principal be treated once the single particle resonance spectra are known.

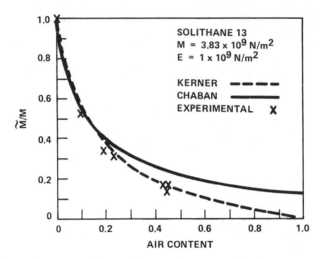

Figure 4. Comparison of Kerner's and Chaban's Theory with
Experiment

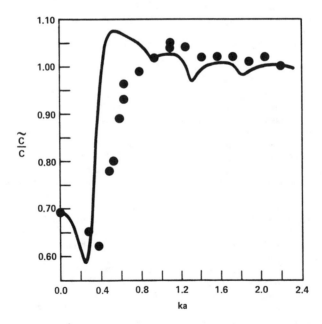

Figure 5. Comparison of Measured and Calculated Wave Speeds;
15% Pb in Epoxy. (Reproduced with permission from ref. 39.
Copyright 1983 American Institute of Physics.)

Literature Cited

1. Ferry, J., Viscoelastic Properties of Polymers, John Wiley and
    Sons, New York, 1970, p. 314.
2. Madigosky, W. and Lee, G., "Automated Dynamic Young's Modulus
    and Loss Factor Measurements," J. Acoust. Soc. Am., 1979, 66,
    pp. 345-349.
3. Madigosky, W. and Lee, G., "Improved Resonance Technique for
    Materials Characterization," J. Acoust. Soc. Am., 1983, 73,
    pp. 1374-1379.
4. Madigosky, W., "The Effects of Crosslinking on the Dynamic
    Mechanical Properties of Urethanes," Rubber World, 1987, 196,
    No. 6, pp. 27-29.
5. Varadan, V. K., "Multiple Scattering of Acoustic, Electromag-
    netic and Elastic Waves, Acoustic, Electromagnetic and Elastic
    Wave Scattering-Focus on the T-Matrix Approach, Pergamon Press,
    New York, 1980, pp. 103-134.
6. Urick, R. J., "A Sound Velocity Method for Determining the
    Compressibility of Finely Divided Substances," J. Appl. Phys.,
    1947, 18, pp. 983-987.
7. Madigosky, W. M. and Bradley, D. L., "Construction and Focusing
    Properties of a Solid Spherical Acoustic Lense," NOL TR 68-145,
    Silver Spring, MD, Nov 1968.
8. Kerner, E. H., "The Elastic and Thermo-elastic Properties of
    Composite Media, Proc. Phys. Soc. London, B, 1956, 69, pp. 808-
    813.
9. DeRuntz, T. A., Jr., "Micromechanics and Macromechanics of
    Syntactic Foams," LSMC Report 6-78-68-44, Lockheed Palo Alto
    Research Laboratories, Sep 1968.
10. van der Pol, C., "On the Rheology of Concentrated Dispersions,
    Rehol. Acta, 1958, 1, pp. 198-203.
11. Dewey, J. M., "The Elastic Constants of Materials Loaded with
    Non-Rigid Fillers," J. Appl. Phys., 1946, 18, pp. 578-581.
12. Christensen, R. M., Mechanics of Composite Materials,
    John Wiley and Sons, New York, 1979, pp. 31-72.
13. Hashin, Z., "The Elastic Moduli of Heterogeneous Materials,"
    J. Appl. Mech., 1962, 29, pp. 143-149.
14. Honig, E. P.; Wierenga, P. E., and van der Linden, J. H. M.,
    Theory of elastic behavior of composite materials, J. Appl.
    Phys., 1987, 62, pp. 1610-1612.
15. van der Linden, J. H. M.; Wierenga, P. E.; Honig, E. P.,
    "Viscoelastic Behavior of Polymer Layers with Inclusions,"
    J. Appl. Phys., 1987, 62, pp. 1613-1615.
16. Chaban, I. A., "Calculation of the Effective Parameters of
    Micro-Inhomogeneous media by the Self-Consistent Field Method,"
    Sov. Phys.-Acoust., 1965, 11, PP. 81-86.
17. Meyer, E.; Brendel, K.; and Tamm, K., "Pulsation Oscillations
    of Cavities in Rubber," J. Acoust. Soc. Am., 1958, 30, pp.
    1116-1124.
18. Gaunaurd, G.; Scharnhorst, K. P., and Uberall, H., "Giant
    Monopole Resonance in the Scattering of Waves from Gas-Filled
    Spherical Cavities and Bubbles," J. Acoust. Soc. Am., 1979, 65,
    pp. 573-594.

19. Kligman, R.; Madigosky, W.; and Barlow, J., "The Effective Dynamic Properties of Composite Viscoelastic Materials," J. Acoust. Soc. Am., 1981, 70, pp. 1437-1444.

20. Kuster, G. T. and Toksoz, M. N., "Velocity and Attenuation of Seismic Waves in Two-Phase Media: Part 1. Theoretical Formulations; and Part 2. Experimental Results," Geophysics, 1974, 39, pp. 587-618.

21. Gaunaurd, G. C. and Uberall, H., "Resonance Effects and Ultrasonic Effective Properties of Particulate Composites," J. Acoust. Soc. Am., 1983, 74, pp. 305-313.

22. Scharnhorst, K. P., "Comments on the Applicability of the Kuster-Toksoz Method to the Derivation of the Dynamic Material Parameters of Inhomogeneous Media," J. Acoust. Soc. Am., 1987, 82, pp. 692-695.

23. Anson, L. W. and Chivers, R. C., "Ultrasonic Propagation in Suspensions -- A Comparison of a Multiple Scattering and Effective Media Approach," J. Acoust. Soc. Am., 1989, 85, pp. 535-554.

24. Foldy, L. O., "The Multiple Scattering of Waves," Phys. Rev., 1945, 67, pp. 107-119.

25. Twersky, V., "On Propagation in Random Media of Discrete Scatters," Proc. Am. Math. Soc. Sym. Stochas. Proc. Math. Phys. Eng., 1964, 16, pp. 84-116.

26. Ishimaru, A., Wave Propagation and Scattering in Random Media, Academic Press, New York, 1978, Vol. 2, Chap. 14.

27. Varatharajulu, V. and Pao, Yih-Hsuig, "Scattering Matrix for Elastic Waves, I. Theory," J. Acoust. Soc. Am., 1976, 60, pp. 556-566.

28. Bostroem, A., "Multiple Scattering of Elastic Waves by Bounded Obstacles," J. Acoust. Soc. Am., 1980, 67, (2), pp. 399-413.

29. Lim, R. and Hackman, R. H., "Fundamental Analysis of a Novel Composite Sound Attenuator Using the T-Matrix Approach," NCSC Technical Report U2120-88-40, 1988.

30. Scharnhorst, K. P.; Hackman, R. H.; and Lim, R., "Damping Via Mode Conversion," Proceedings of the ACS Division of Polymeric Materials: Science and Engineering, 1989, 60, pp. 497-501.

31. Lax, M., "Multiple Scattering of Waves," Rev. Mod. Phys., 1951, 23, pp. 287-310.

32. Waterman, P. C. and Truell, R., "Multiple Scattering of Waves," J. Math. Phys., 1961, 2, pp. 512-537.

33. Varadan, V. V. and Varadan, V. K., "Configuration with Finite Number of Scatterers -- A Self Consistent T-Matrix Approach," J. Acoust. Soc. Am., 1981, 70, pp. 213-217.

34. Domany, E.; Eutin-Wohlman, O.; and Mizrachi, L., "Multiple Scattering Formalism: Application to Scattering by Two Spheres," J. Appl. Phys., 1984, 56, pp. 132-136.

35. Scharnhorst, K. P.; Madigosky, W. M.; and Balizer, E., "Scattering Coefficients and the Absorption Edge of Longitudinal Coherent Sound Waves in Selected Inhomogeneous Materials," NSWC Technical Report 85-196, Silver Spring, MD, Jun 1985.

36. Kinra, V. K., "Dispersive Wave Propagation in Random Particulate Composites, Recent Advances in Composites in the U.S. and Japan," ASTM STP 864, 1985, pp. 309-325.

37.   Madigosky, W. and Warfield, R., "Magnitude of the Ultrasonic
      Volume Viscosity," Acustica, 1984, 3, pp. 3470-3474.
38.   Rogers, L., "Operators and Fractional Derivatives for Visco-
      elastic Constitutive Equations," J. Rheol., 1983, 27, pp. 351-
      372.
39.   Sayers, C. M. and Smith R. L., "Ultrasonic Velocity and
      Attenuation in an Epoxy Matrix Containing Lead Inclusions,"
      J. Phys. D: Appl. Phys., 1983, 16, pp. 1189-1194.

RECEIVED January 24, 1990

Chapter 13

Guided-Wave-Tube Technique for Materials Characterization

R. Harrison

Naval Surface Warfare Center, Silver Spring, MD 20903–5000

The acoustic properties of a layer of visocelastic
material were determined from measurements made in a
water filled guided wave tube. Such a tube consists
of a pipe filled with water, several inches in
diameter, and of a length corresponding to three or
four wavelengths of the lowest frequency for which
measurements are desired. Tone bursts, several cycles
in duration and originating at one end of the tube,
were reflected by the sample and their attenuation
and phase shift measured. A method of obtaining the
phase from the Fourier transform coefficients was
employed with the advantage of working with a small
set of data points rather than a large array. An
extended frequency range was obtained by joining two
pipes of unequal diameters. The propagation
constants were calculated from the complex reflection
coefficient. Results for silicone rubber are given.

In the initial selection of an acoustic absorbing material for an
underwater application, the first considerations are often the
density, and the complex dynamic shear modulus. These quantities can
be measured in the laboratory, requiring only small sample sizes and
hence are useful as a guide to material development.

As the design matures, the direct measurement of the acoustic
properties becomes necessary. These properties include the
longitudinal wave speed, the coefficient of attenuation and the
acoustic impedance, which can be obtained from measurements of the
reflection and transmission of sound by the material. Two acoustic
techniques are available for these measurements, the impedance tube
and the panel test.

In the panel test, a sample of the material is submerged in
water and a sound projector is placed distant from the sample. A
second hydrophone is placed close to the sample so that it can
receive both the transmitted and reflected tone burst. These tone

bursts are usually two or three cycles in time duration of the lowest
frequency to be measured. In order that the sample measurements
approximate those made in free-field conditions, the distance between
the sample and the projector should be such that there is no
curvature to the sound wave and the sample width should be at least
several wavelengths. However, measurements made in a testing tank are
limited by the size of the tank; a sample-projector distance of 170
cm and a sample source distance of 34 cm for a panel measuring 76 cm
by 76 cm are typical of the geometry currently in use. Such a
geometry will have an approximate lower frequency limit of 6 kHz, set
by wave diffraction effects at the sample edges reaching the
receiving hydrophone at the same time as the reflected wave from the
sample.

M. P. Hagelberger and R. D. Corsaro (1) describe the use of an
averaging large planar hydrophone with a rigidly fixed test geometry.
In this case, test data was obtained at frequencies as low as 10 kHz
with only a one foot square sample. Higher frequencies do not present
a measurement problem until their wavelengths become comparable to
geometric irregularities in the sample and test structure at which
point the measurements are no longer characteristic of the entire
sample.

While the panel test has become accepted as a measurement
standard, the acoustic impedance tube (2-5) offers a means of
inexpensively and rapidly surveying a large variety of samples. These
tubes are normally about two 5 cm in diameter and 6 meters long with
the electronic instrumentation similar to that used in a panel test.
The sample is placed at one end and two hydrophones are mounted at
the other end. Signals are generated by one hydrophone and received
by the other. Indeed, a simple system can be constructed from a 6
foot length of two inch diameter steel pipe, filled with water and
having rubber stoppers at the ends. Leads from small hydrophones can
be pushed through holes drilled slightly smaller than their diameters
in the stoppers. Such a device will readily illustrate the
characteristics and utility of a pulse tube. For pulsed operation
with two millisecond pulses, especially evident will be the onset of
non-planar wave propagation at about 16 kHz as evidenced by the
extreme spreading of the echos due to dispersive effects of the
non-planar waves.

METHOD

Knowledge of the propagation constants of a material provides useful
physical insights into the molecular conformation, shear modulus and
density. If the acoustical impedance which is a measure of the
resistance of the sample to penetration by an acoustical wave is
known, then the propagation constants can be obtained through a known
mathematical relationship. One way of obtaining the impedance of a
sample is through the use of an impedance tube. Such a tube consists
of a pipe, filled with water, and of sufficient length to allow
several wavelengths of the sound wave to propagate at the same time
in the water. A sound projector is located at one end of the tube as
is a receiving hydrophone. The sample is placed at the other end.
Sound waves are emitted from the projector and travel through the
water in the tube to the sample. Here they are reflected back torward
the receiving hydrophone where their relative amplitudes and phase

are measured. The complex reflection coefficient, which includes both the amount of sound reflected from the sample and the phase shift experienced by the sound wave is then calculated. The impedance can then be found from this coefficient.

We have developed a water filled impedance tube system for measuring the propagation constants as a function of frequency for polymeric materials including those which exhibit hysteretic or other forms of damping. Fully automated, the system uses a fast Fourier transform procedure to extract the phase angle and calculates the rms value of the coefficient of reflection by a Simpson's rule integration of the digitally acquired data. Calculation of the propagation constants is performed by a small computer code which accepts as input the complex coefficient of reflection. The total time for an analysis of a material over two decades of frequency is less than an hour, which could be further decreased with faster signal processing and instrumentation.

THEORETICAL BACKGROUND

We consider a plane wave propagating along the z-axis in a material characterized by its speed of sound c_m and density ρ. Letting p equal the pressure at point z and time t, we can write:

$$p = p_o e^{-(iKz - i\omega t)}$$

where ω is the angular frequency in radians/sec and K is the propagation constant. For materials which exhibit damping, the propagation constant is complex and iKz is replaced by $(\alpha + ik)z$ yielding

$$p = p_o e^{-\alpha z - ikz - i\omega t}$$

This is the equation of a sinusoidal pressure wave which has an initial amplitude of p_o and decays exponentially with a decay constant of α. α and k are known as the propagation constants and $k = \omega/c$ is also called the wavevector number.

An acoustic wave which is traveling in a medium characterized by an impedance encounters a change of impedance Z_w at the face of the sample. This change causes the wave to be partially reflected with a complex reflection coefficient r, given by

$$r = \frac{Z_m - Z_w}{Z_m + Z_w} \qquad (1)$$

where Z_m is the impedance of the material. For a material which exhibits damping, the acoustic impedance can be shown to be

$$Z_o = i \frac{\omega \rho}{(\alpha + ik)}$$

However, the length of the sample, if it is comparable to the length of the acoustic waves can cause wave interference effects to arise at the sample-water interface and these must be taken into account in a theoretical description. The mathematical treatment of these effects can be found in the literature of transmission line theory which provides the following expression for the impedance of a length l of material

$$Z_m = Z_o * \coth[(\alpha+ik)l]$$

or normalizing to the impedance of the water

$$\frac{Z_m}{Z_w} = \frac{i\omega\rho}{(\alpha+ik)} * \coth[(\alpha+ik)l] \qquad (2)$$

Equation (2) can be combined with equation (1) to yield an expression for r, the complex reflection coefficient.

In practice, we can use the impedance tube to find the complex coefficient of reflection and then vary the two propagation constants in the theory to produce the same complex reflection coefficient. These variations are not easy to perform as the equation is transcendental; however, there are computer programs available to do this (7).

EXPERIMENTAL

The design of a pulse tube is usually dictated by considerations of the maximum working frequency and the sample diameter. Where the sample contains voids or inclusions, the diameter of the tube must be large enough so that the specimen diameter will contain a representative number of voids or inlusions. On the other hand, the diameter of the tube determines an upper frequency of operation. When the tube diameter is small compared to a wavelength of the sound in water in the tube, only plane wave propagation in the tube takes place. However, when the wavelength becomes comparable to the tube diameter, radial modes of propagation are possible making accurate measurements difficult. In this situation, the echos from the sample will have a time duration equal to large multiples of the incident pulse length. This occurs when the frequency of operation is near or exceeds the cutoff frequency given by

$$f_{cutoff} = \frac{.581c}{d}$$

where c is the speed of sound in the tube and d is the diameter of the tube (8-9). In the case where the speed of sound in the tube would have a cutoff frequency of 16 kHz, a 3.5 inch tube would have a cutoff frequency of 9 kHz, and a 7 inch tube a cutoff frequency of 4.6 kHz. These are theoretical limits and the experimentally observed limits will be different. Figure 1 shows this information graphically.

The length of the tube is determined by the requirement that the interrogating pulse completely insonify the sample thus allowing interference patterns arising within the sample to be generated. For

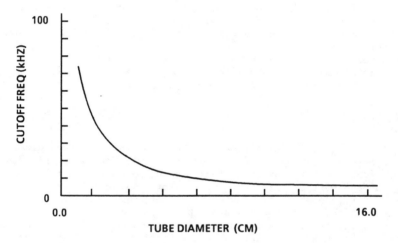

FIGURE 1. Pulse tube cutoff frequency vs. pulse tube diameter. Radial wave formation takes place at frequencies higher than the cutoff frequency.

this reason, use of very short pulses should be restricted to situations where this is not a problem. In general, for the frequency range of 4 kHz to 60 kHz, three cycles will completely insonify a sample several centimeters thick. Thus the tube length must be at least three wavelengths long at the lowest frequency of interest.

Originally, pulse tubes were constructed with a wall thickness equal to their inside diameter. This prevented the tube from bulging as the pressure wave traveled from the transducer to the sample. Recently, in this laboratory and at others, tubes with thin walls of only one or two centimeters have been used. This prevents ringing in thick outer walls but allows bulging as the pressure wave travels the length of the tube. Any such bulging of the tube would lower the propagation velocity of sound in the tube and, as the impedance is the product of the speed of sound multiplied by the density, the impedance of the water to sound. As a difference in impedance between the sample and the water will give rise to a reflection, the velocity of sound in the tube must be accurately known and certainly used in any calculations for sample impedance in any pulse tube (2-3).

The impedance tube developed consists of a 7 foot long section of 2" bore stainless steel tubing coupled to a 6 foot long section of 1/2" bore stainless steel tubing. Connecting the two tubes is an 18" long transition section so designed that it flares from 2" I.D. at one end to 1/2" I.D. at the other end. With the exception of the transition section which has an outside diameter of 6 inches, the 1/2" section of tubing has a wall thickness of 1/4" while the 2" section of tubing has a wall thickness of 1/2". These two wall thicknesses mean that the tube is characterized as a thin wall tube with a velocity of sound in the tube slightly lower than it would be for a corresponding thick wall tube. In our case, the 2" tube permitted a reasonable sample size while the 1/2" tube allowed us to collect data up to 60 kHz. The fifteen foot length of three unequal tubes was chosen to obtain a low frequency limit of 2 kHz and to locate the signal echo in a time span away from other echos. Figure 2 is an illustration of the appearance of the computer display terminal after the acquisition of the data from the waveform recorder. Note in particular how the echo from the sample is located in a quiet region away from other echos which could cause measurement errors. The first signal on the right is leakage through the T-R switch from the transmitted pulse. The next large signal is a reflection from the transition section due to the impedance mis- match of this section. The third signal, located in the gate, is the desired echo.

Observation of the waves arriving at the transducer, will show an initially large wave due to incomplete cancellation of the output pulse in the transmit-receive switch, an echo from the transition section and other echo traveling in the steel walls of the tube. The echoes in the walls of the tube will be found to be superimposed on the sample echo if the length of the tube/tube sections is not chosen with care. The partially cancelled output echo must be blocked out in the preamplifier as its amplitude would hinder accurate measurement of the sample echo. This can be done by introducing a negative gate in the preamplifier to bias off the input transistors.

Any air in the tube or in the water will cause grievous measurement problems as the sound pulses will be reflected from a bubble of air. (This can be checked by observing the echo on an oscilliscope while the tube pressure is varied.) Our tube is

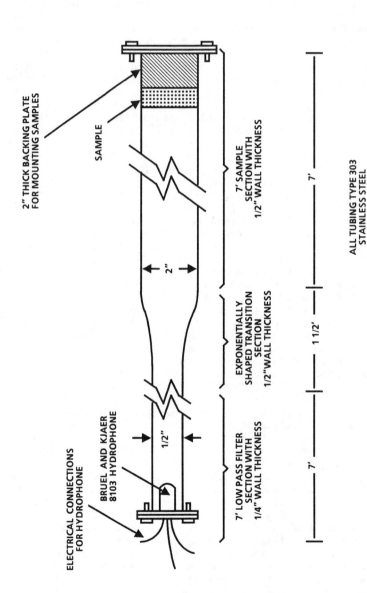

FIGURE 2. The joining of two tubes of unequal size allows operation up to 60 kHz and a two inch diameter sample size.

evacuated and then filled with de-aerated water, produced by either
boiling or by degassing using a mechanical vacuum pump. A small
stainless steel hand pump from Enerpac is used for pressurization;
piston type, air driven, stainless steel pumps could also be used.
For transferring water, a positive displacement, screw type pump is
necessary: centrifugal pumps will introduce cavitation when they have
either a restricted imput or output stream. Even with a screw type
pump, it is advisable to insert a bubble trap between the pump and
the tube. Such a trap consists of a steel cylinder, perhaps a foot
high, in which the water is pumped into near the top and extracted
from the bottom. Bubbles will rise to the top where the air they
carry can be removed by a valve on the cylinder. In addition, perhaps
due to the wall thickness of 1/2", we have found it necessary to
pressurize the tube to 100 psi to obtain accurate data. Bath clear
(VWR Scientific, Inc.) or iodine can be added to the water to control
bacterial growth.

The phase is found by considering only the transform at the
synthesizer frequency and noting that:

Two millisecond pulses of varying frequency were used to
interrogate the sample. The length of time of the pulse is set both
by the need for complete insonnification of the sample and for steady
state conditions to occur. A Bruel & Kjaer Model 8103 hydrophone is
located at one end of the tube and is used for both sending and
receiving. A diode array (Figure 3) is used as a transmit-receive
switch and serves both to short out the preamp during the transmit
time and to disconnect the transmitter during the receiving time.

A Hewlett-Packard synthesizer under computer control generates a
continuous sine wave which is then modulated by General Radio tone
burst generators. A Krohn-Hite 75 watt amplifier provides unusally
clean output signals to the Transmit-Receive switch. The received
signal is filtered by an elliptically designed filter and then
captured by a Biomation waveform recorder where it is digitized and
transmitted to the computer. Because of the need for phase
measurements,accurate timing is crucial to the experiment. It was
found that the best system was to use a differential scanning
analyzer developed for nuclear physics research to obtain a timing
pulse from the positive slope of the modulated signal. This pulse was
used to initiate waveform recording by the waveform recorder.

The acquisition system allowed a gate to be set in the recorded
data as shown in Figure 4. The rms value of the signal was calculated
within this gate. However, the Fourier transform is also taken of the
signal within the gate so that the gate length must be an integral
number of half-wavelengths to avoid leakage. Furthermore, the
frequency of the synthesizer must be set at an integer multiple of
the frequency given by

$$\text{freq} = \frac{1.0}{\text{time spanned by gate}}$$

The phase is found by considering only the transform at the
synthesizer frequency and noting that:

$$\text{Phase} = \frac{\text{Real part of Fourier transform}}{\text{Imaginary part of Fourier transform}}$$

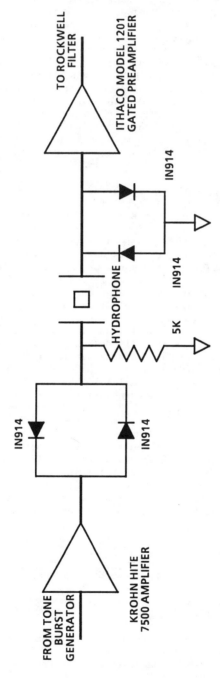

FIGURE 3. The transmit-receive switch allows one transducer to serve as both the sound projector and receiver.

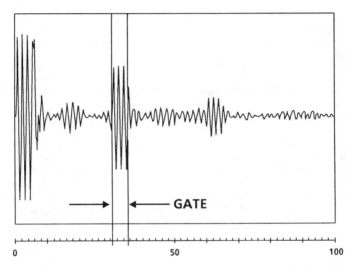

FIGURE 4. The computer calculates the rms value of the signal only within the user defined gate.

For measurement, the sample was mounted on a 15 cm long cylindrical steel base. A calibration cylinder was fabricated whose length was equal to the sample plus its steel base cylinder. This cylinder served to provide both an amplitude and phase reference. The amplitude reference was simply that of a wave being reflected from a steel surface. However, the length of the tube in terms of wavelengths varied from frequency to frequency: thus the phase of the signal as recieved at the hydrophone was constantly changing with respect to the outgoing signal. In order to calculate the phase shift of the reflected signal from the sample, a knowledge of the phase shift from the calibration cylinder is essential. Two runs for each sample were made, the first being for calibration data and the second for sample data. Separate data for each frequency was collected as the sharp filtering by the elliptic filter also introduced some dispersion, making it necessary to reset the gate for each frequency.

RESULTS AND CONCLUSIONS

A short computer program was used to find the propagation constants
from the impedance tube data using suggestions from (6-7). Data for a
silicone rubber sample appears in the table below and a plot of the
data is shown in Figure 5.

TABLE OF FREQUENCY VS ATTENUATION AND SOUND SPEED FOR
SILICONE RUBBER

FREQ(kHz)	α	c(m/sec)	FREQ(kHz)	α	c(m/sec)
15.6	2.4	1530	34.1	21.2	1550
24.4	9.4	1500	39.1	13.0	1570
26.3	15.6	1530	46.8	23.0	1570
29.9	37.6	1570	55.6	37.6	1590
34.1	21.2	1550	58.6	41.8	1400

For the boundary conditions in an acoustic impedance tube, the
measurement corresponds to the dilational modulus, K +4/3 G, where K
is the bulk modulus and G is the shear modulus. This can be
calculated from measurements of α and c through the relations
$M' = \rho c^2 (1-r^2)/(1+r^2)^2$, $M'' = 2\rho c^2 r/(1+r^2)^2$, where M' and M'' are the
real and imaginary components respectively of the dilational modulus
and r= $\alpha c/\rho$ is the density of the sample in kg/m^2, c is the
measured sound speed and α is the attenuation per unit length.

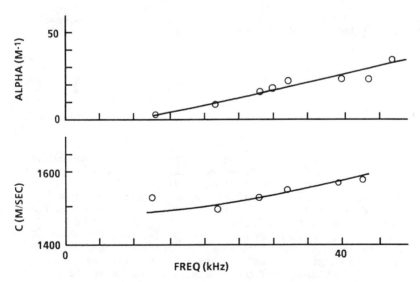

FIGURE 5. The attenuation constant and the speed of sound vs
frequency for a silicone rubber sample.

REFERENCES

1. Hagelberg, M. P. and Corsaro, R. D. A Small Pressurized Vessel for Measuring the Acoustic Properties of Materials, J.Acoust. Soc. Am. 77, 1222-1228 (1985).
2. Bobber, R. J. , Underwater Acoustic Measurements;Peninsula Publishing: Los Altos, 1988, p. 306.
3. Eynck, J. J. Characterisitics of Wave Guides Used To Evaluate Materials Intended for Underwater Acoustic Applications, Rubber Laboratory, San Franciso Bay Naval Shipyard Report 165-50,1967.
4. Sabin, G. A. Acoustic Impedance Measurements at High Hydrostatic Pressures , J. Acoust. Soc. Am.40, 1345-1353 (1966)
5. Meyer, E. et al. Sound Absorption and Sound Absorber in Water, NAVSHIPS 900,166, U. S. Department of the Navy, Washington, D. C. 1950.
6. Cramer, W. S. and Bonwitt, K. S. Pulse Tube for Acoustic Measurements, U. S. Naval Ordance Laboratory Report NAVORD 2257, 30 April 1952.
7. Lastinger, J. L. and Sabin, G. A. A PDP-8 Fortan Program for Reduction of Acoustic-Impedance Data, Naval Research Laboratory, Underwater Sound Reference Detachment, Rept 6906, 28 March 1969.
8. Skudrzyck, E. The Foundations of Acoustics, Springer, New York, 1971 p 431.
9. Morse, P. M. and Ingard, K. U. Theoretical Acoustics, McGraw-Hill Book Company: New York, 1981; p 123.

RECEIVED January 24, 1990

Chapter 14

Numerical Simulation of Acoustic Response of Discontinuous Viscoelastic Fluids

B. C. H. Wendlandt

DSTO, Materials Research Laboratory, P.O. Box 50, Ascot Vale, 3032, Victoria, Australia

Acoustic response of viscoelastic fluids may be simulated by Lagrangian finite element or Eulerian finite difference schemes. Both techniques have been used to predict the behaviour of discontinuous viscoelastic layers which may contain inserts such as air cavities and which are backed by metal plates. The advantages and disadvantages of applying these numerical techniques to the study of the acoustic response of fluid composites are discussed.

The acoustic response of resonant viscoelastic fluid structures to a pressure wave may be simulated by a four-dimensional calculation, three dimensions in space and one in time. The Lagrangian, primitive finite element and Eulerian finite difference schemes form the basis for two models presented in this paper which are able to simulate a wide range of fluid structures containing inclusions of arbitrary spacing, shape and composition.

The Lagrangian model subdivides the fluid into a large number of small elements whose size and shape is influenced by the passage of an acoustic wave. This model is especially suitable for the simulation of the acoustic response of the fluid when the material properties are discontinuous or significant deformations of sections of the fluid occur. The Lagrangian model is, however, computationally expensive, particularly when multidimensional problems are considered. The Eulerian model solves the mathematical equations governing the acoustic response of the fluid rather than simulating the viscoelastic responses of the fluid in detail which simplifies the algorithms but reduces the flexibility of the model.

Both models can readily compute the passage of a pressure pulse or sinusoidal wave through a multilayer, multicavity or multi-inclusion fluid such as coatings used to quieten metal structures. For ease of interpretation, the calculation examples are limited to two space dimensions only. The results of

0097–6156/90/0424–0260$06.00/0

representative calculations show the acoustic response of square, air filled columns and elastic and viscoelastic cylindrical inserts embedded in viscoelastic and elastic fluids. The inserts are irradiated with sound which is incident normal to the long axis of the columns and cylinders.

Model

Propagation Model. The propagation of a pressure or stress wave through a compressible medium is described by the laws of conservation of mass and momentum and the equation of state which relates the pressure or stress in the medium to the strain and its material properties (1).

Numerical difference schemes able to simulate discontinuous media in a simple fashion require the basic laws of physics governing propagation of acoustic waves to be carefully transformed into numerical analogues. Hence these basic physical principles (1) will be briefly reconsidered.

Conservation of Mass. The law of conservation of mass for a compressible medium is usually expressed in an Eulerian framework as, "the time rate change of mass density at any point is equal to the negative divergence of the momentum density at that point."

The conservation of mass equation can be expressed in a frame of reference which moves with the fluid, the Lagrangian frame. In mathematical notation the conservation law becomes in this frame, (2),

$$\frac{d\rho}{dt} = -\rho \, \nabla \cdot \underset{\sim}{v}$$
(1)

where ρ is the density, $\underset{\sim}{v}$ the velocity vector of an element of the medium and t is the time variable.

Conservation of Momentum. The law of conservation of momentum can be expressed for a fluid in tensor notation and in terms of material stresses and its velocity components as (2)

$$\rho \, \frac{\partial v_i}{\partial t} = \frac{\partial \, \sigma_{ij}}{\partial x_j}$$
(2)

where σ_{ij} are the stresses in the material as described below. Equations (1) and (2) are recognised as the usual equations of motion in the Lagrangian frame.

Stress-Strain Relations as Equations of State. Simple theory of elasticity assumes that the material is isotropic and that induced stresses and strains are linearly related to each other as long as they are small. The theory further assumes that the stress and the strain tensors always have the same axes. Poisson's ratio and

Young's modulus describe the extent to which an element of the
material responds when stress or strain is applied along one
direction.

If the stresses and strains along three orthogonal axes are
considered, then the general stress–strain relation can be written
as,

$$\sigma_{ij} = \lambda \epsilon_{mm} \delta_{ij} + 2\mu \epsilon_{ij} \qquad (3)$$

where the ϵ_{ij}'s are the strains and parameter $\epsilon_{mm} \delta_{ij}$ represents the
sum of the orthogonal strain; λ and μ are known as Lame's
constants (3). The parameter μ is also called the modulus of
rigidity and measures the resistance of the substance to
distortions. These constants are related to Young's modulus E and
Poisson's ratio ν by $\lambda = \nu E/(1+\nu)(1-2\nu)$ and $\mu = E/2(1+\nu)$.

Stress–Strain Relations for a Fluid. The medium is hydrostatic
when the direct stresses in three orthogonal directions are equal
and the shear stresses are zero; $\sigma_{ii} = \sigma_{jj} = (3\lambda + 2\mu)\epsilon_{ii}$.

Stress–Strain Relations for Viscoelastic Materials. The
viscoelastic behaviour of an elastomer varies with temperature,
pressure, and rate of strain. This elastic behaviour varies when
stresses are repeatedly reversed. Hence any single mathematical
model can only be expected to approximate the elastic behaviour of
actual substances under limited conditions (2).

The response of a typical elastomer to small pressure
excitation can be expressed by a linear relationship between the
stresses in the material and the strains caused by the pressure
excitation (2).

The simplest general linear relationship between a change in
stress and associated strain can be written in tensor notation as

$$\sigma_{ij} + \beta_0 \frac{\partial \sigma_{ij}}{\partial t} = \lambda(1 + \beta \frac{\partial}{\partial t}) \epsilon_{mm} \delta_{ij} + 2\mu(1 + \gamma \frac{\partial}{\partial t}) \epsilon_{ij}$$

$$(4)$$

where β, β_0 and γ are characteristic relaxation times of the
material.

The linear model of Equation (4) gives a representation of
damping of vibrations by internal friction (2). When a steady
sinusoidal excitation is involved, the internal friction causes a
phase delay in the transmission of signals through the material
which can be expressed as a loss–tangent, tan δ, which is related
to β_0, β, γ and the frequency ω, of the signal by
$\tan \delta_\lambda = \omega(\beta - \beta_0)/(1 + \omega^2 \beta_0 \beta)$ for direct or normal stresses and
$\tan \delta_\mu = \omega(\gamma - \beta_0)/(1 + \omega^2 \beta_0 \gamma)$ for shear stresses.

The angle δ measures the lag of strain behind stress and is known as the loss angle of the material and provides a measure of the internal damping of stress waves. A simpler model of viscoelasticity, the Kelvin–Voigt model places $\beta_o = 0$, $\tan \delta_\lambda = \omega\beta$ and $\tan \delta_\mu = \omega\gamma$.

Wave Equation. The propagation of the displacement vector $\underset{\sim}{u}$ of an acoustic excitation can be directly expressed in terms of materials parameters from the law of conservation of momentum and Equations (3) and (4) as a wave equation ($\underline{4}$). In vector notation,

$$\rho \; \frac{\partial^2 \underset{\sim}{u}}{\partial t^2} = \nabla \; \{\lambda \; \nabla\cdot\underset{\sim}{u} + \nabla\cdot\mu*\underset{\sim}{u}\} + \nabla.\mu*\nabla \; \underset{\sim}{u} \tag{5}$$

Here the Kelvin–Voigt model is assumed to adequately describe the viscoelastic properties of the elastomer and the Lame constants can be written to include the characteristic relaxation times of the material. They become the operators $\lambda^* = \lambda(1 + \beta \frac{\partial}{\partial t})$ and $\mu^* = \mu \, (1 + \gamma \frac{\partial}{\partial t})$, where β and γ are viscoelastic relaxation times.

The wave equation may be separated into two equations which describe torsion and dilatation. Dilatation is defined by $\theta = \nabla.\underset{\sim}{u}$ and is related to pressure by $p = \lambda^*\theta$ through the equation of continuity.

The wave equation which describes the propagation of dilatational waves can be expressed as

$$\rho \frac{\partial^2\theta}{\partial t^2} = \nabla^2\{(\lambda* + \overline{\mu}*)\theta\} + \overline{\mu}*\nabla^2\theta - (\frac{\nabla\rho}{\rho})\cdot\{\nabla(\lambda*+\overline{\mu}*)\theta+\overline{\mu}*\nabla\theta\}+(\nabla\overline{\mu}*).\nabla\theta \tag{6}$$

and is linked by sink and source terms, not shown, which arise where material parameters vary with position, to a similar equation describing the propagation of torsion waves. Terms in Equation (6) which involve ($\nabla\rho$) and $\nabla\overline{\mu}*$ describe the modifying influence discontinuous densities and shear moduli have on the propagation of the dilatational wave. The $\overline{\mu}*$ represents the average value of $\mu*$ over a computational cell and is derived from the integral definition of its divergence. The present study assumes as ($\underline{5}$) that the transformation of dilatation to torsion waves can be neglected for the material response of interest and the source and sink terms are zero.

The acoustic response of many elastomeric fluids can be adequately described by assuming that the contribution of $\mu*$ to the propagation of acoustic waves is negligible ($\underline{5}$). Under this assumption

$$\frac{\partial^2\theta}{\partial t^2} = \frac{1}{\rho} \nabla^2 \{\lambda*\theta\} - \frac{(\nabla\rho)}{\rho^2} \cdot \nabla \{\lambda*\theta\} \tag{7}$$

Lagrangian Numerical Scheme. The Lagrangian approach defines cells of whose corners, and hence boundaries, move with the local velocity. Cell corner movement is used to update densities, strains and stresses in the fluid. The velocity is then updated to complete one step in the time evolution of fluid acoustic response.

Considering motion in cartesian coordinates, and illustrating that in the x direction, the position of a cell corner at time $t + \delta t$, $x_{ijk}(t + \delta t)$ is related to its previous position $x_{ijk}(t)$ at time t by

$$x_{ijk}(t + \delta t) = x_{ijk}(t) + \int_{t}^{t + \delta t} v_{x_{ijk}}(s)\, ds \qquad (8)$$

where ijk are the cell corner indices in the three orthogonal directions of the cartesian coordinate system.

This expression can be approximated by

$$x_{ijk}^{n+1} = x_{ijk}^{n} + v_{x_{ijk}}^{(n + 1/2)} \delta t \qquad (9)$$

where index n indicates a particular instant of time via $t = n\delta t$. Corresponding expressions can be derived for the y and z coordinate of the cell corner. The examples to be discussed are limited to x and y coordinates only. Hence in the time interval n-1 to n, the density variation of the cell designated ij and bounded by the corners (x_{ij}, y_{ij}), (x_{i+1j}, y_{i+1j}), (x_{ij+1}, y_{ij+1}) and (x_{i+1j+1}, y_{i+1j+1}) is,

$$\rho_{ij}^{n} = \rho_{ij}^{n-1} \frac{|(x_{i+1j}^{n-1}, y_{i+1j}^{n-1}; x_{ij}^{n-1}, y_{ij}^{n-1}) \times (x_{ij+1}^{n-1}, y_{ij+1}^{n-1}; x_{ij}^{n-1}, y_{ij}^{n-1})|}{|(x_{i+1j}^{n}, y_{i+1j}^{n}; x_{ij}^{n}, y_{ij}^{n}) \times (x_{ij+1}^{n}, y_{ij+1}^{n}; x_{ij}^{n}, y_{ij}^{n})|} + \qquad (10)$$

$$\frac{|(x_{ij+1}^{n-1}, y_{ij+1}^{n-1}; x_{i+1j+1}^{n-1}, y_{i+1j+1}^{n-1}) \times (x_{i+1j}^{n-1}, y_{i+1j}^{n-1}; x_{i+1j+1}^{n-1}, y_{i+1j+1}^{n-1})|}{|(x_{ij+1}^{n}, y_{ij+1}^{n}; x_{i+1j+1}^{n}, y_{i+1j+1}^{n}) \times (x_{i+1j}^{n}, y_{i+1j}^{n}; x_{i+1j+1}^{n}, y_{i+1j+1}^{n})|}$$

where, for example, $(x_{i+1j}, y_{i+1j}; x_{ij}, y_{ij})$ defines the vector linking the positions x_{i+1j}, y_{i+1j} and x_{ij}, y_{ij}.

The strains ϵ at the cell centres are related to the displacements at the cell corners by, (3),

$$\epsilon_{xy_{ij}}^n = \frac{1}{4} \{ \frac{u_{y_{i+1j}}^n - u_{y_{ij}}^n}{x_{i+1j}^n - x_{ij}^n} + \frac{u_{y_{i+1j+1}}^n - u_{y_{ij+1}}^n}{x_{i+1j+1}^n - x_{ij+1}^n}$$

$$+ \frac{u_{x_{ij+1}}^n - u_{x_{ij}}^n}{y_{ij+1}^n - y_{ij}^n} + \frac{u_{x_{i+1j+1}}^n - u_{x_{i+1j}}^n}{y_{i+1j+1}^n - y_{i+1j}^n} \} \tag{11}$$

and

$$\epsilon_{xx_{ij}}^n = \frac{1}{2} \{ \frac{u_{x_{i+1j}}^n - u_{x_{ij}}^n}{x_{i+1j}^n - x_{ij}^n} + \frac{u_{x_{i+1j+1}}^n - u_{x_{ij+1}}^n}{x_{i+1j+1}^n - x_{ij+1}^n} \} \tag{12}$$

where $u_{x_{ij}}^n = x_{ij}^n - x_{ij}^0$ and $u_{y_{ij}}^n = y_{ij}^n - y_{ij}^0$. A similar expression can be derived for $\epsilon_{yy_{ij}}^n$.

The viscoelastic stress–strain equation, Equation (4) can be expressed in finite element formulation which relates the stress tensor σ_{ij}^n at time index n and cell centre (ij) to the corresponding strain tensor arising from the movement of the adjoining cell corners. Using backward differences for the time step, at time index n.

$$\sigma_{ij}^n = \frac{\beta_o \sigma_{ij}^{n-1}}{\beta_o + \delta t} + \lambda \left[\{ \frac{\delta t + \beta}{\delta t + \beta_o} \} \epsilon_{mm}^n - \frac{\beta}{\delta t + \beta_o} \epsilon_{mm}^{n-1} \right] \delta_{ij}$$

$$+ 2 \mu \left[\{ \frac{\delta t + \gamma}{\delta t + \beta_o} \} \epsilon_{ij}^n - \frac{\gamma}{\delta t + \beta_o} \epsilon_{ij}^{n-1} \right] \tag{13}$$

The velocities of the cell corners are, using time centered differences for the acceleration,

$$v_{x_{ij}}^{n+\frac{1}{2}} = v_{x_{ij}}^{n-\frac{1}{2}} + \frac{2\delta t}{\sum_{k=i-1}^{i} \sum_{\ell=j-1}^{j} \rho_{k\ell}^n} \{ \sum_{\ell=j-1}^{j} \frac{\sigma_{xx_{i\ell}}^n - \sigma_{xx_{i-1\ell}}^n}{X_{i\ell}^n - X_{i-1\ell}^n}$$

$$+ \sum_{k=i-1}^{i} \frac{\sigma_{xy_{kj}}^n - \sigma_{xy_{kj-1}}^n}{Y_{kj}^n - Y_{kj-1}^n} \} \tag{14}$$

where X_{ij} and Y_{ij} are the the coordinates of the centre of the cell (ij).

The above system of equations is solved in sequence to provide a complete description of the response of a material to a stress wave.

The von Neumann stability condition must be satisfied for stability (7), which in the present case, is approximately

$$\delta t < \frac{(\delta x)^2 \omega}{(\frac{\lambda + 2\mu}{\rho})(1 + \tan^2 \delta)} \qquad (15)$$

where δx is the average width of the cell and $\tan \delta$ is the larger of $\tan \delta_\lambda$ and $\tan \delta_\mu$.

Eulerian Numerical Scheme.   Equation (7) assumes that the material parameters are differentiable with respect to the spatial coordinates.   The scheme presented here will approximate discontinuites in media to second order accuracy in space.   A somewhat similar scheme has been published for a wave equation able to describe the propagation of torsion waves (6).   That work considers a wave equation after the assumptions of material continuity have removed the material modulus from the second order spatial differentiation and was unable to successfully model the problem considered in the present paper.

Equation (7) can most readily be solved by an explicit finite difference scheme which steps forward in time across the spatial grid.   The value of θ is updated at each spatial grid point in turn.   When all of the spatial grid has been updated a solution at that point in time has been calculated for the problem considered.

Using centered differences for the time derivative of θ and limiting the present considerations to two space dimensions (x,y) only

$$\frac{\partial^2 \theta^n}{\partial t^2} ij = \frac{\theta_{ij}^{n+1} - 2\theta_{ij}^n + \theta_{ij}^{n-1}}{(\delta t)^2} \qquad (16)$$

where δt is an increment in time and θ_{ij}^{n+1}, θ_{ij}^n, θ_{ij}^{n-1} are dilatations at three consecutive time instants at the spacial node ij.

If centered differences are also used for the spatial derivatives in Equation (7) and backward differences are used to represent the time derivative describing the viscous Kelvin-Voigt effect (2) then the first term on the RHS of Equation (7), in two space dimensions (x,y) become,

$$\frac{\nabla^2}{\rho}\{\lambda(1 + \beta\frac{\partial}{\partial t})\theta\}_{ij}^n \approx \frac{1}{\rho_{ij}(\delta x)^2}\{\lambda_{i+1j}[(1 + \frac{\beta_{i+1j}}{\delta t})\theta_{i+1j}^n -$$

$$\frac{\beta_{i+1j}}{\delta t}\theta_{i+1j}^{n-1}] - 2\lambda_{ij}[(1 + \frac{\beta_{ij}}{\delta t})\theta_{ij}^n - \frac{\beta_{ij}}{\delta t}\theta_{ij}^{n-1}](1 + [\frac{\delta x}{\delta y}]^2) +$$

$$\lambda_{i+1j}[(1 + \frac{\beta_{i-1j}}{\delta t})\theta_{i-1j}^n - \frac{\beta_{i-1j}}{\delta t}\theta_{i-1j}^{n-1}] + (\frac{\delta x}{\delta y})^2\{\lambda_{ij+1}[(1 + \frac{\beta_{ij+1}}{\delta t})$$

$$(17)$$

$$\theta^n_{i\,j+1} - \beta_{i\,j+1}\,\theta^{n-1}_{i\,j+1}] + \lambda_{i\,j-1}\,[(1+\beta_{i\,j-1})\,\theta^n_{i\,j-1} - \beta_{i\,j-1}\,\theta^{n-1}_{i\,j-1}]\}\}$$
$$\tfrac{}{\delta t} \qquad\qquad\qquad\qquad \tfrac{}{\delta t} \qquad\qquad \tfrac{}{\delta t}$$

and

$$(\tfrac{\nabla\rho}{\rho^2}) \cdot \nabla\{\lambda(1 + \beta\tfrac{\partial}{\delta t})\,\theta\,\}^n_{ij} \approx \tfrac{(\rho_{i\,j+1} - \rho_{i\,j-1})}{4\rho^2_{ij}\,(\delta y)^2}\,\{\lambda_{i\,j+1}\,[(1 + \beta_{i\,j+1})}{\delta t}$$

$$\theta^n_{i\,j+1} - \beta_{i\,j+1}\,\theta^{n-1}_{i\,j+1}] - \lambda_{i\,j-1}\,[(1 + \beta_{i\,j-1})\,\theta^n_{i\,j-1} - \beta_{i\,j-1}\,\theta^{n-1}_{i\,j-1}]\}$$
$$\tfrac{}{\delta t} \qquad\qquad\qquad\qquad \tfrac{}{\delta t} \qquad\qquad \tfrac{}{\delta t}$$

$$+ \tfrac{(\rho_{i+1\,j} - \rho_{i-1\,j})}{4\rho^2_{ij}(\delta x)^2}\,\{\lambda_{i+1\,j}\,[(1 + \beta_{i+1\,j})\,\theta^n_{i+1\,j} - \beta_{i+1\,j}\,\theta^{n-1}_{i+1\,j}] - \lambda_{i-1\,j}}{\delta t \qquad\qquad \delta t}$$

$$[(1+\beta_{i-1\,j})\,\theta^n_{i-1\,j} - \beta_{i-1\,j}\,\theta^{n-1}_{i-1\,j}]\} \tag{18}$$
$$\tfrac{}{\delta t} \qquad\qquad \tfrac{}{\delta t}$$

where δx, δy are spatial intervals between nodes.

The dimensional variables of Equation (18) may be transformed into the following dimensionless variables,

$$\alpha_{ij} = \beta_{ij}/t_1\,,\quad \tau = t/t_1,\quad \xi = x/l,\quad \eta = y/l,\quad \gamma_{ij} = \rho_{ij}/\rho_0,$$

$$a_{ij} = \lambda_{i+1\,j}/\lambda_{ij},\quad b_{ij} = \lambda_{i-1\,j}/\lambda_{ij}\,,\quad C_{ij} = \lambda_{i\,j+1}/\lambda_{ij},$$

$$B_{ij} = t^2_1\,\lambda_{ij}/l^2\rho_{ij},\quad d_{ij} = \lambda_{i\,j-1}/\lambda_{ij}, \tag{19}$$

where l, t_1 and ρ_0 are length, time and density characteristic of the calculation of interest. Hence, Equation (7) can be written,

$$\theta^{n+1}_{ij} = 2\theta^n_{ij} - \theta^{n-1}_{ij} + B_{ij}\,(\tfrac{\delta\tau}{\delta\xi})^2\,\{a_{ij}[(1 + \alpha_{i+1\,j})\,\theta^n_{i+1\,j} - \alpha_{i+1\,j}}{\delta\tau \qquad\qquad\qquad \delta\tau}$$

$$\theta^{n-1}_{i+1\,j}] + b_{ij}[(1 + \alpha_{i-1\,j})\theta^n_{i-1\,j} - \alpha_{i-1\,j}\,\theta^{n-1}_{i-1\,j}] - 2\,(1 + [\tfrac{\delta\xi}{\delta\eta}]^2)}{\delta\tau \qquad\qquad\qquad \delta\tau}$$

$$\{[(1 + \alpha_{ij})\,\theta^n_{ij} - \alpha_{ij}\,\theta^{n-1}_{ij}] + (\tfrac{\delta\xi}{\delta\eta})^2\{C_{ij}[(1 + \alpha_{i\,j+1})\,\theta^n_{i\,j+1}}{\delta\tau \qquad \delta\tau}$$
$$\tag{20}$$

$$- \alpha_{i\,j+1}\,\theta^{n-1}_{i\,j+1}] + d_{ij}\,[(1 + \alpha_{i\,j-1})\theta^n_{ij} - \alpha_{i\,j-1}\,\theta^{n-1}_{ij}]\} - \tfrac{(\gamma_{i+1\,j} - \gamma_{i-1\,j})}{4\gamma_{ij}\rho_0}}{\delta\tau \qquad\qquad\qquad \delta\tau}$$

$$\{a_{ij}[(1+\alpha_{i+1\,j})\,\theta^n_{i+1\,j} - \alpha_{i+1\,j}\,\theta^{n-1}_{i+1\,j}] - b_{ij}[(1+\alpha_{i-1\,j})\theta^n_{i-1\,j} -}{\delta\tau \qquad\qquad\qquad \delta\tau}$$

$$- \alpha_{i-1\,j}\,\theta^{n-1}_{i-1\,j}]\} - \tfrac{(\gamma_{i\,j+1} - \gamma_{i\,j-1})}{4\gamma_{ij}\rho_0}\,(\tfrac{\delta\xi}{\delta\eta})^2\{C_{ij}[(1+\alpha_{i\,j+1})\,\theta^n_{i\,j+1} -}{\delta\tau}$$

$$\underset{\delta\tau}{\underline{\alpha_{i\,j+1}}}\;\theta^{n-1}_{i\,j+1}] \;-\; d_{ij}\;[(1+\underset{\delta\tau}{\underline{\alpha_{i\,j-1}}})\;\theta^{n}_{i\,j-1}\;-\;\underset{\delta\tau}{\underline{\alpha_{i\,j-1}}}\;\theta^{n-1}_{i\,j-1}])\}$$

The expected local error of computation is of second order in time and material properties. In practice an error of about five percent was observed in runs using values of 0.1 for the nondimensional spatial intervals and a factor of 1000 in variation of material parameters across one grid interval.

Stability of the computation is governed by the von Neumann stability condition (7) for a wave equation.

$$\frac{2B_{ij}(\delta\tau)^2}{[(\delta\xi)^2 + (\delta\eta)^2]}\;(1 + \frac{\alpha_{ij}}{\delta\tau}) \;\leq\; 1 \qquad\qquad (21)$$

When $\frac{\alpha_{ij}}{\delta\tau} \gg 1$, Equation (20) becomes a diffusion equation which obeys a stricter stability criterion (< 0.5) than the wave equation.

Comparison of Lagrangian and Eulerian Schemes. The Lagrangian scheme subdivides the anechoic coating into a large number of small cells, the corners of which are moved during each cycle of computation. The acoustic displacements of a passing wave are calculated from these movements. The Lagrangian scheme permits, in this fashion, the calculation of large as well as small signal responses. The Lagrangian scheme also enables material interfaces to be placed at the interfaces of computational cells. This enables acoustic excitation crossing material discontinuities to be considered in a relatively rigorous and self-consistent manner and is particularly useful when the material properties vary greatly across an insert interface and deformation of inserts, such as air cavities, occurs. However, multidimensional algorithms of the Lagrangian schemes are difficult to implement and are computationally expensive.

The Eulerian finite difference scheme aims to replace the wave equations which describe the acoustic response of anechoic structures with a numerical analogue. The response functions are typically approximated by series of parabolas. Material discontinuities are similarly treated unless special boundary conditions are considered. This will introduce some smearing of the solution (6). Propagation of acoustic excitation across water-air, water-steel and elastomer-air have been computed to accuracies better than two percent error (8). In two-dimensional calculations, errors below five percent are practicable. The position of the boundaries are in general considered to be fixed. These constraints limit the Eulerian scheme to the calculation of acoustic responses of anechoic structures without, simultaneously, considering non-acoustic pressure deformations. However, Eulerian schemes may lead to relatively simple algorithms, as evident from Equation (20), which enable multi-dimensional computations to be carried out in a reasonable time.

Sample Computation Results. The Lagrangian scheme of Equation (9) to (15) was used to compute the steady state response to sinusoidal excitation of an anechoic coating glued on a steel plate shown in Figure 1. Symmetry considerations permit the calculations to be limited to the regions between the dotted lines, the unit cell.

Steady state acoustic response of the unit cell occurred for the composites considered after the passage of some five acoustic oscillations. Patterns of direct stress, and shear stress as shown in Figures 2 and 3 were obtained. As expected, the corners of the cavities concentrated the stresses. Viscoelastic energy loss calculations, not discussed here, also show that the corners of the cavities are concentrations of energy losses.

The acoustic response of the coating region above the cavities varied from that of the region above the intercavity regions. The results showed that the regions between the cavities and steel backing plate contributed little to the acoustic response of the coating. The overall response pattern of the coating is of course frequency dependent, but the results are representative for frequencies from 10 to 100 kHz. The material was assumed to have a viscoelastic loss tangent of about 0.5 for Youngs modulus and the first and second Lame constants. This assumption leads to relatively high loss tangents for the bulk modulus which is not representative of all elastomers. The model is able to consider the acoustic response of a single insert as well as a number of closely interacting inserts or cavities which modify the acoustic field in the composite.

The Eulerian scheme of Equation (20) was used to examine various aspects of the response predicted by the Lagrangian scheme. The transmissibility of a steel plate coated with an elastomer and immersed in water to a sinusoidal pulse was computed and is shown in Figure 4, (8).

The acoustic pulse response of an air filled row of square columns to an incident Gaussian excitation is shown in Figure 5. Here a Gaussian pulse incident from the left has passed a row of square, air filled cavities set in a loss-less elastic fluid. The reflected pulse is seen moving to the left. The air in the cavities is responding to the pulse even though the pulse had already passed the cavity. This is expected as the velocity of sound in air is only 1/5 that in the loss-less fluid considered. The surfaces of the cavities showed oscillations after the pulse had passed the cavities. This phenomenon can also be explained by surface wave effects (5).

Equation (20) was also used to compute the acoustic response of fluid cylinders immersed in water and insonified normal to their axis with a sinusoidal wavepacket. The examples shown here can be considered by other techniques (5) but serve as appropriate tests for the accuracy of the model which can then be used to compute the acoustic responses of systems which cannot be readily treated by other methods. The material properties of the cylinder are shown in Table 1 and were chosen to enable the calculated echo structure of the cylinders to be compared with previously published analytical work (5).

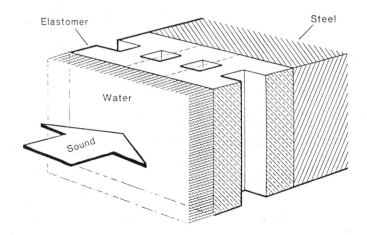

Figure 1.   Generic anechoic coating.

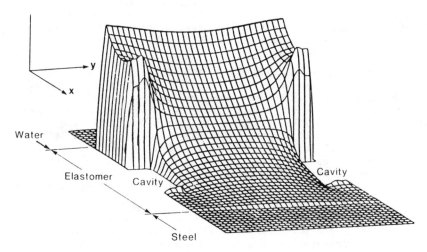

Figure 2.   Amplitude of direct stress.

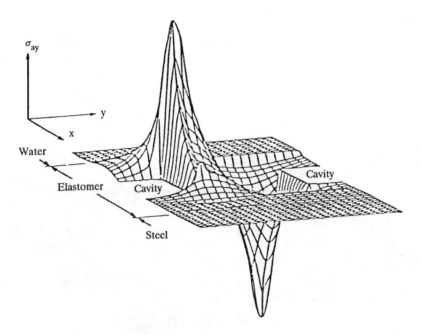

Figure 3.   Amplitude of shear stress.

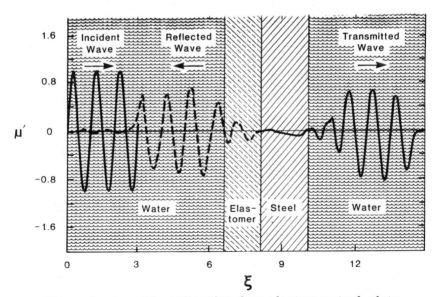

Figure 4.   Acoustic reflection from elastomer-steel plate systems.

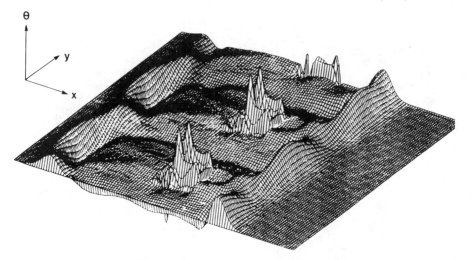

Figure 5.   Impulse response of air cavities.

Table 1   Material Properties of Fluids

	Density (kg/m^3) ρ	Speed of Sound (m/s) $\sqrt{\lambda/\rho}$	λ (GPa)	Loss Tangent $\tan\ \delta=\omega\beta$
Cylinder A	1500	1000	1.5	0
Cylinder B	2000	1000	2.0	0
Cylinder C	1500	1000	1.5	0.5
Water	1000	1500	2.25	0

The acoustic responses of the lossless fluid cylinders are shown in Figures 6-8, where I denote the incident wavepacket, R the wavepackets reflected by the cylinders and T labels the wavepackets transmitted by the cylinder along the axis of advance of the incident packet, normal to the cylinder axis. The acoustic wave patterns at various stages of time evolution of the interaction between the incident wave and cylinder are shown in the same figure for comparison. The subscripts indicate relative time. The incident wave pattern is displayed for reference purposes also in the figures.

The calculations show that the specular reflection from a fluid cylinder whose acoustic impedance equals that of water, cylinder A in Table 1, is very small, or negligible, in agreement with simple impedance theory, Figure 6. The specular reflection is labelled R_1 and its computed magnitude falls within the error of computation. The wave which is transmitted first by the cylinder is labelled T_1. The dominant echo from such a cylinder appears to be caused by waves travelling or creeping around the outside of the cylinder and is generated by the cylinder at a later time after the specular reflection has left the cylinder surface. It is shown as R_2 in Figure 7. At this time a second wavepacket is emitted from the other side of the cylinder and it is labelled T_2. The specular reflection from an elastomer cylinder whose acoustic impedance is different from that of water (cylinder B in Table 1) is shown in Figure 8. The amplitude calculated by the present numerical scheme agrees to within 10% with previous work (5) which was based on Kirchhoff approximation theory.

The amplitude of acoustic excitation within one half of a lossy cylinder C is shown in Figure 9. Symmetry considerations permit calculations to be limited to one half of the cylinder. The power losses are proportional to the square of the amplitude (4). Figure 9 shows amplitude information to bring out details of the acoustic response at the rim. The losses are concentrated at the back of the cylinder where focussing within the cylinder is expected from ray theory. Losses also appear at the rim of the cylinder which are due to complex internal reflections, or due to circumferential waves moving at and close to the surface of the cylinder. The present study has not isolated circumferential or creeping waves from multiple internal reflection phenomena but shows the total local disturbance moving through and along the rim of the cylinder. The observed calculated patterns are caused by the five cycle incident wavepacket and should be enhanced and perhaps be somewhat different for long wavetrains, hence the θ_0 scale is arbitrary.

Conclusion

Two models have been presented which are based on the Lagrangian and Eulerian difference schemes respectively. These models were able to calculate in detail the interaction of acoustic waves with inclusions in discontinuous fluids. The Lagrangian model was able to consider the transfer of normal to shear stresses at fluid discontinuities. The Eulerian model is limited to the study of dilatation waves but its computational efficiency enables it to

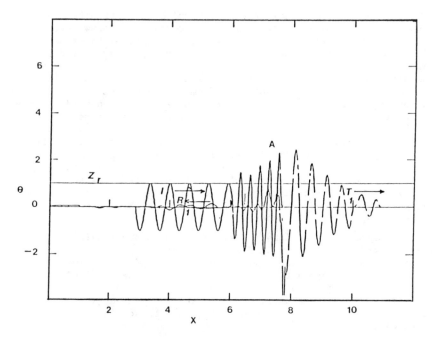

Figure 6. Specular echo generated by fluid cylinder A. The
echo is normalised with respect to the amplitude of the incident
wave packet. The circular cross section is located as shown in
Fig. 3, between x = 6 and x = 8.3, where the unit of x is three
wavelengths of the incident wave packet. I - incident wave
packet, R_1 - first echo or specular echo generated by cylinder,
T_1 - first transmitted wavepacket, Z_r - impedance relative to
impedance of water.

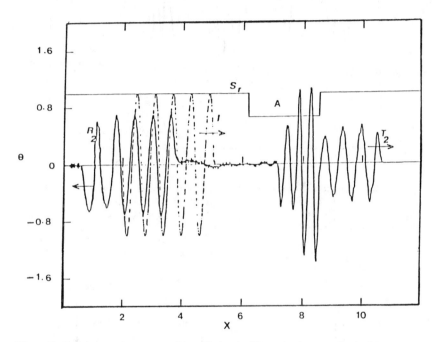

Figure 7. Dominant echo generated by cylinder A. The echo is normalised with respect to the amplitude of the incident wave packet. The distance X is normalised to three wavelengths of the incident wave packet. I is the incident wave packet (shown for reference only); R_2 is the second, or dominant, echo generated by the cylinder; T_2 the second transmitted wave packet; S_r the ratio of the speed of sound in cylinder fluid to the speed of sound in surrounding water.

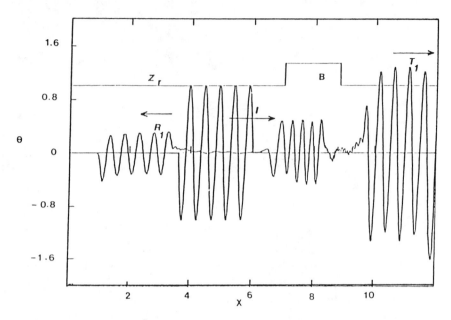

Figure 8. Specular echo generated by cylinder B. The echo is normalised with respect to the amplitude of the incident wave packet. The distance x is normalised to three wavelengths of the incident wave packet. I - incident wave, R_1 - first or specular echo generated by cylinder, T_1 - first transmitted wave packet, Z_r - ratio of acoustic impedance of cylinder fluid to acoustic impedance of water.

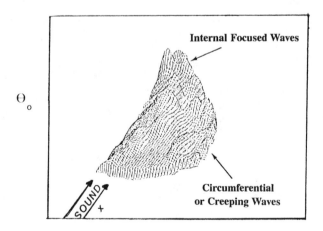

Figure 9. Amplitude θ_o of acoustic excitation in cylinder C. The x and y scales are normalised to the radius of the cylinder. The θ_o is arbitrary. Because the acoustic excitation pattern is symmetric about the diameter of the cylinder, only the θ_o in   half of the cross-section of the cylinder is displayed.

study the time evolution of multiple reflection and refraction at curved media interfaces in a reasonable time on desk top computer installations, such as a VAX 3100. Together, these models provide useful tools for the study of a wide range of acoustic problems covering anechoic coatings, machinery mounts and fabric/skin interaction.

The choice between Lagrangian and Eulerian numerical schemes for the prediction of the acoustic response of viscoelastic fluid systems depends on the particular problem to be solved. While the Lagrangian scheme provides a more precise simulation of discontinuous or composite viscoelastic fluids rather than a numerical approximation to the governing differential equation, the relative simplicity of the Eulerian algorithms suggests their use whenever possible. Only when great precision or uncertainities about the extent of deformations of the coating are significant are Lagrangian schemes preferable to Eulerian schemes.

Literature Cited

1. Temkin, S. Elements of Acoustics, John Wiley, New York, 1981.
2. Jaeger, J.C. Elasticity, Fracture and Flow, Methuen, London, 1956.
3. Jeffrey, H. and Jeffrey, B.S. Methods of Mathematical Physics, University Press, Cambridge, 1962.
4. Sommerfeld, A. Mechanics of Deformable Bodies, Academic Press, New York, 1964.
5. Davies, C.M. et al. J. Acoust. Soc. Am. 63, pp 1694–1698, 1978.
6. Brown, D.L. Mathematics of Computation, 1984, 42, pp 369–391.
7. McCracken and Dorn, W.S. Numerical Methods and Fortran Programming, Wiley, London, 1964.
8. Wendlandt, B.C.H. "The Acoustic Properties of Layered Coatings", MRL-R-1034, 1988.

RECEIVED January 24, 1990

POLYMERIC MATERIALS

POLYMERIC MATERIALS

Many polymer systems are presently used for sound and vibration damping. These include polyurethanes, acrylics, silicones, dienes, butyl, etc. These can be used as homo-polymers, co-polymers, or blends, and are frequently plasticized or compounded. Important features include the width and shape of the glass transition, the effect of variable mobility of components, molecular weight ranges, etc.

In this section, **Duffy** et. al. considers steric effects in the polyurethane system, and relate these structural variations to changes in the measured damping and dynamic modulus. **Roland and Trask** describe the behavior of polydiene blends, with particular emphasis on ensuring miscibility.

Chapter 15

Dynamic Mechanical Properties of Poly(tetramethylene ether) Glycol Polyurethanes

Effect of Diol-Chain Extender Structure

James V. Duffy, Gilbert F. Lee, John D. Lee, and Bruce Hartmann

Polymer Physics Group, Naval Surface Warfare Center, Silver Spring, MD 20903–5000

The steric effect that pendant groups present in the diol chain extender have on the glass (T_g) and melting point (T_m) transitions and dynamic mechanical properties of some PTMG polyurethanes was studied. A series of poly(tetramethylene ether) glycol/4,4'-diphenylmethane diisocyanate prepolymers were extended with diols that contained either methyl, ethyl, or butyl pendant groups. Differential scanning calorimetry (DSC) was used to determine the T_g and T_m of the soft and hard phases. It was found that diol pendant groups increased the amount of phase mixing in the soft segment while preventing crystallization in both the soft and hard segments. The effect of diol pendant groups on the shear modulus and loss factor of these polyurethanes was determined by a resonance technique. The size of the pendant group had no apparent effect on either T_g, T_m, or the dynamic mechanical properties of these polymers.

Polymers are often used in sound and vibration damping areas. In general, one of the most important polymer properties for these applications is the glass transition. At the glass transition, a polymer is most efficient in converting sound and mechanical vibrational energy into heat which results in absorption. By tailoring the polymer structure so that the glass transition is in the required temperature and frequency range, the polymer becomes an effective damper.

For the work presented here, the polymers considered
are in the general class of materials known as
polyurethanes. Polyurethanes are particularly attractive
for a study of the effect of chemical structure on
damping since it is possible to change their T_g's over a
wide range of temperatures (>100°C). This corresponds to
a damping peak location that spans more than 10 decades
of frequency. In addition, changes in polyurethane
structure can be used to produce a transition that can
vary from narrow to broad. To take advantage of these
desirable properties, one must understand the dependence
of T_g on the chemical structure of polyurethane polymers.
     Polyurethanes are alternating block copolymers made
of soft segments derived from polyester or polyether
diols and hard segments which come from the diisocyanate
and diol chain extender (1). Since the soft and hard
segments are chemically dissimilar, they tend to be
incompatible and separate into different phases as shown
in Figure 1. Thus, hard segment domains can separate or
be dispersed in a soft segment matrix. Separate glass
transitions can occur in each phase and either one or
both of the phases can be crystalline (2-3).
     Dynamic mechanical property (DMP) measurements are
used to evaluate the suitability of a polymer for a
particular use in sound and vibration damping. Since the
dynamic mechanical properties of a polyurethane are known
to be affected by polymer morphology (4), it is important
to establish the crystallization and melting behavior as
well as the glass transition temperature of each polymer.
Differential scanning calorimetry (DSC) was used to
determine these properties and the data used to interpret
the dynamic mechanical property results.
     The specific objective of this work was to determine
the effect that diol chain extender structure had on
soft/hard phase mixing, morphology, and the dynamic
mechanical properties of a series of poly(tetramethylene
ether) glycol (PTMG) polyurethanes. Information
regarding the influence of PTMG molecular weight and hard
segment content on these properties was also obtained.

Synthesis

Prepolymers were prepared from poly(tetramethylene ether)
glycol (PTMG) having nominal molecular weights of 650,
1000, 1430, 2000 and 2900, and 4,4'-diphenylmethane
diisocyanate (MDI). The polyol, which was added to MDI
at 45-50°C, produced an exothermic reaction that raised
the temperature to 80°C. The mixture was held at 75-80°C
for 1-2 hours and the mixture was then degassed, cooled,
and sealed under nitrogen. The percent free isocyanate
was determined using ASTM method D1638. The prepolymers,
which had MDI/PTMG molar ratios which varied from 2:1 to
6:1, were chain extended with either 1,4-butanediol (1,4-
BDO), 1,3-butanediol (1,3-BDO), 2,2-dimethyl-1,3-

propanediol (DMPD), 2-ethyl-2-methyl-1,3-propanediol (EMPD), 2,2-diethyl-1,3-propanediol (DEPD), or 2-butyl-2-ethyl-1,3-propanediol (BEPD). Figure 2 is an idealized structure of a linear polymer that was synthesized using one mole of PTMG, three moles of MDI, and two moles of 1,4-BDO. At the bottom of Figure 2, five other chain extender structures are shown, namely: 1,3-BDO, DMPD, EMPD, DEPD, and BEPD. Stoichiometry of the chain extender was adjusted, in all cases, so that the isocyanate index was 1.05, which means that a 5% excess of isocyanate was used beyond the stoichiometric ratio required for complete reaction. This insured that these polymers would be crosslinked through the formation of allophanate bonds during the curing reaction (4). Test specimens were cured at 100°C for 16 hours. The polymers are identified in the following manner: a polymer of 1000 molecular weight PTMG reacted in a molar ratio of 1:3 with MDI and chain extended with DMPD is denoted as PTMG 1000/M3/DMPD, and so forth.

Experimental

Thermal Analysis. A DuPont 9900 computer/thermal analyzer was used in conjunction with a 910 DSC module to obtain thermograms. Samples (10-15 mg) were cut from the DMP test bars and placed in aluminum test pans for analysis. Measurements were made from -170 to 250°C at a scanning rate of 10°C/min, in a nitrogen atmosphere.

Dynamic Mechanical Analysis
Data Collection. The dynamic mechanical apparatus (5) used is based on producing resonance in a bar specimen. Typical length of a specimen is 10-15 cm with square lateral dimensions of 0.635 cm. In brief, measurements are made over 1 decade of frequency in the kHz region from -60 to 70°C at 5 degree intervals. By applying the time-temperature superposition principle, the raw data are shifted to generate a reduced frequency plot (over as many as 20 decades of frequency) at a constant reference temperature. Frequencies greater than 10^7 Hz are not significant in sound and vibration applications, but at these frequencies the effect of variation of chemical structure on glassy modulus can be determined.

As shown in Figure 3, an electromagnetic shaker is used to drive a test specimen at one end while the other end is allowed to move freely. Miniature accelerometers are adhesively bonded on each end to measure the driving point acceleration and the acceleration of the free end. The output signals from the accelerometers are amplified and routed to a dual channel Fast Fourier Transform spectrum analyzer. The analyzer digitizes and displays the measured signals. The signals (amplitude and phase of the acceleration ratio) can be measured over a frequency range of three decades (25 to 25,000 Hz).

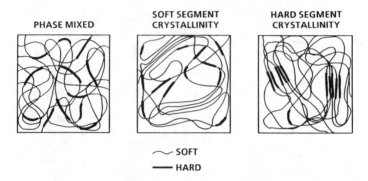

Figure 1.   Microphase separation in polyurethanes.

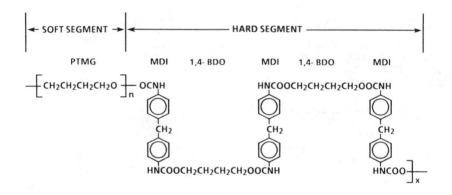

Figure 2.   Chemical structure of a typical linear
polyurethane.

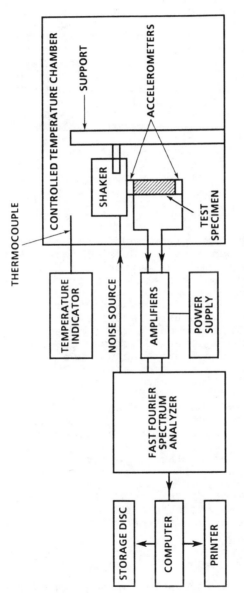

Figure 3. Resonance apparatus.

At certain frequencies, the amplitude of the
acceleration ratio goes through local resonant peaks.
The number of resonant peaks that can be measured is
dependent on the loss factor of the material, but,
typically, there are three to five peaks. As expected,
the resonant peaks appear at higher frequencies in the
glassy state than in the rubbery state. From the
amplitude and frequency of each measured resonant peak,
Young's modulus and loss factor are determined at the
corresponding frequency and temperature. By assuming
Poisson's ratio of 0.5, Young's modulus is converted to
shear modulus. The loss factor in extension is assumed
to equal the loss factor in shear.

In making measurements over a temperature range of
-60 to 70°C, the following thermal cycle was used:
(1) cool the test specimen, which has been mounted in the
test apparatus, to -60°C; (2) allow the specimen to
equilibrate at -60°C for at least 12 hours before making
a measurement; (3) after each measurement, raise the
temperature by 5 degrees; (4) allow 20 minutes to elapse
between each change in temperature to obtain thermal
equilibrium before making the next measurement.

Generation of Master Curves. Modulus and loss factor
data were processed into a reduced frequency plot in the
following manner: modulus curves at different
temperatures were shifted along the frequency axis until
they partially overlapped to obtain a best fit minimizing
the sum of the squares of a second order equation (in log
modulus) between two sets of modulus data at different
temperatures. This procedure was completely automated by
a computer program. The modulus was chosen to be shifted
rather than the loss factor because the modulus is
measured more accurately and has less scatter than the
loss factor. The final result is a constant temperature
plot or master curve over a wider range of frequency than
actually measured. Master curves showing the overlap of
the shifted data points will not be presented here, but a
typical one is found in another chapter of this book
(Dlubac, J. J. et al., "Comparison of the Complex Dynamic
Modulus as Measured by Three Apparatus").

The amount of shift one set of data at a given
temperature has to move along the log frequency axis to
overlap another set of data at a different temperature is
represented by the shift factor, $\log a_T$. The $\log a_T$
versus temperature data is then fitted to the WLF
equation (6)

$$\log a_T = -c_{1_0} (T - T_0) / (c_{2_0} + T - T_0) \qquad (1)$$

where c_1 and c_2 are constants for a given polymer and are
subscripted to indicate the reference temperature T_0 at
which the equation is evaluated. Knowing c_{1_0} and c_{2_0}, the
master curve at T_0 can be replotted at a different

reference temperature using the procedure described by
Ferry (6).
    The lower limit in selecting a reference temperature
is T_g, while the upper limit is about $T_g+100°C$ (6). This
upper limit can change for different polymers. The
limits exist because the WLF equation only applies in the
transition region. All the master curves to be presented
here are referenced to 25°C.
    The WLF constants c_1 and c_2 are determined by
fitting the log a_T versus temperature data only in the
temperature range of the glass transition. It was found
that these constants can vary considerably depending on
the temperature range selected. To avoid subjectivity in
the choice of temperature range, a systematic approach
was devised for this work using hypothetical data to
which a small error was added to each shift factor to
simulate experimental data. Larger and larger
temperature ranges were included in the fitting and shift
constants evaluated for each temperature range. It was
found that both shift constants went through a minimum as
the temperature range increased, and the values at the
minimum were the correct values. Using this approach on
the experimental data, the minimum values were selected
to be the shift constants of the material.

Results and Discussion

Factors Effecting T_g, T_m, and Dynamic Mechanical Properties

Diol Chain Extender Structure. The DSC results for PTMG
650-2900/M3 prepolymers chain extended with 1,4-BDO, 1,3-
BDO, DMPD, EMPD, DEPD, and BEPD are shown in Table I.

Table I. T_g and T_m for Diol Extended PTMG 650-2900/M3
Prepolymers by DSC

PTMG MW	T_g ss (°C)						T_m ss (°C)	T_m hs (°C)
	1,4-BDO	DMPD	EMPD	DEPD	BEPD	1,3-BDO	1,4-BDO	1,4-BDO
650	-24	32	37	31	35	-	-	160
1000	-48	6	9	2	11	0	-	157
2000	-66	-40	-35	-39	-34	-	2	184,197
2900	-71	-58	-54	-56	-54	-56	10	191,198

ss - soft segment   hs - hard segment

Transition temperatures were determined from expanded
versions of the original traces. For PTMG 650/M3, the
soft segment T_g with 1,4-BDO is -24°C while DMPD, EMPD,
DEPD, and BEPD polymers have glass transitions at much

higher temperatures (32, 37, 31, and 35°C). Thus, diols
that contain pendant groups appear to promote phase
mixing as indicated by higher soft segment T_g's (7). A
well defined hard segment melting occurs at 160°C with
1,4-BDO but no hard segment crystallization or melting
occurs in any diol extended polymer that contains pendant
groups (DMPD, EMPD, DEPD, or BEPD). These pendant groups
do not fit into the lattice because of their bulkiness
thereby preventing crystallization. The same type of
results are obtained in the PTMG 1000/M3 series of
polymers. Soft segment T_g's for DMPD, EMPD, DEPD, BEPD,
and 1,3-BDO are located at much higher temperatures (6,
9, 2, 11, and 0°C) than the 1,4-BDO extended polymer
(-48°C) and hard segment melting is seen only in the
1,4-BDO polymer (157°C).

The PTMG 2000 (Table I) and 2900 polymers (Table I
and Figure 4) extended with 1,4-BDO show soft segment
crystallization and melting for the first time (2 and
10°C respectively). This is the result of phase
separation in these higher MW PTMG polymers which allows
the soft segments to aggregate and then crystallize.
Hard segment melting appears as a series of peaks at
higher temperatures (8). The corresponding DMPD, EMPD,
DEPD, BEPD, and 1,3-BDO extended polymers as before
exhibit neither soft nor hard segment crystallization or
melting. T_g's of the PTMG 2000 and 2900 polymer soft
segments are again lower with 1,4-BDO (-66 and -71°C)
than with the corresponding DMPD (-40 and -58°C), EMPD
(-35 and -54°C), DEPD (-39 and -56°C), BEPD (-34 and
-54°C), and 1,3-BDO (-, and -56°C) polymers because they
are less phase mixed.

This major morphological difference between 1,4-BDO
extended polymers (crystalline hard segment) and those
extended with any of the diols containing pendant groups
(non-crystalline hard segment) also yields significant
differences in the shear moduli and loss factors of these
polymer systems.

Shear modulus curves for 1,4-BDO, and DMPD extended
polymers are shown in Figures 5 and 6. The EMPD, DEPD,
BEPD, and 1,3-BDO extended polymers had modulus curves
which were similar to the DMPD results and therefore are
not presented in graphical form. An additional molecular
weight, 1430, is also available for the 1,4-BDO system.
The rubbery modulus for 1,4-BDO polymers is about
1×10^7 Pa whereas DMPD, EMPD, DEPD, BEPD, and 1,3-BDO
polymers had moduli that are lower by about a factor 5
(2×10^6 Pa). This difference can be explained by the
fact that the 1,4-BDO polymers are reinforced by crystals
from the hard segment which act as a filler whereas diols
containing pendant groups yield polymers that have lower
rubbery moduli because they lack hard segment
crystallinity. At high frequencies, these polymers are
in the glassy state and the moduli for all systems have
approximately the same value, 1×10^9 Pa (3).

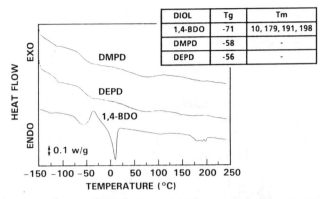

DIOL	Tg	Tm
1,4-BDO	-71	10, 179, 191, 198
DMPD	-58	-
DEPD	-56	-

Figure 4.   DSC thermograms of PTMG 2900/M3/diols.

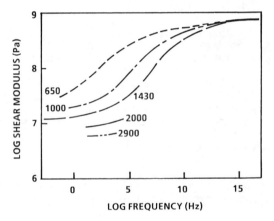

Figure 5.   Shear moduli for PTMG 650-2900/M3/1,4-BDO polymers.

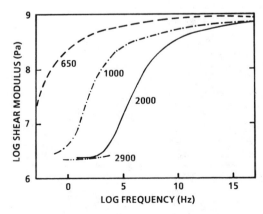

Figure 6.   Shear moduli for PTMG 650-2900/M3/DMPD polymers.

The loss factor shows similar dramatic differences
among the diol chain extenders (Figures 7 and 8) as a
result of hard segment crystallinity. The loss factor
curve, which is a measure of the damping properties of a
polymer, has a maximum value of 0.3 for 1,4-BDO extended
PTMG 650, 1000, and 1430 polymers and is very broad in
nature. The half width is about 8 decades on the
average. The loss factor curves for PTMG 2000 and 2900
are only partially plotted because the data could not be
shifted due to crystallization in the soft segment which
takes place as the temperature is raised. The DMPD,
EMPD, DEPD, BEPD, and 1,3-BDO extended polymers, in
contrast, gave very narrow (half width 3 decades) loss
factor curves and at much higher maxima values of 1.0.
It appears that the lack of any crystallinity in the diol
chain extender allows for increased molecular mobility in
the region of the glass transition which in turn leads to
higher loss factor maxima.

Diol Chain Extender Blends. Some measurements were made
on polymer systems cured with blends of 1,4-BDO and DMPD.
It should be remembered that polymers extended with
1,4-BDO show hard segment crystallization while polymers
extended with DMPD do not crystallize. The hard segment
content is maintained constant at a given level (45-50
percent) while adjusting the 1,4-BDO/DMPD ratio. Thus
PTMG 1000/M3 was chain extended with three different
blends of 1,4-BDO/DMPD (25/75, 50/50, and 75/25), and the
component percentages are shown along with DSC results in
Table II.

Table II.   Composition and DSC Results for PTMG 1000/M3
            Prepolymers Chain Extended with 1,4-BDO/DMPD
            Blends

EQ.WT. RATIO 1,4-BDO/DMPD	% hs	%BDO in hs	%DMPD in hs	T_g ss ($^\circ$C)	T_m hs ($^\circ$C)
100/0	47.0	17.7	0	-48	157
75/25	48.1	13.2	5.1	-30	160
50/50	48.3	8.7	10.1	-4	-
25/75	48.4	4.3	15.1	-3	-
0/100	48.6	0	20.0	6	-

The data shows the shift in soft segment T_g from -48°C
(100 percent 1,4-BDO) to 6°C (100 percent DMPD) as
expected due to phase mixing. As anticipated, hard
segment crystallization occurs when the DMPD content is
less than 50 percent. The shear modulus data (Figure 9)
show a steady decrease in the rubbery modulus as the DMPD
content of the blend increases because the degree of hard
segment crystallinity is decreasing, and these results

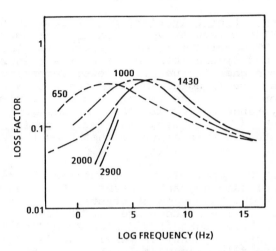

Figure 7. Loss factors for PTMG 650-2900/M3/1,4-BDO polymers.

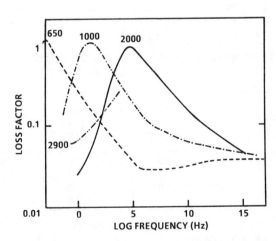

Figure 8. Loss factors for PTMG 650-2900/M3/DMPD polymers.

are consistent with the DSC crystallization results. For
the polymers with DMPD contents of 50 percent or more,
the loss factor reaches a value of 1.0 as shown Figure
10, while the polymers with 0 and 25 percent DMPD have
broader peaks at lower values of 0.3.

Molecular Weight of Polyetherdiol. A summary of the
effect of PTMG molecular weight on T_g is given in Table
I. As the PTMG molecular weight increases, the modulus
(Figures 5 and 6) and loss factor curves (Figures 7 and
8) shift to higher frequencies (lower temperatures).
These trends are in agreement with the DSC data for T_g.
The position of the loss factor peak can be moved within
a frequency range of 10^{-3} to 10^{7} Hz by changing the MW of
the PTMG, while the rubbery and glassy moduli remain
fairly constant within a series that uses the same chain
extender.

Hard Segment Content. It was of interest to determine
the effect of hard segment content on DSC and DMP for a
system where the hard segment crystallizes and for one
that does not. The systems are: PTMG 1000/M3/1,4-BDO
(crystalline hard segment) and PTMG 2000/M3/DMPD (non-
crystalline hard segment) of varying hard segment
content. The diisocyanate/polyetherdiol ratio, for the
PTMG 1000/1,4-BDO polymers varied from 2:1 to 6:1, which
corresponds to hard segment content from 37 to 66
percent, respectively. The DSC results show that the
soft segment T_g's (-44, -48, and -43°C) do not change
very much as the hard segment content increases from 2:1
to 3:1 to 4:1 as presented in Table III.

Table III. Effect of Hard Segment Content and Diol
Structure on Transition Properties

MDI/1,4-BDO RATIO	PTMG 1000/M/1,4-BDO hs CONTENT(%)	T_g ss (°C)	T_m hs (C)
2:1	37	-44	156
3:1	48	-48	155
4:1	56	-43	169
6:1	66	-18	174,196

MDI/DMPD RATIO	PTMG 2000/M/DMPD hs CONTENT(%)	T_g ss (°C)
3:1	32	-40
4:1	39	-21
6:1	50	2

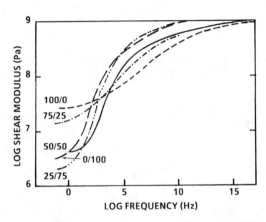

Figure 9. Shear moduli for PTMG 1000/M3 chain extended with blends of 1,4-BDO and DMPD.

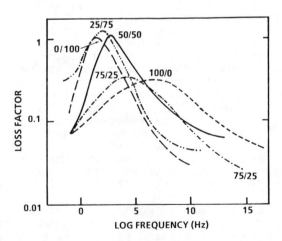

Figure 10. Loss factors for PTMG 1000/M3 chain extended with 1,4-BDO/DMPD blends.

At a ratio of 6:1, however, T_g jumps dramatically to
-18°C as a result of phase mixing. In this system (6:1),
the hard segment melts at a higher temperature which is
an indication that the purity of the phase and crystal
size is increasing as the hard segment content increases.
     The DMP results correlate well with the DSC results.
Shear modulus versus frequency for the various ratios are
plotted in Figure 11. The rubbery shear modulus is
increasing due to the reinforcing effect of the
crystalline hard segment on the amorphous soft segment.
Loss factor versus frequency for the various ratios are
plotted in Figure 12. The positions of the loss factor
peaks for the 2:1, 3:1, and 4:1 ratios are the same. At
a ratio of 6:1, the position of the peak shifts to lower
frequency, corresponding to a high T_g, which implies a
greater degree of phase mixing. The peak height also
decreases from 0.6 to 0.1 and the peak changes from
narrow (half width 5 decades) to broad (half width 11
decades) as the hard segment increases, because the
degree of crystallinity increases. A similar result has
been found for semi-crystalline polyesters (9-10).
     The diisocyanate/polyetherdiol ratio, for the PTMG
2000/DMPD polymers, was varied from 3:1, to 4:1 and 6:1
which raised the hard segment content from 32 to 50
percent. The DSC thermograms for these polymers contain
no crystallinity in either the soft or hard segments.
The soft segment T_g, however, steadily increased (-40,
-21, 2°C) as the ratio increased, indicating more phase
mixing of the amorphous phases.
     The DMP results reflected those obtained in DSC.
The rubbery shear modulus appears to increase only
slightly (Figure 13), since no hard segment crystallinity
is apparent in any of these polymers. The amorphous hard
segment is not as effective in reinforcing the modulus as
crystalline hard segment. The loss factor peaks are high
(about 1.0) and the half width is about 3 decades as
shown in Figure 14. The peaks shift to lower frequencies
(higher T_g's) as the ratio increases due to increase
phase mixing. It should be noted that as hard segment
content increases so does the percent DMPD in the polymer
and because DMPD contains bulky methyl groups that do not
fit into the lattice, crystallization fails to occur.

Fractional Free Volume and Coefficient of Thermal
Expansion. The shift constants c_1 and c_2 from the WLF
equation are not only fitting parameters that describe
the frequency-temperature relation of a given polymer,
but they are also related to chemical structure. Ferry
has shown (6) that these constants can be related to the
fractional free volume and coefficient of thermal
expansion of the free volume, which have physical meaning
in terms of the polymer structure. One can define the
free volume at the glass transition divided by the total
volume as f_g and the coefficient of thermal expansion of

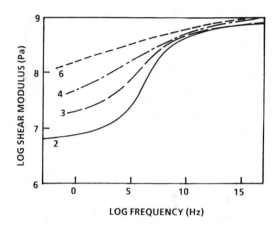

Figure 11. Shear moduli for PTMG 1000/M/1,4-BDO of various MDI mole ratios.

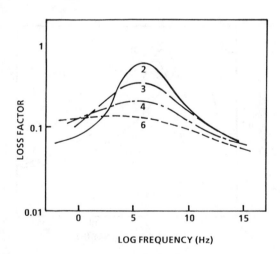

Figure 12. Loss factors for PTMG 1000/M/1,4-BDO of various MDI mole ratios.

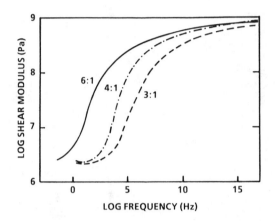

Figure 13.   Shear moduli for PTMG 2000/M/DMPD of various MDI mole ratios.

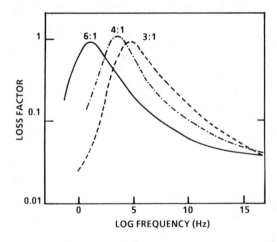

Figure 14.   Loss factors for PTMG 2000/M/DMPD of various MDI mole ratios.

the free volume as the increase of thermal expansion coefficient at the glass transition, α_f.

While the shift constants can be determined at any arbitrary reference temperature, if they are referenced to T_g, then

$$f_g = B / (2.303 \ c_{1g}) \tag{2}$$

where B is the Doolittle constant ([11]) and can be taken to be equal to 1 ([6]). The coefficient of thermal expansion of the free volume is expressed as

$$\alpha_f = B / (2.303 \ c_{1g} \ c_{2g}) \tag{3}$$

Values of f_g and α_f are tabulated in Table IV and are in the range of values as reported by Ferry ([6]) on polymers of widely different structures. While the uncertainty in the values of f_g and α_f, 10 and 50 percent respectively, is comparable to the variations due to changes in chemical structure, the systematic procedure used to evaluate f_g and α_f here is believed to give results that can at least indicate trends in the data.

As shown in part A of Table IV, f_g increases slightly with PTMG MW, while α_f decreases. A similar trend was observed with low molecular weight polymers of poly(vinyl acetate) and polystyrene ([6]). It is reasoned that the molecular weight dependence of f_g arises because longer soft segment length requires more volume to accommodate the increased number of conformations available to such a chain.

Phase mixing at the lower molecular weights (650 and 1000) causes higher f_g values. As hard segments phase mix into soft segments, both f_g and α_f increase (parts B and C) because the hard segment contains bulky, rigid MDI phenyl groups as well as pendant groups like methyl and ethyl from certain chain extenders. The soft segments are forced apart to accommodate these groups, thus creating more free volume.

The size or the number of the pendant groups do not influence either f_g or α_f (parts B - D). This is in agreement with the T_g and DMP results.

Hard segment crystallinity tends to decrease f_g and α_f, at least for the lower molecular weights. As shown in parts A thru C, f_g and α_g for 650 and 1000 MW polymers of 1-4,BDO are less than for DMPD and DEPD, because of crystalline hard segments. The free volume of a crystalline material is less than the free volume of the same material when it is amorphous. Since the soft segments are chemically bonded to the hard segments, the soft segment chains are drawn toward the hard segment domains when crystallization occurs. Thus, the free volume in the soft segment domains is reduced when crystallization occurs. The 2000 MW is evidently the limiting MW where the soft segment is no longer affected

Table IV. WLF Shift Constants, Fractional Free Volume,
and Thermal Expansion Coefficient

PTMG MW	T_g (°C)	c_{1g}	c_{2g} (deg)	f_g x 10^2	α_f x 10^4 (deg^{-1})	DENSITY at 25°C (g/cm^3)
		A.	PTMG 650-2900/M3/1,4-BDO			
650	-24	25	79.6	1.7	2.2	1.174
1000	-48	23.7	126	1.8	1.5	1.139
1430	-57	23.5	121	1.9	1.5	1.105
2000	-66	13.5	185	3.2	1.7	1.080
2900	-71	14.9	289	2.9	1.0	1.056
		B.	PTMG 650-2900/M3/DMPD			
650	32	11.4	36.1	3.8	10.6	1.157
1000	6	13.8	64	3.2	4.9	1.123
2000	-40	16.0	54.7	2.7	5.0	1.074
2900	-58	21.5	133	2.0	1.5	1.033
		C.	PTMG 650-2900/M3/DEPD			
650	31	11.6	50.6	3.8	7.4	1.140
1000	2	11.6	55.3	3.7	6.8	1.073
2000	-39	13.7	49.3	3.2	6.4	1.072
2900	-56	15.8	126	2.7	2.2	1.043
		D.	PTMG 1000/M3/1,3-BDO			
1000	0	11.7	48.6	3.7	7.7	1.116

MDI/PTMG RATIO	T_g (°C)	c_{1g}	c_{2g} (deg)	f_g x 10^2	α_f x 10^4 (deg^{-1})	DENSITY at 25°C (g/cm^3)
		E.	PTMG 1000/M/1,4-BDO			
2:1	-44	17.2	82.8	2.5	3.0	1.107
3:1	-48	23.7	126	1.8	1.5	1.139
4:1	-43	41.7	244	1.0	0.4	1.160
6:1	-18	23.5	85.3	1.8	2.2	1.161
		F.	PTMG 2000/M/1,4-BDO			
3:1	-40	16.0	54.7	2.7	5.0	1.074
4:1	-21	13.1	54.9	3.3	6.1	1.092
6:1	2	14.6	72.4	3.0	4.1	1.108

* A T_0 of 35°C was used rather than the steepest descent
of the modulus vs frequency data, due to the presence of
crystal melting.

by the hard segment crystallinity, since both f_g and α_f are about the same at this MW for all three chain extenders.

As the hard segment content increases, both f_g and α_f decrease due to crystallinity for 1000 MW polymers as shown in part E. However, at a molar ratio of 6:1, they increase because of the greater degree of phase mixing. Neither f_g and α_f are influenced by the hard segment content at 2000 MW as presented in part F. Again, the 2000 MW is a limiting MW where the hard segment does not have any effect on the soft segment.

Conclusions
Based on the thermal and dynamic mechanical measurements reported here, the following specific conclusions were reached:
1. Phase mixing increases in polymers that are chain extended with diols which contain pendant groups. This raises the soft segment T_g while suppressing crystallinity in both the hard and soft segments.
2. The size of the pendant group does not effect the glass transition, DMP results, f_g, or α_f values.
3. The rubbery shear modulus is higher when hard segment crystallinity is present because of the filler effect.
4. Loss factor peaks are higher and narrower in the non-crystalline polymers.
5. When the hard segment is non-crystalline both f_g and α_f decrease as the PTMG MW increases because the degree of phase mixing decreases.

Acknowledgments
In appreciation, the authors would like to thank Mary T. Treusdell, who performed the DSC measurements and prepared some of the dynamic mechanical specimens. This work was sponsored by the NSWC Independent Research Program and the Office of Naval Research.

Literature Cited
1. Cooper, S. L.; Tobolsky, A. V. J. Appl. Polym. Sci. 1966, 10, 1837.
2. Huh, D. S.; Cooper, S. L. Polym. Eng. Sci. 1971, 11, 369.
3. Hartmann, B.; Duffy, J. V.; Lee, G. F.; Balizer, E. J. Appl. Polym. Sci. 1988, 35, 1829.
4. Lelah, M. D.; Cooper, S. L. Polyurethanes in Medicine; CRC Press, Inc.: Boca Raton, Florida, 1986.
5. Madigosky, W. M.; Lee, G. F. J. Acoust. Soc. Am. 1983, 73, 374.
6. Ferry, J. D. Viscoelastic Properties of Polymers; John Wiley & Sons, Inc.: New York, 2nd. ed., 1970.

7.   Sung, C. S. P.; Schneider, N. S. J. Mater. Sci.
     1978, 13, 1689.
8.   Leung, L. M.; Koberstein, J. T. Macromolecules
     1986, 19, 706.
9.   Boyd, R. H.; Aylwin, P. A. Polymer 1984, 25, 330.
10. Ibid, 340.
11. Doolittle, A. K.; Doolittle, D. B. J. Appl.Phys.
     1957, 28, 901.

RECEIVED January 24, 1990

Chapter 16

Damping in Polydienes

C. Michael Roland and Craig A. Trask[1]

Naval Research Laboratory, Washington, DC 20375–5000

Cis 1,4-polyisoprene and 1,2-polybutadiene form thermodynamically miscible, nearly ideal mixtures; consequently, the respective chain subunits are statistically distributed in space apart from the constraints of chain connectivity. This segmental mixing and the resulting uniformity of access of the segments to the free volume, however, does not result in the onset of liquid like mobility for all segments at precisely the same temperature or frequency. The differing free volume requirements to make the transition from glass to liquid behavior for chain units within the blends cause the transition to occur over a very broad region in certain compositions. Since the glass transition is associated with a high level of damping, such damping can be obtained therefore over a broad range of frequency or temperature with these materials.

Recent studies of blends of polyisoprene (PIP) with polybutadiene (PBD) have revealed a number of remarkable features [1-5]. Non-polar hydrocarbon polymers such as PIP and PBD are not expected to exhibit miscibility given the absence of specific interactions. When the polybutadiene is high in 1,2 microstructure, however, it has a remarkable degree of miscibility with PIP. This miscibility is the consequence of a close similarity in both the polarizability and the expansivity of the two polymers [3,4]. Their mixtures represent a very unusual instance of miscibility between chemically distinct, non-reacting homopolymers. As its 1,4- content increases, both the polarizability and the thermal expansivity of the PBD diverge from that of PIP, resulting in a reduced degree of miscibility. This effect of PBD microstructure on miscibility with PIP can be seen in the data in Table I [3].

The critical degree of polymerization N^*, above which phase separation is expected for a mixture containing equal volume fractions of components of equal N, increases from 830 to over 10,000 as the 1,2- content of the PBD increases from 8 to 97%. Actually, phase separation has never been observed in blends of PIP with high 1,2-PBD, with 10,000 representing a lower limit for N^* [4]. Only the miscible region of the phase diagram has been accessible for this system.

[1]Current address: Allied–Signal Inc., P.O. Box 31, Petersburg, VA 23804

It is often desirable that a material intended for the attenuation of mechanical or sonic energy exhibit high damping over a wide frequency regime. Since the glass to liquid transition region is associated with such energy dissipation, a material in which this transition transpires over a broad range will be an attractive candidate for certain damping applications. The glass transition temperature of 1,2-PBD is about 0°C and that of high cis 1,4-polyisoprene roughly -63°C. Their miscible mixtures exhibit intermediate transition temperatures, of course, since the homogeneous morphology implies share equivalent free volumes for the components. The research described herein was directed toward probing the transition behavior of mixtures of 1,2-PBD and PIP in order to explore their suitability as damping materials.

Table I. Mixtures[a] of PIP with PBD [3,4]

% 1,4 units in PBD	α_{PBD}[b]	X[c]	N*[d]
92	1.8×10^{-3}	2.4×10^{-3}	830
74	1.6×10^{-3}	1.7×10^{-3}	1200
59	1.5×10^{-3}	0.7×10^{-3}	2900
3	9.2×10^{-4}	$<0.2 \times 10^{-3}$ [e]	>10000

[a] containing the critical concentration of components at 27°C
[b] thermal expansion coefficient ($\alpha_{PIP} = 9.3 \times 10^{-4}$)
[c] Flory-Huggins interaction parameter
[d] critical degree of polymerization (= 2/X)
[e] phase separation not observed

Experimental

The polybutadienes and the cis-1,4-polyisoprenes used in this work were synthesized in house or obtained from commercial sources. The 1,2-polybutadienes were atactic. Blend compositions, prepared by dissolution in cyclohexane, precipitation into methanol, and vacuum drying, are listed by volume in Table II, along with the weight average degree of polymerization of the component.

Table II. Blend Compositions

Blend	Component (Nw)	Component (Nw)	% 1,2-PBD
IA	PIP (23000)	97% 1,2-PBD (7700)	0.64
IB	PIP (90)	97% 1,2-PBD (2600)	0.75
IC	PIP (4500)	97% 1,2-PBD (2600)	0.32
IIA	PIP (6700)	95% 1,2-PBD (3200)	0.75
IIB	PIP (6700)	95% 1,2-PBD (3200)	0.50
IIC	PIP (6700)	95% 1,2-PBD (3200)	0.25
IIIA	1,4-PBD (100)	97% 1,2-PBD (2600)	0.16
IIIB	1,4-PBD (100)	97% 1,2-PBD (2600)	0.75
IIIC	1,4-PBD (440)	97% 1,2-PBD (2600)	0.29

A Perkin-Elmer DSC-2 with thermal analysis data station was employed for calorimetry measurements. Typically cooling rates from 5 to >200 degrees per minute with a constant heating rate of 20 degrees per minute were utilized, over temperature ranges of from -125° to +30°C. Sample weights were between 5 and 10 milligrams. Linear thermal expansivities were measured with a Perkin-Elmer TMA 7 using 0.2 cm thick samples. A heating rate 1 degree per minute was employed.

Dynamic mechanical testing was conducted in uniaxial extension with an Imass Corp. Dynastat Mark II instrument. Specimens were typically 35 mm in length and 12 mm wide. The measurements were made over a frequency range of from 0.01 to 100 s⁻¹ at various temperatures down through the glass transition temperature of the sample.

An inter-diffusion sample was prepared by annealing plied films of PIP and deuterated PBD at 52°C in vacuo for 162 hours, at which time the specimen was quenched and maintained at 0°C. The respective film thickness was 0.5 mm, which greatly exceeds the distance over which the species could spontaneously mix in that time period. Small angle neutron scattering from the plied films was measured subsequent to the annealing at the National Institute of Standards and Technology. The development of scattering contrast in the sample reflects the extent of inter-diffusion, since only regions of mixed isotopic species would possess a scattering contrast significantly above the background. An incident neutron wavelength of 6Å and a sample to detector distance of 3.6m in conjunction with a converging collimation system yielded SANS measurements down to 0.01 Å^{-1}. Background and detector sensitivity corrected data were converted to absolute intensity units using a secondary silica standard.

Results

Glass Transition Behavior.   Blends of high molecular weight PBD (of 85% 1,2-microstructure) and PIP exhibit single glass transitions suggestive of a segmentally dispersed phase morphology [1]. The glass transitions of these blends are relatively sharp, although miscible polymer mixtures in general have glass transitions somewhat broader than those of the pure components (see Figure 1). In blends of 1,4-PBD with PIP, narrow glass transitions are similarly observed; however, the less favorable interaction energy limits miscibility to components of low molecular weights (Table I).

More interesting glass transition behavior is found when PIP is blended with PBD of very high (>92%) 1,2 microstructure. Displayed in Figure 2 are differential scanning calorimetry results for three blends of varying relative concentrations of PIP and 1,2-PBD. Although only a single transition is observed for these mixtures, in two cases the transition can be seen to occur over a broad temperature range. As judged by the temperature difference between the onset and midpoint of the heat capacity change at T_g, the pure components had transition breadths of only a few degrees and for blends the transition was typically several degrees broad (congruous with the results in Figure 1 for blends with PBD of higher 1,4 content). However, in mixtures that had a high concentration of high 1,2-PBD the breadth of T_g was over 20 degrees. This broad nature of the transition was unaffected by the rate at which the temperature of the calorimeter was scanned or by the direction in temperature from which T_g was approached. Annealing samples at various temperatures did not influence subsequent measurements of T_g. The invariance of the anomalous transition breadths to the thermal treatment imposed suggests that they are not an artifact of non-equilibrium.

A broadening of the glass transition range can of course be associated simply with the onset of phase segregation, as has been observed, for example, in poly(vinyl methyl ether) mixed with polystyrene [6]. The blend in Figure 2 most proximate to the spinodal line demarcating the thermodynamic stability limit (as judged by the magnitudes of X^*, the spinodal value of the Flory interaction parameter [1,3,4]) is the one designated IA. When the molecular weight of both the PBD and the PIP are significantly reduced (sample IB), resulting in a more stable blend, the transition remains broad. Contrarily, a mixture (sample IC) of lower miscibility but containing a lower concentration of the 1,2-PBD has

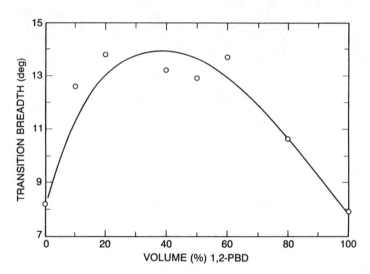

<u>Figure 1</u> - The breadth of the glass to liquid transition in blends of PIP and 85% 1,2-PBD as reflected by the temperature range over which the loss tangent had a magnitude greater than 50% of its maximum value [1].

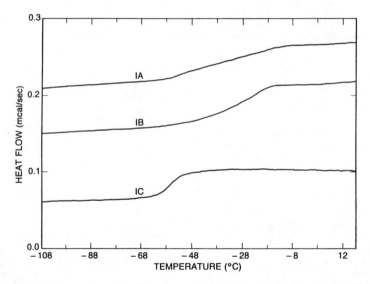

<u>Figure 2</u> - Calorimetry results for three miscible blends of PIP and PBD (sample designations in this and the ensuing figures are defined in Table II). The uppermost curve has the lowest spinodal interaction parameter value (1.7×10^{-4}). Broad glass transitions are observed in both it and the curve for blend IB, for which $X^* = 7.6 \times 10^{-3}$.

a fairly sharp glass transition. Clearly the transition breadths, while correlating with the concentration and microstructure of the PBD, are unrelated to the relative phase stability of these compositions.

In calorimetric measurements, a glass to liquid transition is established from the abrupt increase in sample heat capacity. This transition will similarly be accompanied by an increase in the thermal expansion coefficient of the material. The thermal expansion coefficients for pure 1,2-PBD and PIP have a difference that marginally exceeds the error of such measurements, indicating the similarity in their liquid state structure [3]. The consequent absence of equation of state effects is in fact necessary for the extreme miscibility of these polymers. In Figure 3 the thermal expansion behavior of two mixtures of PIP and 1,2-PBD are displayed. As this figure illustrates, in blends with an abundance of PBD high in 1,2 microstructure, the increase in the thermal expansion coefficient at T_g transpires over a broader temperature range than for other compositions. If the PIP was higher in relative concentration, or if the 1,2- content of the PBD was reduced to 85% or less, the change of the thermal expansion coefficient at T_g became relatively abrupt. The discontinuity in the temperature dependence of the volume (Figure 3) is seen to mirror that of the enthalpy (Figure 2) with respect to the anomalous broadening.

Since the appearance of a single glass transition in a blend only evidences homogeneity on a certain level (perhaps 100 Å [7,8]), thermodynamic miscibility can only be proven by other means. While the results exemplified in Figures 1 to 3 are consistent *in toto* only with miscibility in the 1,2-PBD/PIP blends, corroboration of homogeneity in samples exhibiting broad T_g is nonetheless desirable. Toward this end the occurrence of any inter-diffusion between the PIP and deuterated 1,2-PBD was assessed from measurement of small angle neutron scattering after sheets of the two polymers had been brought into contact. Mixtures of these polymers show the same anomalous glass transition behavior (Figures 3 and 4) as blends containing protonated PBD. The SANS experiment serves as an unambiguous test of miscibility since inter-diffusion between the contacted polymers will take place only if they are thermodynamically compatible. The magnitude of the small angle neutron scattering contrast measured from the layered polymer sheets was found to be significantly greater than the incoherent background intensity which arises independently of any inter-diffusion (Figure 5). The increase in coherent scattering intensity with contact time indicates that inter-diffusion has occurred. The differential scattering cross-section, I(q), is given by [9,10]

$$I(q) = (b_{PIP} - b_{PBD})^2 \, S(q) \, / \, V \qquad (1)$$

where b refers to the scattering length of the repeat units (equal to 3.33×10^{-13} cm and 6.662×10^{-12} for protonated PIP and deuterated 1,2-PBD respectively). The reference volume is taken to be the root mean square volume of the component segment volumes, giving $V = 1.1 \times 10^{-2}$ cm^3 for a blend in which the volume fractions of the two components are equal ($\phi = 0.5$). The structure factor for a two component mixture is given by [9,10]

$$S(q)^{-1} = \{N_i \phi_i g_i(Q)\}^{-1} + \{N_i \phi_j g_j(Q)\}^{-1} - 2X_{ij} \qquad (2)$$

where g is the Debye scattering function for an ideal chain, and Q the momentum transfer. In the limit of zero angle the intensity becomes proportional to the proximity of the interaction parameter to the spinodal value

$$S(0) = \{2 \, (X^* - X)\}^{-1} \qquad (3)$$

The zero angle differential scattering cross-section for the completely inter-diffused couple is calculated from equations 1 and 3 to be roughly 1100 cm^{-1}. The SANS intensity from the blend will approach this at a rate governed by the respective inter-diffusion

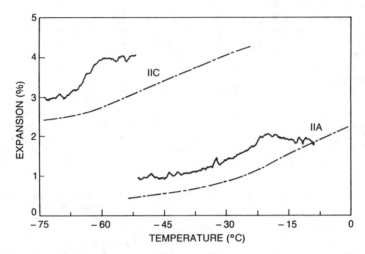

Figure 3 - The linear thermal expansion measured for two blends of PIP and deuterated 1,2-PBD (— -), along with the derivative (—), which is the thermal expansion coefficient, α. As is more clearly seen in the latter, the glass to liquid transition is less abrupt when the 1,2-PBD has the higher relative abundance.

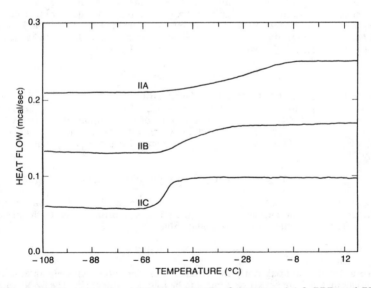

Figure 4 - DSC measurements on three blends of deuterated 1,2-PBD and PIP of varying composition. The glass transition is increasingly broadened as the PBD concentration is increased.

constants. The scattering intensity in Figure 5 indicates that the extent of inter-diffusion is far from complete, as expected given the high molecular weights employed. The rate of inter-diffusion between the polymer species is actually proportional to S(0) [11,12]

$$D = f(\phi) \ (X_* - X) \qquad (4)$$

where $f(\phi)$ depends on the respective tracer diffusion coefficients, D°

$$f(\phi) = 2 \ \phi \ (1 - \phi) \ [D_i^{\ 0}N_i\phi + D_j^{\ 0}N_j(1-\phi)] \qquad (5)$$

The spontaneous mixing of the two polymers will transpire at a rate which reflects the degree of miscibility of the system. As X approaches the critical value for phase separation, "thermodynamic slowing down" of the interdiffusion will occur [12]. The rate of increase of the scattering contrast reflects the proximity of the system to criticality, as well as the strong composition dependence of the glass transition temperature of the blend. Extraction of a value for either the self diffusion constants [13,14] or the interaction parameter is not feasible from the presently available data.

It was observed that the polymer layers used for the SANS experiment could not be separated after contact had been maintained for several hours, the interface between the two sheets having been eliminated by the inter-diffusion. Such an increase in peel adhesion with contact time have in fact been directly measured for these blends [1,5]. Notwithstanding the broad glass transitions, the inter-diffusion revealed by both the SANS and adhesion experiments provides direct evidence of the thermodynamic miscibility of the blend.

The absence of an abrupt glass transition is not unique to mixtures of 1,2-PBD with PIP. Blends of 1,2-PBD and 1,4-PBD are miscible at sufficiently low molecular weights [3] and calorimetry measurements for three such mixtures are displayed in Figure 6. The sample designated IIIA has a critical concentration of the components and thus represents the least stable composition for these particular component molecular weights. Since the molecular weights are low enough to yield a X^* that is less than the interaction parameter for the blend, the blend is miscible; accordingly, phase homogeneity gives rise to a single glass transition. When the stability of the mixture is increased by altering the composition away from that associated with the critical point (sample IIIB), the glass transition broadens. If the blend stability is sufficiently reduced by employing components of higher molecular weights (sample IIIC), two distinct glass transitions are observed, reflecting a heterogeneous phase morphology. These data are analogous to the results described above for PIP/1,2-PBD blends and demonstrate that a gradual change from glass to liquid behavior with elevation of temperature (that is, a broad T_g) is not a result of reduced blend stability or the onset of phase segregation.   Broad transitions are observed in mixtures containing an abundance of PBD which is, moreover, high in 1,2 microstructure.   Parenthetically, copolymers of 1,4- and 1,2-PBD, even when the proportion of the latter is high, exhibit only sharp glass transitions.

The occurrence of broad transitions in homogeneous blends has been reported previously.   Miscible blends of poly(methyl methacrylate) with poly(styrene-co-acrylonitrile) [15] and of poly(2,6-dimethyl-1,4-phenylene oxide) with either poly(p-chlorostyrene-co-o-chlorostyrene) [16] or polystyrene [17] exhibit broad T_g's when the constituent of higher glass transition temperature is present in high concentration. In the present work the anomalous broadening is also associated only with blends containing a high concentration of the higher T_g 1,2-PBD, although this is likely coincidental. From observation of a composition invariance of the distribution of relaxation times, it has been suggested that the anomalously broad transitions in polystyrene/poly(2,6-dimethyl-1,4-phenylene oxide) blends results but from an alteration of the structure of the glass [17]. The nature of such a structural change, or the mechanism underlying it, were unstated. The anomalous breadths of glass transitions observed in the PIP/1,2-PBD blends are invariant to the thermal treatment imposed, demonstrating at least that any peculiarity of the glassy state in these mixtures is probably not an artifact of non-equilibrium.

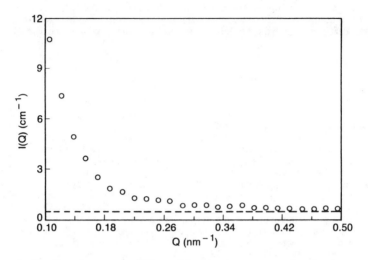

Figure 5 - Small angle neutron differential scattering cross section (ooo) measured from a sample consisting of sheets of PIP (N_w = 23000) and deuterated 1,2-PBD (N_w = 3200) which were in contact for 162 hours at 52°C. The scattering contrast significantly exceeds the incoherent background (---) determined from measurements on the individual polymers, evidencing the thermodynamic miscibility of the blend.

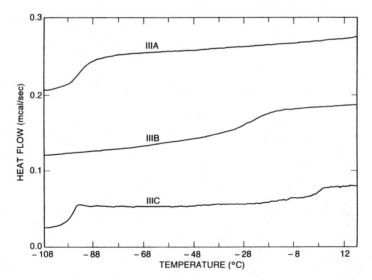

Figure 6 - Calorimetry results for three blends of 1,2-PBD with 1,4-PBD, in which it can be seen that reducing X^* from 6.4 x 10^{-3} (sample IIIA) to 2.4 x 10^{-3} (sample IIIC) effects phase separation. A broad T_g is only observed in the sample (IIIB) with a high concentration of 1,2-PBD, despite the fact that this is a more thermodynamically stable composition than IIIA.

Recent ^{13}C solid state N.M.R. experiments have revealed that, despite compositional homogeneity that extends to the segmental level, upon heating from below T_g the vinyl carbons of the PIP begin to manifest liquid-like motions at temperatures for which the vinyl carbons on the 1,2-PBD remain in the glassy state [18]. Since the blend is miscible, the chain units of the two polymer species experience the same free volume, both on average and with respect to its fluctuations. This, of course, is why only a single T_g, intermediate between the glass transition temperatures of the pure polymers, is measured for the blends. Nevertheless, a greater local free volume is evidently required for liquid like mobility in the vinyl carbons of the PBD. Although the mixtures are homogeneous and thus the components share the same free volume, differences exist in their dynamical response in reflection of a divergence in the free volume requirements for segmental mobility. The bulk T_g measurements are broadened because of these differences in the temperature at which local motion is engendered in the various moieties. Consequently, morphological homogeneity can be accompanied by dynamical heterogeneity [18].

In both calorimetry and expansivity experiments, the anomalously broad transitions were only seen in blends with a high concentration of PBD that was high in 1,2-microstructure. It is not clear why the transition anomaly appears to require both a PBD which is high in 1,2- units, and a relative abundance of the 1,2-PBD itself. Segmental motions in the mixtures are undoubtedly inter-related, particularly with respect to the competition (between deformation induced local motion and fluctuations in the available free volume) governing the glass transition. The details of this interdependence, along with the reasons that manifestations of it become apparent only in certain compositions, remain to be investigated.

Mechanical Damping. For damping purposes, the relevant glass transition behavior is that exhibited during mechanical deformation. The dynamic moduli of a 97% 1,2-PBD, PIP, and their blends were measured in tension over a range of frequencies at a series of temperatures spanning the glass transition region. These low temperature data were expressed as a single function of frequency by time-temperature superpositioning [19]. Previously time-temperature master curves of the dynamic moduli of 1,2-PBD/PIP blends were obtained from higher temperature measurements, encompassing the rubbery and terminal zones of the viscoelastic spectrum [2,5]. A breakdown of superpositioning due to thermal effects on the chemistry of the blend (which could have a different temperature dependence than the friction coefficient) is not of concern with strictly van der Waals mixtures. In miscible blends of poly(ethylene oxide) and poly(methacrylate), a failure of time-temperature superposition has been reported [20]. Consistent with tracer diffusion measurements in polystyrene/poly(2,6-dimethyl 1,4-phenylene oxide) blends [21], this failure suggests a difference in the friction factor for the components of the blends [20]. Nevertheless, the results for PIP/1,2-PBD mixtures [2,5] demonstrate unambiguously the validity of time temperature superposition for this system. For example, distinct local maxima were observed for the loss modulus in the terminal region of the viscoelastic spectrum [2,5], providing a sensitive test of time-temperature shifting. A apparent difference in the friction constant of components of a miscible blend may arise when the extent of chain entanglement (as defined by the ratio of the molecular weight and the critical value of molecular weight) differs significantly for the components [22]. This is the situation in blends described in references 19 and 20, but not for the 1,2-PBD/PIP blends [2,5]. Such an effect would not be operative at frequencies and temperatures for which glassy behavior is approached.

The only fundamental requirement for the superpositioning of viscoelastic data is that the shift factors be a single, smooth function of temperature, as is seen for the present mixtures in Figure 7. The master curves for the storage modulus for a blend containing 75% of the 1,2-PBD, along with the corresponding E' measured for the pure PIP and 1,2-PBD, are presented in Figure 8. Different reference temperatures were used in depicting the curves in order to display the very gradual approach of the blend to a glassy plateau modulus. Its transition to the full glassy state occurs over a very broad

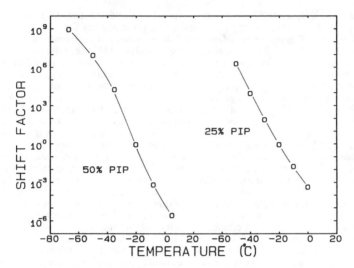

<u>Figure 7</u> - Empirically determined time-temperature shift factors for two blends of 97% 1,2-PBD with PIP. The concentration of the PIP is as indicated.

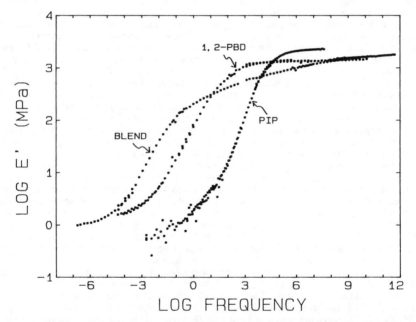

<u>Figure 8</u> - Superpositioned storage moduli for pure 97% 1,2-PBD, pure PIP, and a mixture containing 25% by volume of the PIP. Different reference temperatures (+5°C and -50°C for the pure 1,2-PBD and PIP respectively and -20°C for the blend) were employed to allow depiction of the data on an expanded scale. The transition temperatures of all blends were, of course, intermediate to the pure component T_g's (-63°C and 0°C for the PIP and 1,2-PBD respectively).

range (more than fifteen decades of frequency). The loss moduli for these same materials is shown in Figure 9. The pure 1,2-PBD has a significantly broader mechanical glass transition than does pure PIP, an effect not noticeable in the heat capacity and expansivity data. The 25% PIP blend, however, exhibits an extremely broad transition, with virtually no maximum in evidence in Figure 9. In the blend the peak value of the loss modulus, and perhaps also its integrated area, is lower than for the pure components. It is expected that when a transition is very broad the magnitude of the damping will be decreased since at a given temperature or frequency fewer dissipative mechanisms are active.

The loss tangents for the two pure materials and the 25% PIP blend are shown in Figure 10. Interestingly the width of the damping peak at the half maxima points is almost equivalent for the pure 1,2-PBD and the blend. The pure PIP has a significantly sharper loss tangent peak, which may indicate something about the dynamical free volume requirements for the respective chain constituents. The only manifestation in Figure 10 of a broadened glass transition in the blend is the high frequency tail of its loss tangent, which can be seen to persist many decades of frequency in the glassy region of the spectrum.

As described above, anomalously broad transitions occurred in calorimetry and expansivity measurements only in blends with a high concentration of PBD that was moreover high in 1,2- microstructure. The equivalent effect is found in the mechanical spectra. For example, a blend containing 50% PIP, although having a broader transition region than the pure polymers, exhibits a sharper glass transition than a blend with higher 1,2-PBD content (Figure 11). Similarly an absence of the anomalous broadening was found in the dynamic mechanical spectra of blends having a high concentration of PBD, but in which the latter had only an 85% 1,2-content.

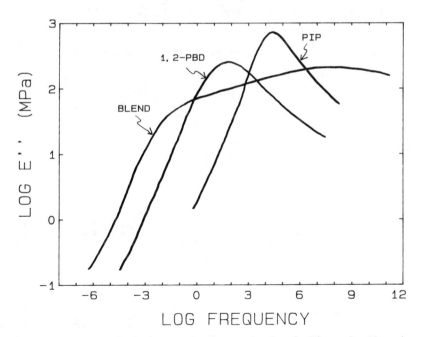

Figure 9 - The loss moduli corresponding to the data in Figure 8. Note that different reference temperatures were chosen in order to allow the curves to be displayed on an expanded scale.

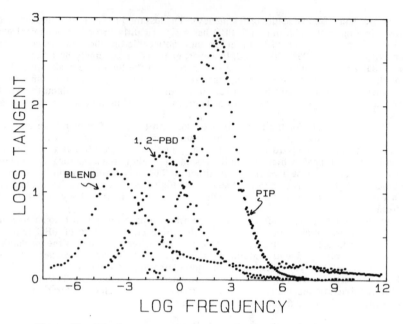

Figure 10 - The loss tangents corresponding to the data in Figure 8.

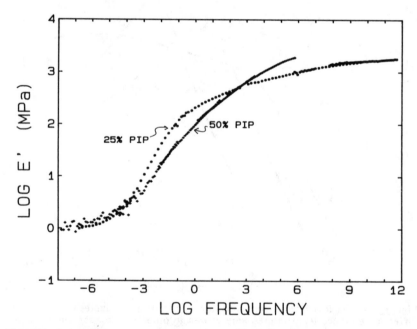

Figure 11. Superpositioned storage moduli for two blends of PBD that had 97% 1,2- units with 50% PIP and 25% PIP. The reference temperature is -20 °C in both cases, with the shift factors given in Figure 7.

Conclusions

Mixtures of cis 1,4-polyisoprene and 1,2-polybutadiene make apparent that the structural homogeneity conferred by thermodynamic miscibility is not always accompanied by equivalence in the dynamics of the components. This heterogeneity of the chain motions gives rise to very broad glass transitions in blends with a high proportion of polybutadiene that is high in 1,2- content. The current absence of any detailed understanding of the phenomenon need not inhibit the exploitation of these blends as novel damping materials.

Literature Cited

1. Roland, C.M.; Macromolecules 1987, 20, 2557.
2. Roland, C.M.; J. Polym. Sci. Polym. Phys. Ed. 1988, 26, 839.
3. Trask, C.A.; Roland, C.M.; Polym. Comm. 1988, 29, 322.
4. Trask, C.A.; Roland, C.M.; Macromolecules 1989, 22, 256.
5. Roland, C.M.; Trask, C.A.; Rub. Chem. Tech. 1988, 61, 866.
6. Bank, M.; Leffingwell, J.; Thies, C.; J. Polym. Sci.: A 1972, 10, 1097.
7. Roland, C.M. In Handbook of Elastomers - New Developments and Technology, Bhowmick, A.K.; Stephens, H.L., Eds.; Marcel Dekker: New York, 1988; chapter 6.
8. Roland, C.M. Rubber Chem. Technol. 1989, 62, 456.
9. DeGennes, P.G.; Scaling Concepts in Polymer Physics, Cornell University Press: Ithaca, N.Y., 1979.
10. Wignall, G.D.; Encyclopedia of Polymer Science and Engineering 1987, 10, 112.
11. Kramer, E.J.; Green, P.F.; Palmstrom, C.J. Polymer 1984, 25, 473.
12. Green, P.F.; Doyle, B.L. Macromolecules 1987, 20, 2471.
13. Bartels, C.R.; Crist, B.; Graessley, W.W.; Macromolecules 1984, 17, 2702.
14. Roland, C.M.; Bohm, G.G.A.; Macromolecules 1985, 18, 1310.
15. Suess, M.; Kressler, J.; Kammer, H.W.; Polymer 1987, 28, 957.
16. Alexanderovich, P.; Karasz, F.E.; MacKnight, W.J.; Polymer 1977, 18, 1022.
17. Prest, W.M.; Roberts, F.J.; Proc. 28th IUPAC Macromol. Symp. 1982, p. 664.
18. Miller, J.B.; McGrath, K.J.; Roland, C.M.; Trask, C.A.; to be published.
19. Ferry, J.D.; Viscoelastic Properties of Polymers, 3rd Edition, Wiley: New York, 1980.
20. Colby, R.H.; Polymer 1989, 30, 1275.
21. Composto, R.J.; Kramer, E.J.; White, D.M.; Macromolecules 1988, 21, 2580.
22. K.L. Ngai, private communication.

RECEIVED January 24, 1990

VIBRATION DAMPING

VIBRATION DAMPING

This section contains two tutorial papers on vibration damping, both of which are included to provide essential introductory information about the design and material considerations involved in developing vibration damping systems for practical applications. Of course, it is this aspect of damping which leads to the many commercial applications available today.

Kerwin and Unger first describe the application of materials to vibration damping. He includes basic engineering concepts, summarises the important commercial applications, and describes the material requirements. **Drake** then describes the stepwise approach used in designing a damping system and selecting material components for engineering application.

Chapter 17

Requirements Imposed on Polymeric Materials by Structural Damping Applications

Edward M. Kerwin, Jr., and Eric E. Ungar

Laboratories Division, BBN Systems and Technologies Corporation, 10 Moulton Street, Cambridge, MA 02138

We review the application of polymeric materials in structural damping, giving an introduction to the field as it is seen by the designer of damping treatments. Our objective is to provide an overview of the ways in which visoelastic damping materials are used, and of the resulting material-property requirements, and the corresponding polymer deformation types (extension, shear, or wavebearing). Finally, we show how the requirements of a given damping problem define the visoelastic properties required of candidate polymers, as influenced by the mechanical and environmental conditions that obtain.

The purpose of this paper is to present an introduction to structural damping for those polymer chemists who are not already familiar with the field. We shall treat several subjects, including what structural damping can accomplish, the definition of damping in terms of energy and loss factor, and the important classes of damping mechanisms. Finally, we look briefly at the range of problems in which damping may be applied, and at the resulting requirements imposed on the viscoelastic materials involved.

This approach presents a narrow view of the whole field of structural damping. It considers only those damping mechanisms that involve viscoelastic materials, and covers the wide variety of treatment schemes only generically with regard to the types of viscoelastic-material deformations of import. Thus, we are not able to cover the details of damping-treatment design and engineering, nor the material-property requirements from other areas of acoustics, such as the fields of vibration isolation, shock mitigation, and underwater sound.

0097–6156/90/0424–0317$08.25/0
© 1990 American Chemical Society

Nevertheless, we hope that this special-purpose
presentation will prove useful.

The Functions of Structural Damping

Structural damping is defined as the irreversible removal
of energy from a vibrating system. Damping, therefore,
serves to control "free" responses, i.e., those responses
for which the elastic and inertial forces balance, leaving
the system response controlled by the losses in the system.
In the absence of losses, the responses would be force-
free, and once excited, would continue arbitrarily far in
time or space. The category "free" responses thus includes
both resonances and freely traveling waves. It is these
that are controlled by system damping.
   Thus, damping can accomplish the following:
* limit steady-state, resonant response,
* cause transient responses to decay more rapidly,
  and
* attenuate travelling waves.

   Broader potential results include the reduction of
structural vibration, of radiated sound, and of structural
fatigue.

Damping, Energy, and Loss Factor

If, in a freely vibrating system that contains a loss
mechanism, the system energy is seen to decay at a rate
that is proportional to the instantaneous energy W, i.e.,
$-dW/dt = a*W$, it follows that the decay is exponential, $W = W_o e^{-at}$. (This is the case for a linear system in which the
loss coefficient itself is independent of vibration
amplitude.) The decay coefficient is $a = \omega\eta$, where $\omega = 2\pi f$ is the angular frequency of vibration, and η is defined
as the system loss factor. Since the energy decay rate is
Π_{dis}, the power dissipated, it follows directly that $\Pi_{dis} = \omega\eta W$. Thus, we have the system loss factor η defined in a
very basic way in energy and power terms as follows:

$$\eta = \Pi_{dis}/\omega W$$

   In many cases of interest, the vibrating system
comprises a number of elements, each of which may have its
own loss factor η_i and its own contribution W_i to the
system vibratory energy. These elements might be, for
example, the several layers of a laminated beam or plate.
In such a case, the power dissipation rate is the sum of
the several contributions, and the system energy is the sum
of the several energies. We, therefore, have the familiar
and useful definition of system loss factor as follows (1):

$$\eta = \frac{\Sigma \eta_i W_i}{\Sigma W_i}$$

This is an important result. It shows that in order for a given system element to be effective in damping the system, that element must not only have a significant loss factor, but it must also participate significantly in the total energy of the system.

It is customary in dealing with linear (or linearized) behavior of dynamical systems to describe the properties of viscoelastic materials as complex quantities (2). For example, we write Young's modulus

$$\overline{E} = E' - iE'' = E'(1-i\eta)$$

where
$\overline{E}$ is the complex modulus
E' is the "storage" modulus
E'' is the "loss" modulus
and
$\eta \equiv E''/E'$ is the loss factor of the material

The loss factor η is sometimes denoted as $\tan \delta$, where δ is the phase angle between the storage and loss components of harmonic stress.

Correspondingly, we have for the shear modulus

$$\overline{G} = G' - iG'' = G'(1-i\eta)$$

Other elastic moduli or compliances, e.g., bulk, dilatation, etc., may be described as above. Note that the use of $i = \sqrt{-1}$ in the above expressions implies a time dependence $e^{-i\omega t}$. (Some authors prefer $+j$; the choice is arbitrary.)

The Function of Damping. The function of damping is to control the "free" responses of a system. Such responses are those in which the system kinetic and potential energies are in balance, so that dissipative influences are controlling. Free responses include (a) resonance in frequency and (b) propagation of free waves in space. In the familiar case of a resonance in frequency, the response to a given oscillating force in the vicinity of resonance or natural frequency is peaked, showing a maximum at the resonance frequency and a finite bandwidth of high response. Both features are controlled by the system loss factor: See Figure 1.

The peak amplification factor Q is defined as

$$Q = \frac{1}{\eta},$$

and describes the peak response at resonance relative to
the response that would exist at the resonance frequency if
only the system stiffness (or only the mass) were
controlling. (Consider the simple case of a mass on a
spring.)

The resonance fractional bandwidth is $\Delta f / f_n = \eta$, for η
not too large, and represents the fractional frequency
difference between the "half-power" or "3-dB-down" points
in the response, i.e., the points where the squared
amplitude drops to half its peak value.

The above discussion implies steady-state response in
time. An equivalent reciprocal view of steady-state
resonance response is that in the vicinity of resonance
there is a dip in the force required to maintain a constant
level of response. The force-reduction ratio is Q, and the
fractional bandwidth of the force reduction is η. In
contrast, a truly force-free response of a resonant system
(once excited) would involve the exponential decay of
vibration amplitude with time. As we have mentioned
earlier, decay is also controlled by the system loss factor
as follows:

Temporal decay rate: $\Delta_t = 27.3 \eta f_n$ (dB/sec),

where a vibration level L_v would be expressed in the
customary logarithmic decibel (dB) measure for a response
quantity v, such as velocity,

$$L_v = 10 \log \left(\frac{v}{V_{ref}}\right)^2 \quad (dB \text{ re } V_{ref}).$$

Thus a level change of 10 dB corresponds to a factor 10
change in an energy or intensity-like quantity such as v^2.

The concept of the influence of damping on vibration
decay is very familiar. However, the quantitative
implication of a given value of loss factor may be less so.
To help convey the character of the temporal decay, Table I
shows the following for a range of loss factors: (a) the
number of cycles of vibration required for the vibration
amplitude to decay to one-tenth of its initial value, (b) a
corresponding representative sound, and (c) the clarity or
pitch of the sound. Note that as the loss factor
increases, the decay time decreases, and the frequency
bandwidth increases. The result is that a unique frequency
or pitch for the transient becomes progressively less well
defined. This is in keeping with the "uncertainty
principle" relating time and frequency (3).

Table I.   Dependence of the Character of a Transient Decay
on Loss Factor

Loss Factor η	N: cycles for decay to 0.1 amplitude	Sound—Pitch
≤ 0.001	≥ 730	"Clang"—approaches pure tone
0.01	73	"Bong"—clear pitch
0.1	7.3	"Bunk"—discernable pitch
0.5	1.5	"Thud"—almost without pitch

In a way that is parallel to the temporal decay of
vibration of a resonant system, damping governs the spatial
attenuation of freely propagating waves. In Figure 2, we
see a wave, assumed excited somewhere on the left,
traveling to the right without further excitation. The
wave travels with "phase speed" c at a frequency f, with a
wavelength λ given as

$$\lambda = c/f.$$

As we have noted above, the system loss factor defines
the temporal decay of the system energy. Therefore, in
considering the spatial decay of vibration, we find the
"energy" speed (or "group velocity") c_g involved in the
result:

Spatial decay rate: $\Delta_1 = 27.3 \, \dfrac{c}{c_g} \, \eta$ (dB/wavelength)

[The energy speed is defined (4-5) as $c_g = df/d(1/\lambda)$.]
In non-dispersive systems, such as acoustic waves in a
fluid or simple tension waves on a string, the wave speed
does not vary with frequency. Thus the energy speed c_g is
the same as the phase speed c, so that

$$\Delta\lambda = 27.3 \, \eta \text{ (dB/}\lambda) \text{ (non-dispersive waves)}$$

There are wave types, however, in which the wave speed has
an inherent dependence on frequency. A very important
practical example is the propagation of bending waves on
beams or plates (5) where (when the wavelength is large
with respect to the thickness of the plate or beam) the
energy speed is twice the phase speed: $c_g = 2c$. For such
bending waves, we have

$$\Delta\lambda = 13.6 \, \eta \text{ (dB/}\lambda) \text{ (bending waves)}$$

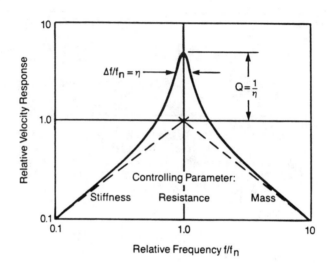

Figure 1.   Resonance Bandwidth and Amplification
            Determined by Loss Factor

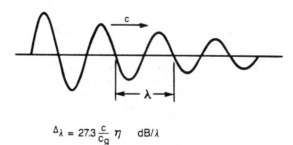

$$\Delta_\lambda = 27.3 \frac{c}{c_g} \eta \quad \text{dB}/\lambda$$

$$= 27.3 \, \eta \qquad \text{dB}/\lambda \text{ for nondispersive waves}$$

$$= 13.6 \, \eta \qquad \text{dB}/\lambda \text{ for bending waves}$$

Figure 2.   Free Wave Spatial Attenuation Due to Damping

Damping Configurations that Utilize Viscoelastic Materials

Viscoelastic materials have found application in vibration
damping in a myriad of configurations. Some of these are
very much specific to a particular application or device or
structure, while others are broadly useful. In this
section, it is our purpose to indicate the general ways in
which viscoelastic materials are applied by describing two
widely used types of damping treatments, outlining their
modes of operation, the dependence of their performance on
material properties and system geometry, several
elaborations on the basic configurations, and a few
examples of applications. In this way, we hope to convey
some understanding of why certain parameters and properties
of viscoelastic materials are important in damping
applications.
      The two general treatments that we shall explore are
the "free" viscoelastic layer, and the constrained
viscoelastic layer. (These are sometimes referred to as
"extensional" and "shear" damping, respectively.) These
are sketched in Figure 3, which shows each treatment as
applied to a plate (or beam), and also shows the type of
deformation that the viscoelastic layer undergoes due to
bending of the base plate. Bending vibrations are very
important in the fields of noise and vibration control, and
much of our discussion will concern the damping of bending
waves. In addition, in considering the effectiveness of
damping treatments, one finds that there are typically one
or more geometric parameters of the structure and the
treatment, and one or more elastic or stiffness parameters
that government the system performance.

The Free Viscoelastic Layer. As we see in Figure 3, the
free viscoelastic layer is bonded to the plate to be
damped. As the plate vibrates in bending, the viscoelastic
layer is deformed principally in extension and compression
in planes parallel to the plate surface. Such damping
layers have long been known, and at first were applied
more-or-less empirically. In the early 1950s Oberst (6,7)
and Liénard (8,9) published analyses describing
quantitative analytical models of free layer behavior.
      The damping performance of a free layer treatment for
plate bending waves is shown in Figure 4 (10,11). This
chart, which is Oberst's result, gives the system loss
factor η relative to η_2, the loss factor of the
viscoelastic material, as a function of the thickness ratio
H_2/H_1 (viscoelastic layer to plate). Each of the several
curves corresponds to a particular value of the relative
Young's storage modulus E_2/E_1 (viscoelastic layer to
plate).
      This description of the damping effectiveness of the
free layer is clear and simple: Its features are that
system loss factor increases

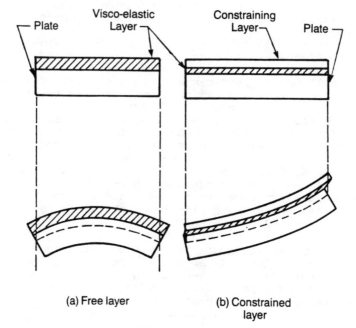

(a) Free layer   (b) Constrained
                                       layer

Figure 3.   Free and Constrained Viscoelastic-Layer
              Damping Treatments

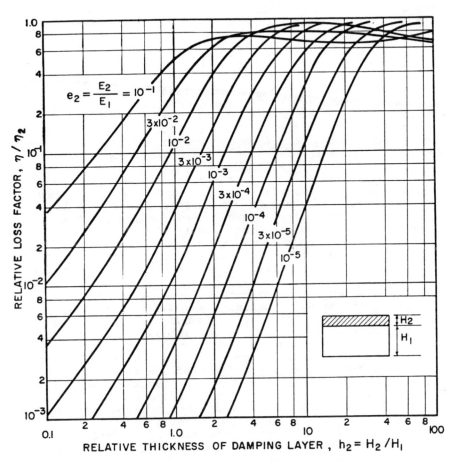

Figure 4. Damping Performance of a Free Viscoelastic
Layer (η_2 and η: Loss Factors of Viscoelastic
Layer and Composite) (Adapted from ref. 6)

a) with layer thickness
b) with layer modulus, and, of course,
c) with the loss factor of the viscoelastic layer
    (the resultant system loss factor is proportional
    to η_2, the loss factor of the viscoelastic layer.)

In the central region of Figure 4 a particularly
simple result applies: the loss factor ratio η/η_2 is
proportional to $(H_2/H_1)^2$. Thus, over a useful range

$$\eta \doteq A \, \frac{E_2}{E_1} \left(\frac{H_2}{H_1}\right)^2 \eta_2$$

Here the influences of a geometric parameter, H_2/H_1, and of
an elastic parameter, E_2/E_1, are clear. This analytical
model for free-layer damping makes clear the desirable
properties of the viscoelastic material. Specifically, the
free-layer damping material should have both large storage
modulus E_2 and large loss factor η_2. Over much of the
range, it is fair to say that the system loss factor is
maximized by maximizing the loss modulus, that is, the
product $E_2\eta_2$. However, at the upper end of performance
shown in Figure 4, a higher value of loss factor η_2 is
somewhat more effective than a higher modulus E_2 (M.L.
Drake, personal communication, 1989).

We note here that for simplicity, we have implied that
"best performance" means highest system loss factor. This
is the case for a number of damping applications, but not
for all cases. D. J. Mead (12,13) has noted the
quantitative importance of other system parameters
(stiffness and mass) in optimizing a damping treatment in
cases where maximum loss factor is not the criterion of
best performance (e.g., minimizing stress, acceleration,
etc.). These considerations are particularly important in
controlling structural fatigue and equipment malfunction.

An interesting characteristic of the free viscoelastic
layer is that it is, to first order, "locally reacting".
That is, the strain in the layer is proportional to the
local curvature of plate, so that the energy dissipated in
the viscoelastic layer is proportional locally to the
elastic energy in the plate. Therefore, for full coverage
by the viscoelastic layer, the damping performance (as
system loss factor) is independent of the mode shape of the
vibration. This fact simplifies the specification of a
treatment for wide frequency coverage (as there is no
inherent frequency or mode-shape dependence). It is,
however, possible to optimize partial coverage for a
particular mode or a limited class of modes.

The Constrained Viscoelastic Layer. The second of our two
general damping treatments is the constrained viscoelastic
layer shown in Figure 3(b). The complete constrained-layer
configuration is a three-layer laminate comprising base
layer to be damped, viscoelastic layer, and constraining

layer. As Figure 3 suggests, the principal deformation of
the viscoelastic layer occurs as shear as the composite
undergoes transverse (bending) motion. (The behavior of
the constrained-layer system is described in a number of
references. See, for example, References 2, 14-19.)

In bending vibration, this three-layer system exhibits
an inherent dispersion from "coupled" bending for large
wavelengths (low frequencies) to "uncoupled" bending for
small wavelengths (high frequencies). In these low- and
high-frequencies regimes, the elastic energy of deformation
lies almost entirely in the bending and extension of the
base and constraining layers. Thus, in these limits of
very low and very high frequency, the viscoelastic layer
cannot bring about significant damping of the system. (But
see the later discussion of segmented constraining layers.)

[The meaning of the terms is as follows: In coupled
bending, the wavelength is long enough so that the shear
stiffness of the viscoelastic layer (acting over a length
of half of the bending wavelength between stress reversals)
is able to cause the base and constraining members to
stretch and compress rather than allowing significant shear
strain in the viscoelastic central layer. At sufficiently
large wavelengths, the composite bends essentially as
though it were "perfectly" bonded together by a central
layer with "infinite" shear stiffness.]

[In contrast, for short wavelengths (high frequencies)
the shear stiffness of a half wavelength of the
viscoelastic layer is small relative to the extensional
stiffness of the base and constraining layers. In the
limit of very short wavelengths, the laminate behaves as
though the two outer layers were joined by a central layer
of vanishing shear stiffness, but which maintains the
spacing between layers. Thus the "uncoupled" bending
stiffness of the laminate approaches simply the sum of the
bending stiffness of the outer layers, i.e., a lesser
stiffness than that of the layers tightly coupled at large
wavelengths.]

However, in the mid-frequency region, where the
elastic energy in the viscoelastic layer can represent a
useful fraction of the total, the system loss factor rises
and passes through a maximum. Figure 5 shows this behavior
with the wavelength (λ) dependence represented by a "shear
parameter" γ defined as follows:

$$\gamma = \left(\frac{1}{2\pi}\right)^2 \alpha^2 = \left(\frac{1}{2\pi}\right)^2 \frac{G_2}{H_2} \left(\frac{1}{K_1} + \frac{1}{K_3}\right)$$

where the reciprocal of the parameter α defines a
characteristic distance in which a shear perturbation in
the laminate relaxes to $1/e$ of its initial value (14, 17,
20). (See Figure 6.) Therefore, the shear parameter
measures the square of the ratio of the bending wavelength
to this characteristic length. Clearly α involves simply
the product of the shear stiffness G_2/H_2 of a unit length

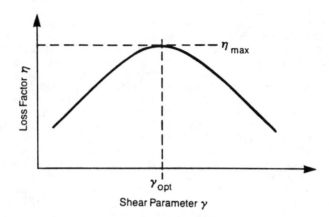

Figure 5. Characteristic Damping Performance of a
            Continuous Constrained Viscoelastic Layer

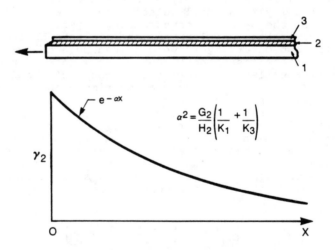

Figure 6. Shear Relaxation in a Constrained Visco-
            elastic Layer Treatment

and width of the viscoelastic layer, and the summed
extensional compliances $1/K_1$, $1/K_3$ of unit length and width
of the base and constraining layers.

Since, for bending of a simple beam or plate,
frequency and wavelength are related as $f \propto 1/\lambda^2$, it
follows that the shear parameter γ is inversely
proportional to frequency.

The loss factor peak which occurs at γ_{opt} can be
located in frequency by the appropriate choice of the
parameters that define γ (see below). The maximum loss
factor η_{max} at the peak is governed by η_2 the loss factor
of the viscoelastic layer and by a "stiffness" parameter Y
simply defined as (17)

$$Y = \frac{B_{coupled} - B_{uncoupled}}{B_{uncoupled}} \text{,}$$

where the coupled and uncoupled bending stiffnesses have
been discussed above. Figure 7 (10, 21) shows η_{max} versus
Y for a range of values of η_2. We see that η is
proportional to Y for smaller value of these parameters,
but increases more slowly as either parameter becomes
larger (say $Y > 0.3$ or $\eta > 0.5$).

The parameter Y depends only on the properties of the
base and constraining layers. Although we have called it a
stiffness parameter, Y does become dependent only on
geometry when the two outside layers have equal elastic
moduli.

Above we referred to the optimum shear parameter γ_{opt},
which defines the location of the peak damping as a
function of wavelength (or frequency). We have (11)

$$\gamma_{opt} = \frac{1}{\sqrt{(1+Y)(1-\eta_2^2)}}$$

Thus, the constrained-layer system parameters of
import are the following:

(1)   The Stiffness/Geometric parameter Y which
       defines the "severity" of the bending-wave
       dispersion characteristic between coupled and
       uncoupled conditions. Y and the loss factor η_2
       of the viscoelastic layer determine the maximum
       system damping for the treatment.

(2)   The Shear Parameter γ which defines the location
       within the dispersion characteristics of a given
       frequency by comparing the corresponding bending
       wavelength with the characteristic shear
       relaxation length in the laminate. γ is a
       combined geometric and elastic parameter. The
       governing property of the viscoelastic layer is
       its shear stiffness G_2/H_2.

The design of a constrained-layer treatment (11, 18, 2)
involves a number of steps and iterations. Typically (but
not always) our goal would be to achieve a given maximum
loss factor η_{max} and to locate this peak performance at a
given frequency for the operating temperature range
expected. Out of the design process will come desired
ranges of values of Y and η_2, hence a value of γ_{opt}, and
eventually a desired range for the shear stiffness G_2/H_2 of
the viscoelastic layer. Thus, the material-property
requirements will be

a)   a usefully large loss factor η_2, and
b)   a selected value of G_2/H_2, with a practically
        achievable H_2.

As a result, one may be able to choose from a set of
materials, adjusting the shear stiffness by adjusting the
layer thickness H_2. (This is something of an
oversimplification, because the value of H_2 influences Y.)
This ability to adapt a viscoelastic material to a
particular application can allow the inclusion of a range
of candidate materials having desirable properties aside
from the dynamic elastic properties G_2 and η_2.

Variations on Constrained Viscoelastic Layer Systems. The
constrained viscoelastic layer treatment, as applied to a
base member, is at its simplest, a continuous 3-layer
laminate. Numerous variations and elaborations have been
explored, and a number have found practical application.
In this section we review a few of these as further
illustrations of the application of viscoelastic materials
in shear-dependent configurations. Examples discussed are
  • Segmented constraining layer
  • Multiple constrained-layer treatments
  • "Fish-scale" treatment: multiple, overlapping
    segmented layers.

Segmented Constraining Layer. As we said in the preceding
section, a characteristic of the continuous constrained
viscoelastic layer treatment is its inherent wavelength
(frequency) dependence. The system loss factor falls off
on either side of a peak value (see Figure 5). Happily,
the damping performance in the low-frequency, large-
wavelength region below the damping peak can be improved
significantly by segmenting (cutting) the constraining
layer periodically. (See Figure 8.)
    This approach, suggested by D. J. Mead (22) and
analyzed by Parfitt et al. (22, 23), and later by Plunkett
and Lee (24) and by Zeinetdinova et al. (25), results in
locally increased strains in the viscoelastic layer in the
regions around the cuts, and therefore results in
additional dissipation. Parfitt's experimental results

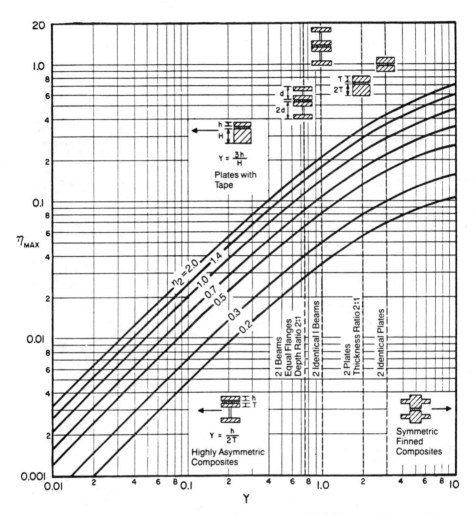

Figure 7.   Maximum Loss Factor versus Stiffness Parameter
for Constrained Viscoelastic Layer Treatments
(Adapted from ref. 10)

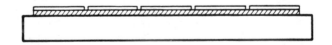

Figure 8.   Segmented Constrained-Layer Treatment

are shown in Figure 9, where increasing numbers of cuts in
the constraining layer of a treatment applied to a beam
cause the low frequency damping to rise, reach a maximum,
and then fall. There is, for a given treatment, an optimum
segment length, which is shown in both theory and
experiment. As one would expect, the optimum length
balances segment length against the shear-relaxation length
for the laminate. When the cuts are too close to each
other (in terms of the shear relaxation length), the
induced strain fields interfere with each other, reducing
the effectiveness of segmenting. As Figure 9 shows, the
optimum design raises the low-frequency branch of the loss-
factor characteristic practically to the level of damping
seen at the peak.

Incidentally, Kerwin and Smith (20) have shown that
essentially the same segment-length optimization yields
best damping of longitudinal or extensional waves in the
base plate. The segmented constrained viscoelastic layer
is one of the few treatments capable of providing useful
damping of such waves, which are troublesome in certain
cases.

*Multiple, Continuous Constrained Viscoelastic Layer
Treatments.* Multiple layers of constrained-layer treatment
are an obvious possibility for consideration. Some
practical advantages might accrue, such as easier handling
and installation on curved surfaces. It might also be
possible to manufacture and stock fewer "sizes" of
treatment, gaining applications flexibility through the use
of multiple layers. Further, design problems may be eased
in finding optimum configurations without being forced to
use very thin layers of a good, but low-modulus,
viscoelastic material.

Several studies of multiple layer treatments have been
made (26, 22, 27, 2). For such treatment comprising
identical layers, there does not appear to be a strong
performance advantage over a single constrained-layer
treatment of equal weight (26, 22).

An important exception is the use of multiple layers
of different properties, especially where damping
performance must be achieved over a broader temperature
range, or equivalently at several widely separated
temperatures. In such an application, as Henderson (27)
points out, an inner constrained-layer treatment contains a
high-temperature viscoelastic damping material, while an
outer treatment uses a lower-temperature damping material.
Such a configuration provides good shear coupling to the
outer treatment in the lower temperature range where the
inner viscoelastic layer is stiff. At higher temperatures,
the inner constrained-layer treatment is, of course, well
coupled to the vibrating structure, and is therefore able
to provide effective damping.

Multiple-Layer Segmented Treatments. Another interesting
damping configuration involves the use of multiple-layer
segmented constrained layers. See Figure 10. This
configuration was patented by Painter (28, 29) and has been
analyzed by Warnaka, Kerwin, and Carbonell (30) and by
Plunkett and Lee (24).

The elastic segments (platelets or "fish scales") take
on the role of a macroscopic filler for the viscoelastic
material, raising E' and therefore E" (at least at filler
concentrations that are not so high as to reduce the
effective loss factor excessively). The result is an
orthotropic composite material that can exhibit the
properties of a very effective "free" layer treatment. As
one would expect, the design process involves optimizing
the increased extensional loss modulus by balancing the
extensional stiffness of the segments and the shear
stiffness of viscoelastic layer. Essentially the same
physical considerations apply as in optimizing the segment
length in a single segmented-constrained-layer treatment.

This platelet-filled treatment, although attractive in
some respects, does not appear to have found wide
application, perhaps because of its complexity.

Spaced Damping Treatments for Bending Waves. A homogeneous
member that vibrates in bending experiences oscillating
curvature of its neutral plane (and of its surfaces). The
resulting extensional strains increase linearly with
distance from the neutral plane. There is, therefore, the
possibility of making geometric manipulations of a damping
treatment to make it more effective. One such concept
known as "spaced damping" is particularly applicable to
free and constrained viscoelastic layers (31-36, 29), and
is illustrated in Figure 11. There we see that the
deformation of the viscoelastic layer in extension or shear
is increased through the use of a spacer that places the
free layer or the constraining layer further from the
neutral plane. The results can be higher loss factor,
reduced treatment weight, or the need for a smaller amount
of viscoelastic material.

The properties desired of an ideal spacer layer are
that it be stiff in shear, but that the spacer itself
contribute minimally to the bending stiffness of the base
structure, shifting the neutral plane as little as
possible. We note that for the spaced constrained layer,
the combined function of the viscoelastic layer and spacer
is to provide a thick, dissipative and appropriately stiff
(in shear) layer between the constraining and base layers.
Therefore the order of the viscoelastic and spacer elements
is arbitrary and they may be subdivided as long as the
desired properties are preserved. These possibilities give
additional freedom in adapting viscoelastic materials for
effective damping.

Other spacer-like geometric manipulations have been
considered for constrained viscoelastic layers. These

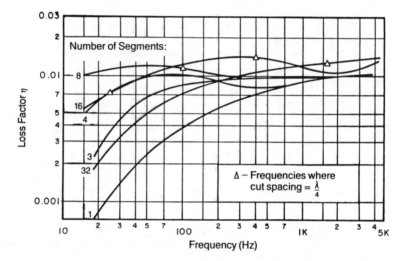

Figure 9. Measured Loss Factor of Segmented Constrained Layer Damping for Various Segment Lengths on a 36-in.-Long Bar (The curve for 3 segments may actually be for 2.) (Adapted from ref. 22)

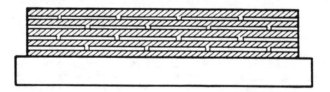

Figure 10. Multiple Segmented Constrained-Layer Treatment

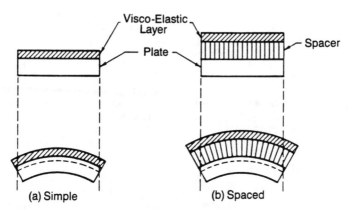

Spacer used to increase extensional strain in free layer.

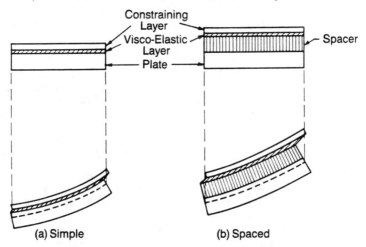

Spacer used to increase shear strain in constrained layer.

Figure 11. Simple and Spaced Free and Constrained
Viscoelastic-Layer Damping Treatments

include Whittier's (35, 29) arched truss, shown in Figure
12(a) and Painter's (36, 29) hat-section of Figure 12(b).

Resonant Damping Treatments. Where damping is required
over a more-or-less limited range of frequency, resonant
damping treatments can be of use (5, 2). Figure 13 shows
examples of resonant or "tuned" dampers that use
viscoelastic materials as the lumped elastic element in
combination with an appropriate mass. The viscoelastic
element can have its principal deformation either in
extension or in shear, depending on the geometry chosen.
    A resonant damper functions as follows: At the
resonance frequency (as determined by the mass and the
stiffness), the input impedance to the base of the damper
becomes resistive and rises to a relatively high value,
which is proportional to the moving mass and inversely
proportional to the loss factor of the viscoelastic
material. Power, therefore, flows into the damper and is
dissipated.
    The performance and material requirements for a
resonant damper are different from those for, say, a free
or constrained viscoelastic layer. For a given base
velocity amplitude, the input resistance, and, therefore
the power dissipated at the resonance frequency is
inversely proportional to the loss factor of the
viscoelastic material. However, the bandwidth over which
the high input resistance is seen is directly proportional
to the loss factor. Thus a low loss factor gives a high
damping peak that covers only a narrow frequency range.
Conversely ,a high material loss factor results in a lower
maximum damping, but an increased effective bandwidth.
Obviously, the choice of material properties depends on the
problem at hand: i.e., on the bandwidth and magnitude of
damping required. A further controlling factor is the
expected temperature range because the temperature
dependence of the dynamic properties of the viscoelastic
material will govern the temperature variability of the
damper performance.
    Viscoelastic materials can also be used as wavebearing
members in resonant dampers (37, 38). In this case a
viscoelastic layer applied to a plate or beam becomes
wavebearing through its thickness. Thus, with increasing
frequency, the layer can present a relatively high
resistive impedance to the plate when the layer thickness
becomes an odd multiple of the quarter wavelength of
compressional waves. (The system damping dips at even
multiples of $\lambda/4$.) The damping performance is dependent on
the material thickness, characteristic impedance, and loss
factor (equivalently on its density, wave speed, and loss
factor.)
    An example of the measured damping by such a layer is
shown in Figure 14 (38). Relatively high damping is shown

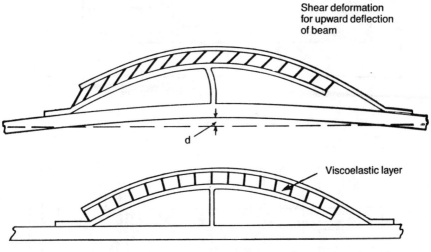

Shear deformation
for upward deflection
of beam

Viscoelastic layer

Figure 12(a). Arched Truss Damping Treatment
(Adapted from ref. 29)

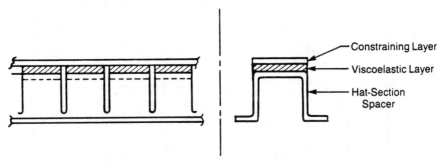

Constraining Layer

Viscoelastic Layer

Hat-Section
Spacer

Figure 12(b). "Hat-Section" Damping Treatment
(Adapted from ref.29)

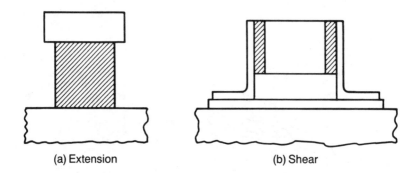

(a) Extension   (b) Shear

Figure 13.   Resonant Damper Configurations

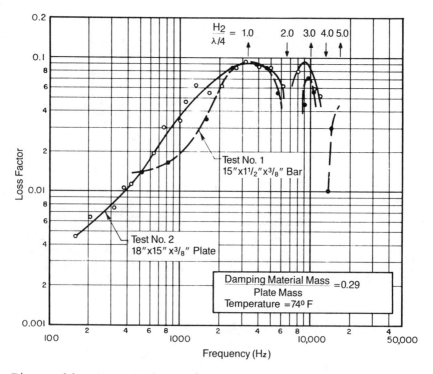

Frequency (Hz)

Figure 14.   Measured Damping Performance of Thickness-
Resonant Viscoelastic Layer (λ is the
wavelength of compressional waves in the
viscoelastic layer) (Adapted with permission
from Ref. 38. Copyright 1964 The American
Institute of Physics.)

over a significant frequency range for a treatment-to-base-member weight ratio of 0.29.

Damping Applications and Treatment Characteristics

In this section we want to indicate the broad range of problems in which damping can be useful, and to present a few specific examples. Further, we list some of the principal properties and characteristics of viscoelastic materials that must be considered in selecting appropriate candidate materials for applications.

Examples of the Applications of Damping. As was pointed out earlier, structural damping can control "free" responses, i.e., (a) resonant response and spatial distribution in steady-state response, and (b) both ringing (temporal decay) and spatial decay in transient response. Useful applications of damping have covered a wide range of vibration issues as indicated in Table II. The list of problem issues range from vibration levels high enough to cause destruction and failure of structure and equipment, through issues of physical harm and interference with personnel performance, to aesthetic considerations and the disruption of precision processes. The corresponding ranges of vibration amplitudes and frequencies cover many decades.

Table II. General Areas of Application of Damping in
Controlling the Effects of Vibration

- Structural Fatigue and Failure
- Damage to Equipment
- Equipment Malfunction
- Vibration (and Noise) Control
  for Personnel Safety and Function
- Comfort, Acceptance, product "Feel"
- Effects on Instruments, Processes, and
  Precision Equipment

In Table III are listed some important areas in which damping is applied. Clearly, wide ranges of parameters (temperature, frequency, amplitude, environment, etc.) are encountered. These issues are discussed further in the following section.

Table III. Examples of Structural Damping Applications

Field	Items treated	Motivation
Transportation		
Automotive	body panels, valve cover, oil pan	noise control
Aircraft	panels, appendages, turbine-engine components, fuselage skin panels	acoustic fatigue, noise control
Ships	hull, decks, machinery piping	noise control
Railroad	car structure, panels, wheels	noise control, wheel squeal
Spacecraft	equipment panels supports, electronic circuit boards, optical equipment, antenna and support structures	acoustic fatigue, malfunction, vibration control
Industry	machinery, e.g., vehicles, drilling rods, grinders circular-saw blades	fatigue and noise control
Commerce	appliances, office machines, computer components, metal cabinets, and furniture	noise and vibration control
Construction	building partitions, multilayer concrete slabs, window panes	noise and vibration control

Requirements on Materials for Damping Application.

Obviously, a polymer or other material to be utilized in a
vibration damping application must possess appropriate
properties over the ranges of temperature, frequency, and
strain amplitude involved. It is also apparent that

consideration must be given cost, weight, durability, stability, etc.

Some of what we see as the principal properties and characteristics of import in the development and selection of viscoelastic materials for structural damping are listed below. Table IV notes "passive" properties and characteristics, that is, those that pertain to the material itself and to its performance in the specified operating conditions. Somewhat in contrast, Table V lists "Active or Interactive" properties and characteristics that concern the interaction of the viscoelastic material with other materials and with the environment of the treatment. These tabulations are surely incomplete, but are intended to suggest the breadth of considerations that may be involved in material selection.

Table IV. Viscoelastic Materials for Damping: Passive Properties and Characteristics

---

- Acoustical (See Table VI)

- Environment-Related

  - temperature   - chemical
  - weather   - bio-degradation
  - moisture   - impact
  - radiation   - fatigue
  - outgassing   - chemical contamination

- Application-Related

  - application procedures and materials
  - cutting, shaping
       e.g., bending, forming of laminates
  - bonding, including self-adhesive properties
  - post-application fastening, welding
  - toughness, durability
  - creep and set
  - appearance, maintainability

Table V. Viscoelastic Materials for Damping: Active-Interactive Properties and Characteristics

---

- toxicity
- odor
- flammability
- outgassing
- compatibility
- electrical and thermal conductivity

In the particular case of the acoustical requirements mentioned in Table IV, we show in Table VI the technical acoustical requirements specifically for free and constrained viscoelastic layer treatments. This table with its notes summarizes the requirements on material dynamic properties that were presented in the earlier discussion of these broadly useful treatment types.

Table VI.   Technical Acoustic Requirements on the Dynamic
            Properties of Viscoelastic Damping Materials

	Free VE Layer	Constrained VE Layer
Principal Deformation:	Extension	Shear
Requirements for max η (Note 1):	Large loss modulus $E'' = E'\eta$ (Notes 2, 3)	• Large loss factor, η • Specified shear stiffness G'/H (Note 2)

1.   Maximum loss factor is not always the design goal. Combined mass, stiffness, and damping criteria often apply.

2.   Good properties are required over specified ranges of frequency and temperature.

3.   Note that at higher loss factors increased η is more effective than increased E' (Personal communication, Michael Drake, 1989)

Summary

Materials that are used in structural damping encounter wide ranges of temperature, frequency, and strain. The pertinent deformations may be in extension-compression, shear, dilatation, or combinations of these. Further, the materials can be subject to large static or transient loads, sometimes of significant magnitude, either in fabrication or in service. Together with the fact that damping treatments can encounter demanding environments, the above conditions make it clear that the selection or development of materials for damping is a complex task.
      The goal of this paper has been to establish the definition and mechanical principles of structural damping as well as the functional types of applications in which damping can be useful.

In addition, we have presented several of the important, practical types of damping treatments and the quantitative role of viscoelastic materials in such treatments. The reader should keep in mind several general principles of structural damping as follows:

1.   Damping operates on free responses:
      • Resonances
      • Free waves

2.   Damping does little or nothing to reduce the amplitude of forced responses.

3.   In an effective damping treatment, the viscoelastic element(s) must participate significantly in the energy of vibration.

4.   Damping may or may not reduce sound radiation.

As an example of principle number 4 above, consider a locally excited large panel on which resonant bending waves account for most of the vibratory response. However, assume that these resonant waves have a wavelength shorter than that of free waves in the surrounding air. In such a case the resonant waves are poorly coupled to the air, and radiate very little sound. What radiation there is can be dominated by non-resonant forced motion around the drive point (and at other discontinuities). As a result, applied damping can reduce the resonant response, but not the forced motion and the radiation of sound.

It is our hope that what we have presented here will be of help to those in the field of Polymer Science and Engineering who have an interest in developing materials for application in structural damping.

Literature Cited

1.   Ungar, E. E.; Kerwin, E. M., Jr. J. Acoust. Soc Am. 1962, 34, 954-57.
2.   Nashif, A. D.; Jones, D. I. G.; Henderson, J. P. Vibration Damping; John Wiley & Sons: New York, 1985.
3.   Morse, P. M. Vibration and Sound (2nd ed.); McGraw-Hill: New York, 1948; p 229. (Reprints available from the Acoustical Society of America, New York)
4.   Strutt, J. W. (Lord Rayleigh). The Theory of Sound (2nd ed.); Dover: New York, 1945; Vol. 1, p 302.
5.   Cremer, L.; Heckl, M. A.; Ungar, E. E. Structure-Borne Sound (2nd ed.); Springer Verlag: Berlin, 1988.
6.   Oberst, H. Acustica 1952, 2, 181-94.
7.   Oberst, H.; Becker, G. W. Acustica 1954, 4, 433-44.
8.   Liénard, P. La Recherche Aeronautique 1951, 20, 11-22.
9.   Liénard, P. Annales des Telecommunications 1957, 12-10, 359-66.

10. Kerwin, E. M., Jr.   In _Internal Friction Damping and_
    Cyclic Plasticity; ASTM Special Technical Publication
    No. 378; American Society for Testing and Materials:
    Philadelphia, 1965; pp 125-149.
11. Ungar, E. E.   In _Noise and Vibration Control_; Beranek,
    L. L., Ed.; McGraw-Hill: New York, 1971; Chapter 14.
12. Mead, D. J.   _Criteria for Comparing the Effectiveness_
    of Damping Treatments; Report No. 125; Department of
    Aeronautics and Astronautics, University of
    Southampton: Hampshire, England, 1960.
13. Mead, D. J.   _Noise Control_ 1961, _7-3_, 27-38.
14. Kerwin, E. M., Jr.   _J. Acoust. Soc. Am._ 1959, _31_, 952-
    62.
15. Ross, D.; Ungar, E. E.; Kerwin, E. M., Jr.   In
    Structural Damping; Ruzicka, J. E., Ed.; Am. Soc.
    Mechanical Engrs.: New York, 1959; Section III.
16. Ruzicka, J. E.   _J. Engineering for Industry_
    (Transactions of the ASME, Series B) 1961, _83-B-4_,
    403-24.
17. Ungar, E. E.   _J. Acoust. Soc. Amer._ 1962, _34,_ 1082-89.
18. Torvik, P. J.   In _Damping Applications for Vibration_
    Control; Amer. Soc. Mechanical Engrs.: New York,
    1980, AMD Vol. 38, pp. 85-112.
19. Soovere, J.; Drake, M. L.   _Aerospace Structures_
    Technology Damping Design Guide; Flight Dynamics
    Laboratory: Wright-Patterson Air Force Base, Ohio,
    1985.
20. Kerwin, E. M., Jr.; Smith, P. W., Jr.   _Vibration_
    Damping 1984 Workshop Proceedings, 1984, pp 27-9.
21. Ungar, E. E.; Kerwin, E. M., Jr.   _J. Acoust. Soc._
    Amer. 1960, _32_, 912(A).
22. Parfitt, G. G.; Lambeth, D.   _The Damping of Structural_
    Vibrations, Aeronautical Research Council "A.1" Report
    (U); PUBLISHER: Ministry of Aviation, U.K., 1960
    (also ASTIA No. AD 253216).
23. Parfitt, G. G.   _Proc. Fourth International Congress on_
    Acoustics, 1962, p 21.
24. Plunkett, R.; Lee, C. T.   _J. Acoust. Soc. Amer._ 1970,
    _48-1_, 150-61.
25. Zeinetdinova, R. Z.; Naumkina, N. I.; Tartakovskii, B.
    D.   _Soviet Physics-Acoustics_ 1978, _24-4_, pp 347-8.
26. Ungar, E. E.; Ross, D.   _Proc. Fourth Conference on_
    Solid Mechanics, 1959, pp 468-87.
27. Henderson, J. P.   In _Damping Applications for_
    Vibration Control; Amer. Soc. Mechanical Engrs.: New
    York, 1980; pp. 145-158.
28. Painter, G. W.   U.S. Patent 3 079 277, 1963.
29. Trapp, W. J.; Bowie, G. E.   In _Damping Applications_
    for Vibration Control; Amer. Soc. Mechanical Engrs.:
    New York, 1980; pp 1-26.
30. Warnaka, G. E.; Kerwin, E. M., Jr.; Carbonell, J. R.
    J. Acoust. Soc. Amer. 1965, _37_, 1215(A).

31. Kerwin, E. M., Jr.   U.S. Patent 3 087 565, 1963.
32. Kerwin, E. M., Jr.   U.S. Patent 3 087 571, 1963.
33. Kerwin, E. M., Jr.   Proc. Third International Congress on Acoustics, 1959, 1961, pp 412-15.
34. Whittier, J. S. M.S. Thesis, University of Minnesota, 1958.
35. Whittier, J. S. The Effects of Configurational Additions Using Viscoelastic Interfaces on the Damping of a Cantilever Beam; WADC TR 58-568, ASTIA Doc. 214381, 1959.
36. Wallerstein, L., Jr.; Painter, G. W. U.S. Patent 3 078 971, 1963.
37. James, R. R. Progress Report No. 5, Development of Damping Treatments for Destroyer Hulls; Report No. 94-30, Rubber Laboratory: Mare Island Naval Shipyard, 1961.
38. Ungar, E. E.; Kerwin, E. M., Jr. J. Acoust. Soc. Amer. 1964, 36-2, 386-92.

RECEIVED January 24, 1990

Chapter 18

General Approach to Damping Design

Michael L. Drake

University of Dayton Research Institute, Dayton, OH 45469

The following chapter covers several important
steps in the design of a damping system. Designer
must 1) verify that there is a problem which is
indeed the result of resonant vibration, 2) define
the dynamic characteristics of the structure
under consideration and 3) define the environmental
conditions in which the structure operates. These
parameters are required because damping material
properties are dependent on both the frequency
and the temperature at which the problem occurs.

At this point, the problem is completely defined,
i.e., the dynamics which cause the problem are
known, the dynamic characteristics of the
component are known, and the operational environment
is established. With this data, the designer can
make logical choices of damping polymers and damping
configurations to develop the final design(1-3).

There are several important steps to consider in the design of
a damping system. They are as follows:

1. Verify that the problem is resonant vibration induced.

2. Dynamic Analysis of the system to determine resonant
    frequencies, mode shapes, and damping.

3. Define the environmental conditions in which the system
    operates.

4. Define the system damping required to eliminate the problem.

0097–6156/90/0424–0346$06.00/0
© 1990 American Chemical Society

5. Select the appropriate damping materials and basic damping
    configuration.

6. Develop the required design from the data collected.

Dynamic Problem Identification

The proper initial step in solving any problem is to completely
define that problem. This is very true when solving a vibration
induced problem using damping technology.   Therefore, the first
step in a damping design is to verify that the problem is indeed
resulting from a structural resonance.
    In case of a new design, the designer must obtain the
anticipated force input, i.e., excitation environment, for the
structural system and correlate the frequency content of this
information with the results of a natural frequency analysis of the
structure. If natural frequencies occur in the frequency band of
excitation, the potential of dynamic problems exists and should be
addressed.
    If a problem develops in an existing part, the designer might
choose one of the following approaches to identify the cause of the
problem.
    In the case of a cracked component a crack analysis should be
run to verify that the crack is a high cycle fatigue failure.   An
instrumented operational test of the component will identify the
frequencies of high vibration levels causing the problem.
    If the problem under consideration is high noise radiation, an
operational evaluation should be made to determine both the
frequencies and magnitudes of the noise being radiated and the
source of radiation.   An unacceptable vibration level environment
problem should be attacked in the same basic manner as the noise
problem using vibration measurements instead of acoustic
measurements.
    As a result of the above investigations, the designer has
determined the operational dynamic cause of the problem, verified
that the cause is resonant vibration, and defined the resonant
frequencies with the high dynamic response.

Dynamic Characteristics

A successful damping design can only be developed from a complete
understanding of the dynamic behavior of the structural system to be
damped. Generally, a frequency range over which this dynamic this
information is needed is defined from the analysis completed during
the first step. The dynamic range can be defined from operational
testing or can be determined from knowledge of the part under
consideration, i.e., a problem in a component where the excitation
forces are known to be engine-order related; low frequency
excitation from road roughness to the suspension; or acoustic
excitation to aircraft fuselage components due to jet engine
exhaust.   Once a frequency range is chosen, a complete dynamic
investigation must be conducted.   One must accurately determine all

the resonant frequencies, corresponding structural mode shapes, and
inherent modal damping values in the required frequency range. If a
tuned absorber is to be properly applied, the modal mass and
stiffness are also needed. This data can be obtained analytically
or experimentally.

In the early design stages where a prototype is available, the
optimum solution for data acquisition is to use experimental
analysis on the prototype structure to refine an analytical model
which can then be used for damping design(4).

Often, when a damping application is used as a redesign
approach, the necessary dynamic characterization can be acquired
efficiently through the use of modern experimental methods.
Experimental methods can quickly determine the data needed for
highly complex structural system; however, measurements on
operational systems can be extremely difficult and costly. The
Fourier analyzer is the most powerful experimental tool currently
available to do the experimental work; however, holographic methods
for determining mode shapes and standard sine sweep methods for
resonant frequencies and modal damping values are extremely
useful(5-7). The Designer must choose the most expedient method to
develop the required data.

Environmental Definition

Important data still required to design a damping application are
the operational environment in which the design must operate. This,
at first thought, might seem to be a rather simple task but the
importance of accurate environmental data cannot be over stressed.

A broad brush approach to temperature such as the standard
temperature range for operation of many aircraft components of -
65°F to 250°F is not the answer. This may be the maximum range seen
by the component; however, it will not generally be necessary to
provide high damping over this entire range. The engineer must
determine over what specific temperature range the damage is
occurring and design his application for that range while
maintaining an awareness of the required survivability temperature
range. Time related recordings of vibration and temperature data
from operational tests can be used to determine the temperature
range over which damaging vibrations occur. Operational tests can
also supply the necessary maximum temperature limits to be used in
the design. If temperature data from a large number of different
operational tests are available, a statistical study of the data
will reveal the temperature range in which the majority of
operational time is spent(8). An example of this type of data for
an aircraft is shown in Figure 1 where minimum and maximum
temperatures are shown along with percent of total operational time
spent in each temperature range. It is easy to see the value of
this type of data, particularly if vibration level and temperature
data cannot be simultaneously obtained for operational conditions.

In the early stages of structural system design, complete
operational temperature data may not be available. In such a case,
data from similar systems should be reviewed and the best estimates
of temperature should be developed and used in the damping design
procedure.

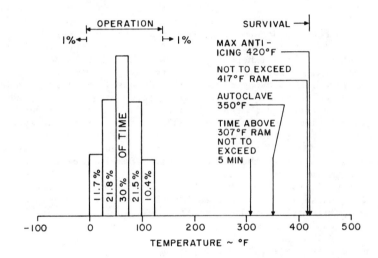

Figure 1.   Statistical Temperature Data.

Temperature is not the only environmental factor which must be considered. The engineer must know if the application will come in contact with contaminates such as salt water, gasoline, jet engine fuel, hydraulic fluid, or any other substance which might affect the performance or longevity of the candidate damping materials(2,4,9).

Required Damping Increase

The remaining question to be answered before a damping design can be started is "How much damping is needed to eliminate the problem?" In the "fix-it" damping business, the general approach found in the literature is to design a damping system with a high damping level and test it in service. If the failures are eliminated the problem is solved. In reality, the designer should use the minimum value of system damping which will eliminate the vibration problem. If the damped design accomplishes just the minimum required damping using an optimum damping system, the design should also be optimized from a weight, size, and cost standpoint.

The method for determining the minimum required system damping will depend on the problem to be solved. From the dynamic characterization, the inherent system damping has been determined. The corresponding vibration problem (high dynamic stress, noise level radiated, high dynamic amplitude response, etc.) is directly related to the inherent damping. Quick calculations can be made to determine the required increase in system damping to eliminate the vibration problem. Basically, if a 20 percent decrease of system response is needed, then the system damping needs to be increased 20 percent. If an analytical model has been developed, an analysis can be conducted to verify the value of system damping needed to eliminate the vibration problem.

In the literature, most of the successful damping systems currently in use were designed to achieve near maximum damping from a given configuration without regard to eliminating the problem with the least required amount of damping.

Damping Material Selection and Application Design

To this point the primary function of the designer has been to develop an accurate and complete definition of the resonant vibration problem. The frequencies of the component at which the problem exists during operation are defined along with the associated dynamic characteristics. It now becomes a simple matter to determine which resonant modes of the component are creating the vibration problem which in turn defines the frequencies at which damping is needed and the corresponding resonant mode shapes, the undamped modal loss factors, and the required damped modal loss factor. This complete set of dynamic data combined with the environmental conditions provide the designer with all the data necessary to begin analysis and evaluation of damping materials and damping design configurations. The designer must first look at the temperature range for which damping is needed and the survivability temperature limits to see if either of these temperature ranges eliminates a particular method of damping approach because no available materials meet the temperature requirements. An example here would be a damping requirement over the temperature range of 150°F to 250°F with a survivability to 600°F. This would eliminate most typical polymeric damping materials such as acrylates and vinyls used in constrained layer or free layer damping designs.

If the temperature range conditions can be met, the next consideration is the mode shapes of the resonant frequencies which require damping. Tuned dampers require displacements of some magnitude while constrained and free layer applications require large areas of localized bending which can deform the damping material(10-12). Highly localized strain distributions will negate the effectiveness of a layered treatment.

From the temperature conditions and the dynamic characteristics, the designer can choose the appropriate class of damping polymers and the appropriate type of damping configuration for the starting point to design the specific application for the structure having the vibration problem. The basic principles of free layer and constrained layer damping applications and tuned dampers and analysis techniques are discussed in References 10, 11 and 12.

Various design analysis methods are often appropriate for problems; however, it is necessary to obtain all the basic information discussed previously to be successful regardless of the analysis procedure used. A design flow chart appropriate for any of the design analysis techniques is seen in Figure 2. The dynamic and temperature data is the input and the output is the structural loss factor. The chart loops are continued until the proper η_s is achieved at which time the damping design is complete. Design techniques are discussed in references 10, 11, and 12.

Summary

Restating the importance of the problem definition is appropriate at this point. Inaccurate temperature range formulation will eliminate any beneficial effects of the damping material. This can be seen in Figure 3 (Figure 3 is dynamic modulus data for 3M ISD-112) where a temperature shift of 100°F causes a significant reduction in the loss factor. If the survivability temperature limits are incorrect, the damping application may well provide the necessary reduction in the vibration levels but will be destroyed by an over-temperature condition(13). Guesses at temperature data will invariably lead to the failure of a damping design.

The other major area where accurate data are necessary is the dynamic characteristics of the system under consideration. The placement of a layered damping design on a portion of the structure which will not undergo major motion in a particular mode is as ineffective as placing a tuned damper on a node line of the mode you wish to control.

As with any design project, successful results require accurate information upon which to base the design. Temperature and dynamic characteristics are the two prime factors which must be meticulously measured to obtain good damping design results. All the steps required in a damping design are summarized on page 346.

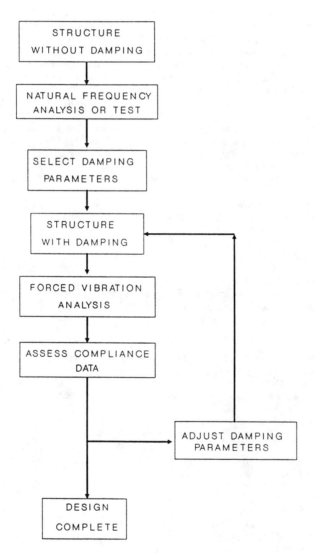

Figure 2.   Basic Flow Chart.

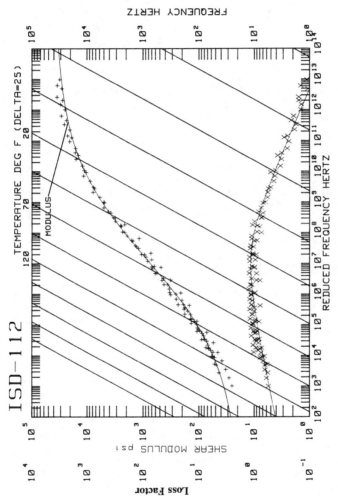

Figure 3.   Reduced Temperature Nomogram.

Literature Cited

1. Sharp, J.D. and M.L. Drake, "Elimination of Resonant Fatigue
    Problems for Major Maintenance Benefits," ASME Publication
    Number 77-DET-135.

2. Drake, M.L. and J.D. Sharp, "An Example of Additive Damping
    as a Cost Savings Alternative to Redesign," ASME Publication
    Number 77-WA/GT-2.

3. Drake, M.L. and M.P. Bouchard, "On Damping of Large Honeycomb
    Structure," Journal of Vibration, Acoustics, Stress, and
    Reliability in Design, Vol. 107, pp. 361-366, October, 1985.

4. Flora eta, "Dynamic Analysis and Testing of Damped
    Intermodule Plates for the Sigma Laser Device," ASIAC Report
    No. 1182.1A, November 1982.

5. Brown, Dave, Lecture notes - short course, "Modal Analysis
    Theory and Measurement Techniques," sponsored by the
    University of Cincinnati and Hewlett Packard.

6. Ramsey, K.A., "Effective Measurements for Structural
    Dynamics Testing," Sound and Vibration, November 1975, pp 24-
    35.

7. Drake, M.L. and J.P. Henderson, "An Investigation of the
    Response of a Damped Structure Using Digital Techniques,"
    Shock and Vibration Bulletin 45, Part 5, 1975.

8. L.C. Rogers, and M.L. Parin, "Additive Damping for Vibratory
    Stress Reduction of Jet Engine Inlet Guide Vanes," presented
    at the 47th Shock and Vibration Symposium, Albuquerque, NM,
    1976, published in the 47th S&V Bulletin.

9. Henderson, J.P. and M.L. Drake, "Investigation of the Effects
    of Damping Treatments on the Response of Heated Fuselage
    Structure," NoisEXPO, National Noise and Vibration Control
    Conference, New York, New York, March, 1976.

10. Jones, D.I.G., J.P. Henderson, and 1/Lt. G.H. Bruns, "Use of
    Tuned Viscoelastic Dampers for Reduction of Vibrations in
    Aerospace Structures," Air Force Materials Laboratory,
    presented at the 13th Annual Air Force Science and
    Engineering Symposium at Arnold Air Force Station, Tennessee,
    September 27-29, 1966.

11. Nashif, Ahid D., David I. G. Jones, and John P. Henderson,
    "Vibration Damping," John Wiley & Sons, 1985.

12. Soovere, J., and M. L. Drake, "Aerospace Structures
     Technology Damping Design Guide, Volume I - Technology
     Review, Volume II - Design Guide, Volume III - Damping
     Material Data," Technical Report AFWAL-TR-84-3089, Dec. 1985.

13. Jones, D.I.G. and C.M. Cannon, "Control of Gas Turbine Stator
     Blade Vibrations by Means of Enamel Coatings," Journal of
     Aircraft, Vol. 12, No. 4, pp. 226-230, April 1975.

RECEIVED January 24, 1990

ADVANCED MATERIALS (IPNs)

ADVANCED MATERIALS (IPN s)

One of the more exciting new areas are polymer systems forming interpenetrating polymer networks (IPN's). These systems offer the potential of allowing more control over the dynamic properties than previously attainable, including broadening the damping peak by a select amount and reducing the temperature dependence of Young's modulus. In some measure, the IPN materials provide a critical need for the acoustic engineer: that is, a slowly varying storage modulus.

In this section, **Fox** et. al. first introduces this area, providing a definition of what constitutes an IPN and the general requirements of the material system selected. **Ting** et. al. describe a urethane/epoxy IPN system, with particular emphasis on acoustical properties. **Sorathia** et. al. also describe results on a urethane/epoxy IPN system with particular emphasis on marine applications. **Greenhill** et. al. review the formation of latex dispersions and their use in damping. **Fay** et. al. use a variety of techniques to study the poly(vinyl methyl ether)/polystyrene system. **Yamamoto and Takahashi** consider the vinyl - phenolic system and the width of the damping peak. Finally, **Tabka** et. al. discuss the influence of kinetics of the IPN formation process.

Chapter 19

Interpenetrating Polymer Networks

An Overview

R. B. Fox[1], J. J. Fay[2], Usman Sorathia[3], and L. H. Sperling[4]

[1]Naval Research Laboratory, Washington, DC 20375–5000
[2]Department of Chemistry, Lehigh University, Bethlehem, PA 18015
[3]David Taylor Research Center, Annapolis, MD 21402–5067
[4]Department of Chemical Engineering, Lehigh University,
Bethlehem, PA 18015

The field of interpenetrating polymer networks,
IPNs, is introduced. General relationships
between synthesis, phase separation, morphology
and properties of IPNs are developed
preliminary to the following chapters.

For many investigators in the sound and vibration damping
field, interpenetrating polymer networks, or IPNs, are a
new class of polymers capable of exhibiting relatively
broad-band damping properties. The purpose of this
chapter is to provide an introduction (1,2) to these
materials, what they are, how they are made, the kind of
morphologies that may be expected, and why they have the
properties they have.

What is an IPN? Operationally, an IPN is an intimate
mixture of two (or more) network polymers in which at
least one of the networks has been formed in the presence
of the other (3). Ideally, bonding between the networks
is absent, interpenetration of the network chains in the
bulk mixture is complete on a molecular scale, and a
single phase results. On the molecular level, chains of
one network are threaded through those of the second
network; they are catenated. The total crosslink density
of the composite material includes a contribution from the
physical crosslinks as well as the chemical crosslinks
created during polymerization. Such an ideal, albeit
complex, system has seldom, if ever, been realized.

Phase Separation

The rarity of the perfect IPN lies in thermodynamics.
Mixing two kinds of long polymer chains produces little
change in entropy simply because the monomeric units are
coupled in a polymer chain, and chains can't cut through
each other. Heats of mixing of two polymers are usually

0097–6156/90/0424–0359$06.00/0

positive unless some specific interaction, such as
hydrogen bonding, occurs between them (4). In most cases,
the free energy of mixing will be positive and phase
separation will exist. This is fortunate for the cause
of sound and vibration damping, which is associated with
mechanical relaxations in a viscoelastic material, since
a broadened region for damping will result if the glass
transition temperatures of the components are different.

One of the advantages of IPNs in damping applications
is their ability to damp over broad temperature ranges.
This is brought about through the development of a
microheterogeneous morphology, in which very small domains
are formed on the order of about 100 to 200Å in size.
Actually, they are comprised mostly of interphase
material, no clearly identifiable "domains" are observed.
Since the smallest unit of polymer chain involved in
damping motions contains 10-50 backbone atoms, these
microheterogeneous domains contain only a few such units.
With the statistics of small numbers, the composition of
each unit of space is highly variable, producing a
multitude of regions exhibiting wide variations in the
glass transition temperature within the same material.
Of course, the loss modulus and the loss tangent are also
correspondingly broad, providing the necessary conditions
for outdoor or variable temperature damping.

While IPN technology is not the only way of
synthesizing the microheterogeneous morphology, the
systematic introduction of crosslinks helps to control the
domain size. The role of thermodynamics in producing
these morpohologies must also be emphasized. Like other
compositions of matter, multicomponent polymer systems
have phase diagrams and free energies of mixing. If the
system has a negative free energy of mixing and meets a
few other requirements, a single phase material will be
synthesized, with properties similar to a statistical
copolymer with the same composition. A material of this
type would then exhibit good damping over narrow
temperature ranges only. When the material has a positive
free energy of mixing, and exists clearly within the phase
separated region of the phase diagram, damping peaks will
be observed which correspond to the individual phases.
However, if the free energy of mixing is nearly zero and
much of the material exists near the phase boundary, a
microheterogeneous morphology will arise as the stable
structure. In this final case, if the glass transition
temperatures of the individual polymers are widely
separated, damping may be achieved with the multicomponent
material over a broad range of temperatures spanning the
region between the glass transition temperatures of the
component polymers.

Another important aspect of developing the
microheterogeneous morphology is the kinetics of phase
separation. If the polymers gel before phase separation
takes place, then the crosslinks will restrict phase

separation. However, if phase separation precedes
gelation, the domains will be large and the crosslinks
will stabilize the phase separated morphology. Obviously,
if the thermodynamics are favorable, phase separation will
occur late in the polymerization, producing the desired
morphology.

Highly phase separated IPNs may very well have
interpenetration only at the interface of the separate
domains, and/or have interpenetrating phases. However,
to the extent that interpenetration exists, adhesion
between domains must still be affected and therefore,
mechanical properties are altered as well.

How and why will IPNs and semi-IPNs differ from
linear blends? All of these mixtures are subject to the
same thermodynamic phenomena. In IPNs, the constituent
networks are catenated or interpenetrated; physical, non-
bonded, crosslinks are present. The components are
inseparable without breaking covalent bonds. In semi-
IPNs, in which only one of the component materials is
crosslinked and the other is linear, there are similar
entanglements, but disentanglement is possible. The
attainment of equilibrium in which linear polymer chains
have separated from the network can be very slow,
particularly where one component is rigid or glassy. In
principle, the linear component can be extracted from the
network component. Blends of two kinds of linear polymer
chains, on the other hand, can be likened to a mixture of
cooked white and green spaghetti. There is no
interlocking that prevents disentanglement. Rather, all
of the thermodynamic push to separate the two kinds of
chains into discrete phases will be effective and the
polymers may be separated.

The extent to which the IPN ideal is approached in
these materials is highly dependent on the manner in which
the mixtures are made. Unless there is some specific
interaction between the unlike chains, phase separation
can occur up to the point at which crosslinking fixes the
final morphology of the material. Often, the result will
be a combination of the phase separated components with
additional mixed phases of varying composition. A further
complication arises when specific interactions lead to
partial miscibility between the components.

Types of IPNs

Because the unique properties of IPNs arise from the
intimate mixing of the component polymer systems, the
synthetic methodology used to produce these materials is
critical. Presently, there are three main routes utilized
to produce IPNs: simultaneous, sequential and latex. The
method employed is determined by the component polymers
selected, polymerization mechanisms, miscibility and the
anticipated end use of the IPN.

Simultaneous IPNs. A representative simultaneous IPN, or SIN, synthesis is illustrated in Figure 1. In this schematic, the "hard" and "soft" polymer segments are combined in a homogeneous solution containing all the necessary components (monomers, crosslinkers, initiators and accelerators) to produce the final IPN. Two non-competing polymerization methods, such as addition and condensation, are employed to polymerize the components in the presence of the other to produce the intimate mixing characteristic of IPNs. The main advantages of SINs arise from their liquid uncured state which makes them well suited for casting and injection molding applications. In addition, variation of the hard and soft polymer segments enables tailoring of the physical and dynamic properties of the end product (5). The main disadvantage of SINs is also a result of the liquid uncured state. Very few polymer/polymer mixtures are soluble in each other over a wide composition range. It is this degree of immiscibility that determines the properties of the final product. Because only a relatively few systems are significantly miscible, the number of possible combinations for use in SINs is limited.

In SIN formation, both timing and rates of polymerization to form the two networks are important. With an acrylate-epoxy system, it was found that simultaneous gelation produced materials with poorer properties than those formed by slightly mismatched polymerization rates (6). In another instance (7), polyurethane-poly(n-butyl methacrylate) SINs in which the acrylate was initiated photolytically at various times after the onset of polyurethane formation produced a series of materials, presumably with the same chemical composition, with an average particle size that decreased as the delay time to acrylic initiation increased. Damping properties of these materials changed systematically across the series.

Sequential IPNs. The second type of IPNs are known as sequential IPNs and, as the name implies, are generally made in a two step process in which a mobile monomer/crosslinker/initiator phase is swollen into and polymerized inside a previously crosslinked three dimensional polymer, as illustrated in Figure 2. This process allows production of systems not possible by the simultaneous route with the ability to vary widely the bulk properties as well as the properties through the thickness of the final structure, effectively acting as a post treatment for a preformed crosslinked polymer system. Much of the work on sequential IPNs has been directed towards the theoretical aspects of clarifying or demonstrating true IPN structure in these materials mainly through the use of small angle neutron scattering or electron microscopy.

Latex IPNs. Latex IPNs are the third type of IPNs and are manufactured according to the general schematic illustrated in Figure 3. Latex IPN synthesis involves the initial synthesis of a crosslinked seed polymer, usually in the form of an aqueous latex. The seed latex is then swollen with a second monomer/crosslinker/initiator system which is then polymerized "in situ" to form an aqueous IPN emulsion. Materials of this type are best suited to coating applications as illustrated by the development of "Silent Paint" by Sperling et al (8). However, latex IPNs are limited to water emulsifiable monomer/polymer systems, most of which have fairly low service temperatures, less than 150°C.

These synthetic routes rely on a knowledge of reactions and reactants to provide, in a giant step, an assumed knowledge of the nature of the product. Side reactions such as graft polymerization are known, of course, but for the most part are admitted to exist and are then neglected. Subsequent characterization often reveals the inadequacy of such assumptions, and frequently properties of the materials are unlike those expected without considering side reactions. Again, it must be realized that these systems are quite complex. Variations in the general synthetic methodology, including the sequence and rates of polymerization, miscibility, viscosity of the mixture at the time of phase separation, overall composition, and crosslink densities of the constituent networks can all act to modify the physical properties of the resultant IPN.

Damping

Although the performance of any sound damping system is ultimately determined by its application, relative performance evaluation and ranking for materials of interst can be obtained from dynamic mechanical analysis techniques designed to provide viscoelastic (vibration damping) property information. Specifically, these dynamic mechanical testing techniques provide storage modulus, E' (elastic response), loss modulus, E" (viscous response) and loss tangent, tan δ (E"/E') as a function of temperature and/or frequency. All polymer systems exhibit a maximum value for tan δ, and hence maximum damping efficiency, at their glass transition temperature, T_g. Whereas homopolymers and statistical copolymers damp effectively over narrow temperature ranges, typically 20-30°C, IPNs constitute a class of multicomponent polymeric materials that are capable of damping over broad temperature ranges in excess of 100°C.

Conclusion

This brief introduction to the synthesis and general properties of interpenetrating polymer networks demonstrates that IPNs are complex multicomponent polymer

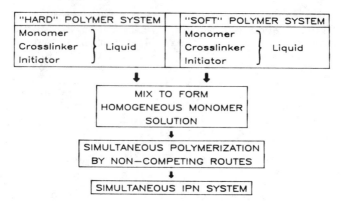

Figure 1. Simultaneous IPNs - schematic synthesis.

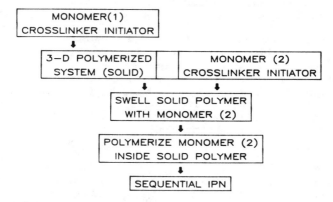

Figure 2. Sequential IPNs - schematic synthesis.

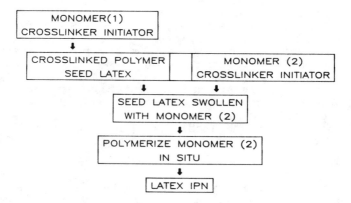

Figure 3. Latex IPNs - schematic synthesis.

materials in many ways. Each of the following papers show
how different systems are synthesized and characterized
in order to yield fundamental information regarding
damping or to evaluate new materials for damping
applications. While there are still many questions to be
answered regarding damping with IPNs, the ensuing papers
serve to present the "state of the art" of IPN research
pertaining to sound and vibration damping.

Literature Cited

1.   Sperling, L. H. In Comprehensive Polymer Science;
     Allen, G., Ed.; Pergamon Press: New York, 1989; Vol.
     6, p 423.
2.   Klempner, D.; Berkowski, L. In Encyclopedia of
     Polymer Science and Engineering; Mark, H. F.;
     Bikales, N. M.; Overberger, C. G.; Menges, G., Eds.;
     John Wiley and Sons: New York, 1987; Vol. 8, p 279.
3.   Sperling, L. H. Interpenetrating Polymer Networks
     and Related Materials; Plenum Press: New York, 1981.
4.   Kim, H.; Kwei, T. K.; Pearce, E. M. Polym. Prepr.
     1987, 28(2), 177.
5.   Xiao, H. X.; Frisch, K. C.; Kordomenos, P. I.; Ryan,
     R. A. In Cross-linked Polymers: Chemistry, Properties
     and Applications; Dickie, R. A.; Labana, S. S.;
     Bauer, R. S., Eds.; ACS Symposium Series No. 367,
     American Chemical Society: Washington, DC, 1988.
6.   Tousaent, R. E.; Thomas, D. A.; Sperling, L. H. in
     Toughness and Brittleness of Plastics; Deanin, R. D.;
     Crugnola, A. M., Eds.; Adv. Chem. Ser. No. 154,
     American Chemical Society: Washington, DC, 1976.
7.   Fox, R. B.; Binter, J. L.; Hinkle, J. A.; Carter, W.
     Polym. Eng. Sci. 1985, 25, 157.
8.   Sperling, L. H.; Chiu, T. -W.; Gramlich, R. G.;
     Thomas, D. A. J. Paint Technol. 1974, 46, 4.

RECEIVED January 24, 1990

Chapter 20

Acoustical Properties of Some Interpenetrating Network Polymers

Urethane—Epoxy Networks

R. Y. Ting[1], Rodger N. Capps[1], and D. Klempner[2]

[1]Underwater Sound Reference Detachment, U.S. Naval Research Laboratory, P.O. Box 568337, Orlando, FL 32856–8337
[2]Polymer Technologies, Inc., University of Detroit, Detroit, MI 48221

Interpenetrating polymer networks (IPNs) composed of polyurethanes and epoxies have been synthesized by the simultaneous polymerization technique. The objective was to utilize the high glass transition temperature (Tg) of the glassy epoxy and the low Tg of the elastomeric polyurethane to form a new material that are able to absorb vibrational energy over a broad temperature and frequency range. The dynamic mechanical properties of the IPN samples were characterized by using a Rheovibron DDV II at 110 Hz and the NRL–USRD string apparatus at 1-10 kHz. Master curves obtained for each sample by using the time-temperature superposition demonstrated the high and broad loss tangent exhibited in these epoxy/urethane IPNs. The effects of polyol selection, epoxy type, component ratio, urethane chain extenders, and fillers and plasticizers on IPN damping properties were investigated.

In recent years, it has been found that the physical and mechanical properties and the processability of polymeric materials can be improved by chemically or physically combining two or more structurally dissimilar polymers, (1,2). Among these new polymeric systems, interpenetrating network (IPN) polymers represent a unique new class of blends that are composed of two crosslinked polymers. They are essentially intimate mixtures of the crosslinked networks with no covalent bonds or grafts between them. The chains of one polymer are completely entangled with those of the other, and this entanglement is of a permanent nature. While normal polymer blends usually result in a multiphase morphology due to thermodynamic incompatibility of the polymers, true IPNs are homogeneous mixtures of the component polymers; and the morphology of the resulting mixture can be controlled to exhibit at worst only limited phase

0097–6156/90/0424–0366$06.00/0
© 1990 American Chemical Society

separation. Such a unique chemical structure suggests that IPNs may be prepared to make them potentially very useful for sound dampening applications.

IPN polymers may also exhibit varying degrees of phase separation depending on the compatibility of the polymer components used in their synthesis. With highly incompatible polymers, the thermodynamic forces leading to phase separation are so powerful that it occurs substantially before the kinetic ramifications can prevent it. In these cases, little phase mixing would be gained. On the other hand, more compatible polymers may lead to cases where phase separation can be almost completely circumvented, (3). By controlling the compatibility, the size of the dispersed phase domains of an IPN polymer can be varied from a few microns to a few tens of nanometers (4). For semi-compatible situations, polymer systems having broad transitions over a wide range of temperature and frequency have been shown to be possible, (5). We were interested in developing IPNs which contain polymer components having a low and a high glass transition temperature but with a semi-compatible behavior. Polyurethane/epoxy networks were prepared via the one-shot, simultaneous polymerization technique, (6). The effects of chain extenders, epoxy type, catalyst and urethane/epoxy ratio on the low-frequency dynamic mechanical properties of these IPNs (as obtained on a Rheovibron) were studied, (7). The effects of various fillers on the acoustical properties of the IPN polymers were also investigated. The results from these studies are summarized here.

Experimental

The IPNs prepared were composed of a rubbery polyurethane and a glassy epoxy component. For the polyurethane portion, a carbodiimide-modified diphenyl-methane diisocyanate (Isonate 143L) was used with a polycaprolactone glycol (TONE polyol 0230) and a dibutyltin dilaurate catalyst (T-12). For the epoxy, a bisphenol-A epichlorohydrin (DER 330) was used with a Lewis acid catalyst system (BF_3-etherate). The catalysts crosslink via a ring-opening mechanism and were intentionally selected to provide minimum grafting with any of the polyurethane components. The urethane/epoxy ratio was maintained constant at 50/50. A number of fillers were included in the IPN formulations. The materials used are shown in Table I.

These IPNs, prepared by the mixing of two components, were essentially simultaneous interpenetrating polymer networks (SIN). One component contained the isocyanate and epoxy resin, and the other component contained the polyols, chain extenders, catalysts, fillers, and plasticizers. The two components were then mixed together for 30 seconds (at room temperature) using a high speed mechanical stirrer at 2000 rpm. The mixture was then quickly poured into a pre-heated mold and pressed on a laboratory platen press at 100°C. The sample was removed from the press and demolded 30 minutes after it gelled. The gel time was approximately 3-6 minutes. The post curing condition was for 5 hours at 120°C. Figure 1 shows schematically this preparation procedure for the IPN

Table I: Materials Used For Preparing IPN Polymers

Material	Chemical Composition	Supplier
Isonate 143L	Carbodiimide-modified diphenyl methane diisocyanate, functionality=2.1	Upjohn Co.
TONE polyol 0230	Poly(caprolactone) glycol	Union Carbide
DER 330	Bisphenol A epichlorohydrin	Dow Chemical Co.
T-12	Dibutyltin dilaurate (catalyst)	M&T Chemical Co.
BF_3-etherate	Lewis acid complex (catalyst)	Eastman Chemical
Sundex Oil 750T	Plasticizer	Sun Products
Ethylene homopolymer	Filler	Allied Corp.
$Al_2O_3 \cdot 3H_2O$	Alumina trihydrate (filler)	Great Lakes Minerals Co.
Suzorite Mica 60-S International LTD	Filler	Manetta Resources
Freon 11	Blowing agent	Dupont

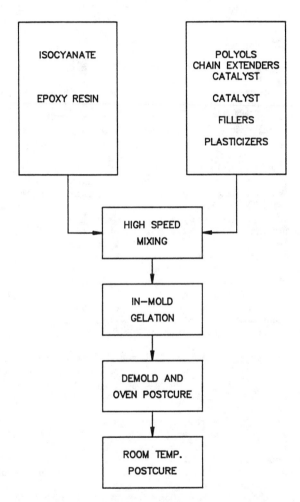

Figure 1   A schematic representation of the procedure for preparing IPN samples.

samples, which can be readily used in a fast-curing reaction injection molding (RIM) operation.

The acoustical properties for all samples were evaluated by using either a Toyo DDV-II Rheovibron viscoelastometer at 110 Hz, or a string apparatus developed at NRL-USRD, (8). In this latter instrument, dynamic Young's moduli and loss tangent were measured in the frequency range of 1-10 kHz, and master curves were obtained by using the time-temperature superposition technique, (9).

Results and Discussion

Polyol Selection. Pure polyurethanes were prepared by using Isonate 143L and T-12 as the catalyst, but without any chain extenders. Different polyols were used, and the dynamic mechanical properties of these polyurethanes were measured to determine which type of polyol produced the highest loss tangent values. At 110 Hz, the results were: poly(oxy-propylene) glycol > poly(oxypropylene-oxyethylene) triol with 21% graft acrylonitrile > poly(1,4-oxybutylene) glycol > poly(caprolactone) glycol.

The final polyol determination was based on more than just the tan Δ values, however. A fast-curing RIM (Reaction Injection Molding) IPN was the desired method of synthesis. Therefore, the reactivity of the IPN as well as its loss property (and not just that of the pure polyurethane) all played an important role in the final polyol determination. Even though pure polyurethanes based on poly(oxypropylene) glycol and poly(1,4-oxybutylene) glycol resulted in very high loss materials, they were very slow reacting and thus not suitable for RIM. Also, they did not respond as well in the IPN to dynamic mechanical measurements as they did in the pure polyurethane form. All subsequent formulations were based on polyurethanes composed of Isonate 143L and a poly(caprolactone) glycol (Niax Tone 0230, Union Carbide). Figure 2 shows the Rheovibron data of this polyurethane.

Effects of Epoxy Type and Amount. Three different types of epoxies were investigated, two of them more extensively. The epoxies were: 1) a bis(glycidyl ether) of bisphenol A (DER 330, Dow Chemical), 2) an epoxy novolac (DEN 431, Dow Chemical Co.), and 3) a tetra-functional epoxy, tetraglycidyl ether of 4,4'-methylene bis(aniline), (Araldite MY720, Ciba-Geigy Co.). The tetrafunctional liquid epoxy was eliminated early in the study due to the fact that too much stiffness and therefore very little loss (i.e. low damping) was imparted into the IPN. It was believed that the use of the tetrafunctional epoxy resulted in too high a crosslink density in the IPN.

In a 60/40 IPN where the epoxy was at 40 weight percent, DEN 431 imparted broader but lower tan Δ peaks, (see Figure 3), while DER 330 gave a higher, although more narrow tan Δ peak. These properties are consistent regardless of the type of polyurethane polyol and chain extender used. DEN 431 also shifted the peaks to a higher temperature (as expected due to its more rigid structure). Therefore, DER 330 was chosen as the epoxy for producing the optimum high loss peaks in the desired temperature range. Figure 4 shows the Rheovibron data of a 50/50 PU/DER 330 IPN sample. The PU/epoxy

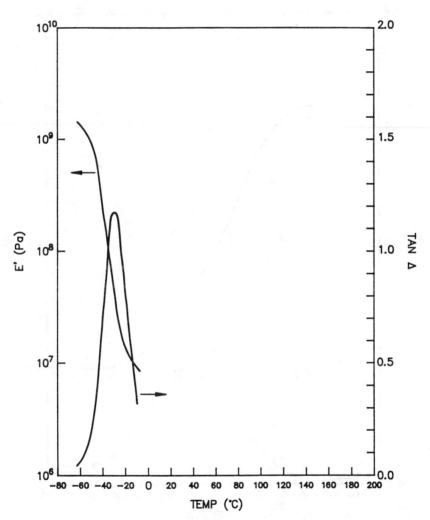

Figure 2   The Rheovibron result of an Isonate 143L/Tone 0230
                polyurethane sample at 110 Hz.

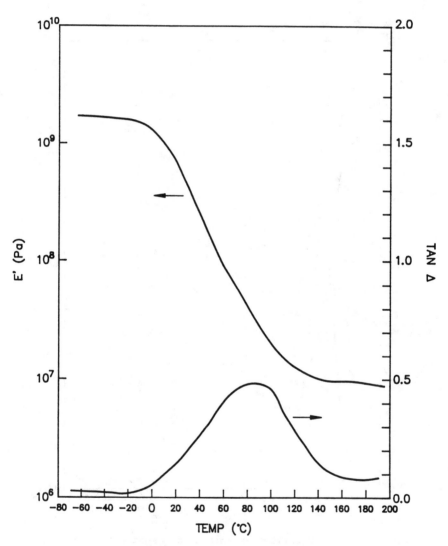

Figure 3   The Rheovibron result of a PU/Novolac DEN 431
                (60/40) sample at 110 Hz.

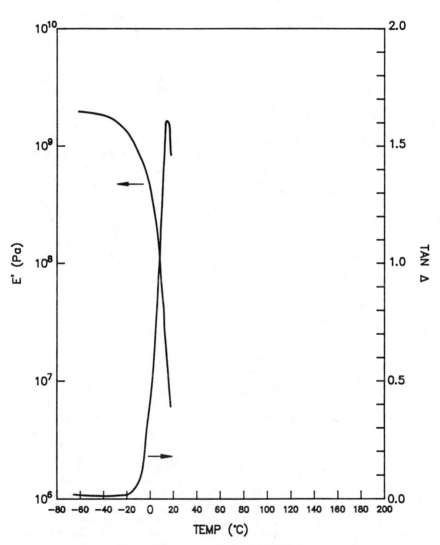

Figure 4   The Rheovibron result of a PU/DER 330 (50/50) sample
at 110 Hz.

ratio was changed to take advantage of the lightly crosslinked
epoxy. In addition to the 60/40 PU/epoxy formulation, others were
prepared at ratios of 50/50, 40/60 and 30/70. The tan Δ values kept
climbing from a maximum of 1.6 in the 50/50 IPN to tan Δ > 1.75 in
the 40/60 and 30/70 mixtures. At 30/70 PU/epoxy, the material
became too soft to obtain dynamic mechanical measurements as the
temperature reached 40°C. The modulus of the material was much less
than 10^9 dynes/cm^2 in the rubbery plateau, and became of little
interest for practical applications.

Chain Extenders. Various weight fractions of 1,4BD (butanediol)
were introduced into the PU/epoxy IPNs as chain extenders for the
PU. The Rheovibron data of these IPNs showed that the degree of
phase separation increased (i.e. the loss peaks broadened) as the
1,4BD content increased from 2 to 15% based on polyols. The results
were independent of the type of epoxy used, DER 330 or DEN 431, and
was most evident when the polyurethane weight percent in the IPN was
50% or greater. This increase in broadening or, in other words,
increase in phase separation between the PU and epoxy was initially
believed to be extremely beneficial for vibration damping, since the
broadening of loss tangent peaks would indicate a broader
temperature range for vibration attenuation. The height of the tan
Δ peak also increased, albeit very slightly. Unfortunately, this
increase in broadening and height coincided with a shifting of these
peaks to a higher temperature range beyond that of our interest,
(see Figure 5). Increasing the 1,4BD shifted the maximum damping
out of the desired range because the low molecular weight of 1,4BD
imparted stiffness into the IPN (hard blocks in the PU). The
increase in peak height and width is not justified when the peaks
are not in the desired temperature range of 0-30°C. When the weight
percent of the polyurethane was decreased to less than 50% of the
IPN, the effect of 1,4BD on broadening the peaks was minimal.
Therefore, 1,4BD was left out in subsequent formulations.
     Addition of trimethylolpropane (TMP -- Aldrich Chemical Co.)
and Isonol N-100 (N,N'-bis(2-hydroxypropyl) aniline, Upjohn Co.) as
chain extenders for the PU greatly broadened the tan Δ peaks in the
following order: Isonol N-100 > TMP > 1,4BD (60/40 PU/epoxy IPN).
The Isonol N-100 and TMP also shifted the peaks to a higher
temperature out of the desired range. They also drastically lowered
the tan Δ peaks. This is because the chain extenders increase the
stiffness of the material by increasing the hard block content of
the PU.

Effect of Fillers. Figure 6 shows the Rheovibron test result at 110
Hz for the urethane/epoxy (50/50) IPN sample (not post-cured)
containing 10% Suzorite mica 60-S filler. At this frequency, the
glass transition temperature is about 20°C, where the loss tangent
shows a peak value of approximately 1.8. The corresponding
acoustical measurement result is shown in Figure 7, where the
dynamic Young's modulus E' and the dynamic loss factor (tan Δ) are
plotted as a function of the reduced frequency $a_T f$ referenced at
10°C. The two sets of measurements agreed very well, providing
confidence in the shifting procedure used for data reduction and for
the generation of the master curve of Figure 7. At 10°C, the loss

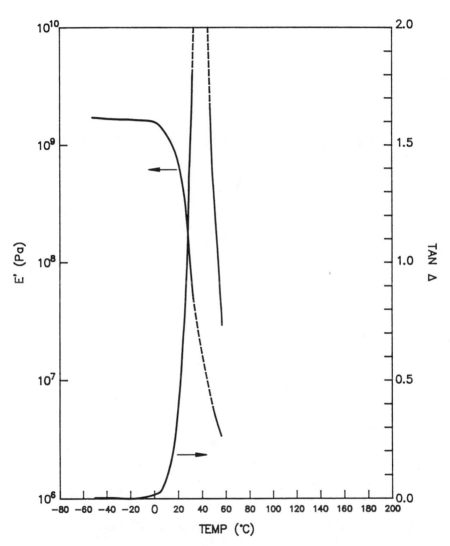

Figure 5   The 110-Hz Rheovibron result of a PU/DER 330 (50/50) sample with 1,4 butanediode chain-extender.

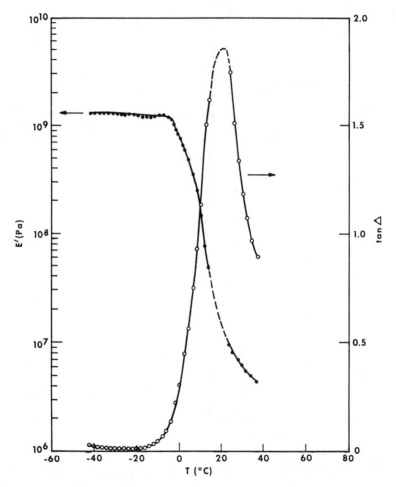

Figure 6   The 110-Hz Rheovibron result of a PU/DER 330 (50/50)
                sample with 10% mica fillers.

tangent curve shows a maximum value of about 1.8 at 10 Hz. The
dynamic Young's modulus at that frequency is about 5×10^7 Pa, typical
for a rubbery material. The glassy modulus reaches 10^9 Pa at
frequencies higher than 10 kHz. Figure 8 shows the effect of 5-hr.
post-cure at 120°C of the same IPN sample. The overall level of E'
is increased, as would be expected. The high value of tan Δ is
still maintained, but shifted to a much lower frequency, ($<10^{-2}$ Hz).
This is an indication of further crosslinking developed in the
interpenetrating networks. Further increase of the mica content
from 10% to 20% reduced the overall tan Δ values by about 25%; there
was a strong accompanying stiffening effect in that the modulus E'
values were nearly doubled over the frequency range displayed, (see
Figure 9).

When 10% by weight of alumina trihydrate was added to the same
urethane/epoxy IPN formulation, the maximum tan Δ value remained at
about 1.8, but appeared at a lower frequency of about 1 Hz when
compared with the result shown in Figure 8 for 10% by weight of
mica. On the other hand, the results presented in Figure 10 suggest
that alumina hydrate fillers did not seem to have as much of a
stiffening effect as mica for the IPN sample. For instance, with
the same post-cure treatment, the dynamic Young's modulus only
reached about 3×10^7 Pa at 10 Hz in this case, whereas the mica-
filled sample showed a E' value of over 4×10^8 Pa.

Freon 11 was also used as a blowing agent to create pores in
the filled IPN samples. The effect of this treatment on the damping
behavior is quite small, except for broadening the loss peak
somewhat toward the infrasonic region. However, it did not increase
the tan Δ peak value any further. The porous nature of the polymer
matrix, however, caused the dynamic Young's modulus to decrease
slightly, (see Figure 11).

An ethylene homopolymer (Allied Corporation) was also used as a
filler for the IPNs. A 10% addition of this material to the 50/50
urethane/epoxy IPN sample had similar effects as those shown by the
10% mica-filled sample of Figures 7 and 8. A clear difference in
this case was a shift of the loss tangent peak to lower frequencies,
resulting in a lower tan Δ value in the frequency range of the
display, (see Figure 12).

A combined filler system of 10% Suzorite mica and 10% alumina
hydrate was further used for the base IPN sample. This total of 20%
by weight of additives did not show the large stiffening effect as
in the 20% mica-filled sample. It seemed to suggest that the
dynamic Young's modulus was not affected by the alumina fillers as
much as by mica. In fact, the E' curve for this sample with
combined fillers was nearly identical to that shown in Figure 8 for
the 10% mica-filled sample, except for a slight stiffening below
10^{-1} Hz. When compared with the loss curve in Figure 8, the loss
tangent maximum was shifted slightly to a higher frequency
(~ 0.05 Hz), with a more rapid drop-off at frequencies higher than
10^5 Hz. This result seems to indicate that alumina trihydrate can
be used as a filler to independently modify the loss behavior of the
urethane/epoxy IPNs with very little effect on the dynamic modulus
E'. If established with further experimental work, this would be
very important in the preparation of damping materials having a

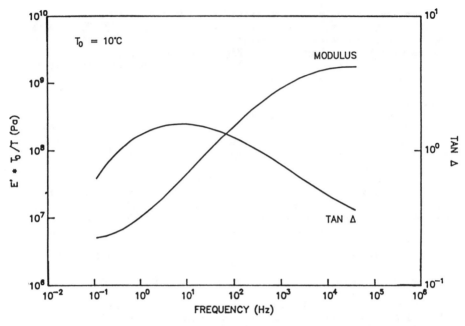

Figure 7   The master curve at 10°C of a PU/DER 330 (50/50)
                sample with 10% mica fillers.

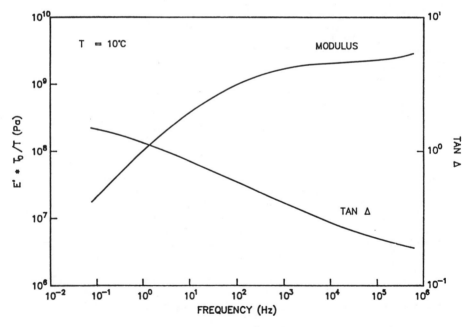

Figure 8   The master curve at 10°C of a post-cured PU/DER 330
                (50/50) sample with 10% mica fillers.

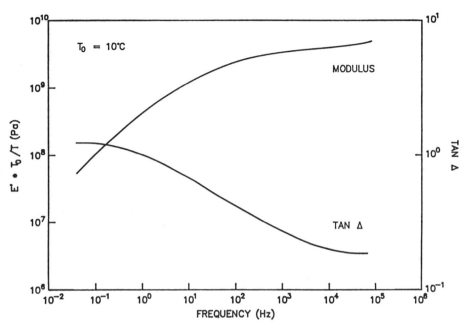

Figure 9   The master curve at 10°C of a post-cured PU/DER 330 (50/50) sample with 20% mica fillers.

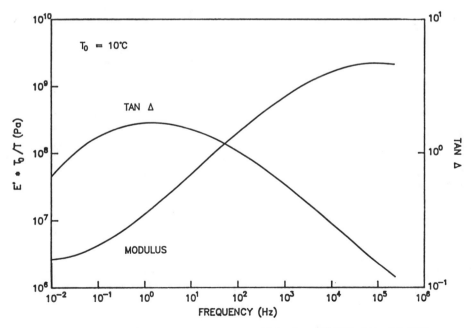

Figure 10   The master curve at 10°C of a PU/DER 330 (50/50) sample with 10% alumina trihydrate.

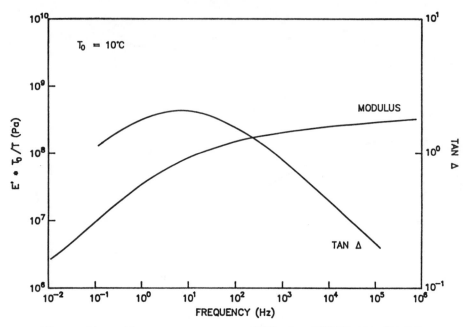

Figure 11   The master curve at 10°C of a PU/DER 330 (50/50)
                sample with blowing agents.

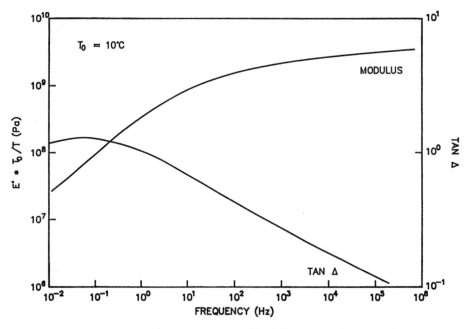

Figure 12   The master curve of a PU/DER 330 (50/50) sample with
                10% mica plus 10% alumina trihydrate.

broad loss tangent peak over an extended frequency range of interest.

Conclusions

It has been demonstrated that IPN polymers exhibiting high loss peaks over a broad temperature range can be produced based on semi-compatible mixtures of polyurethane/epoxy networks. These IPNs (at a 50/50 PU/epoxy ratio) were based on a carbodimide modified, 4,4'-diphenylmethane diisocyanate, a polycaprolactone polyol and small amounts (0-10%) of 1,4- butanediol chain extender for the polyurethane component. A bis(glycidyl ether) of bisphenol-A was used as the epoxy component and dibutyltin dilaurate and BF_3-etherate were used as the urethane and epoxy catalysts, respectively. The polycaprolactone polyol was selected by considering both the damping characteristics of the resulting IPNs and the reactivity of the system for reaction-injection-molding (RIM) applications. Different epoxy resins have the effect of increasing IPN stiffness and glass transition temperature in the order of MY720, Novolac DEN 431 and a bisphenol-A type epoxy, DER 330. By increasing the epoxy content in the PU/epoxy from 60/40 to 50/50, 40/60 and 30/70, there resulted in a consistent increase in the maximum tan Δ peak values. However, the material became so soft that it perhaps would offer very little practical value for most applications. Chain-extenders promoted phase separation, resulting in broadened loss peaks accompanied with a shift of the glass transition to high temperatures. Mica fillers were found to increase the dynamic stiffness of these IPNs, but alumina trihydrate increased the dynamic loss. Non-aromatic plasticizers were also found to decrease the glass transition temperature of the IPN polymers without affecting the tan Δ value or the dynamic Young's modulus, (7).

Literature Cited

1. Bruins, P. F. Bruins, Polymer Blends and Composites, Wiley Interscience, New York (1970).
2. Manson, J. A. and L. H. Sperling, Polymer Blends and Composites, Plenum Press, New York (1976).
3. Frisch, K. C., Klempner, D., Migdal, S., Frisch, H. L., and Ghiradella, H., Polym. Eng. Sci., 15, 339 (1975).
4. Klempner, D. and Frisch, H. L., J. Polym. Sci. B, 8, 525 (1970).
5. Sperling, L. H., Chiu, T. W., and Thomas, D. A., J. Appl. Polym. Sci. 17, 2443 (1973).
6. Klempner, D., Angew. Chem. 17, 97 (1978).
7. Klempner, D., Berkowski, L., Frisch, K. C., Hsieh, K. H., and Ting, R. Y., Rubber World, 192, 16 (1985).
8. Capps, R. N., J. Acoust. Soc. Amer. 73, 2000 (1983).
9. Capps, R. N., Ting, R. Y., and Quinn, M. E., J. Acoust. Soc. Amer. 82, Suppl. 1, S86 (1987).

RECEIVED January 24, 1990

Chapter 21

Advanced Damping Materials for Marine Applications

Usman Sorathia, William Yeager, and Timothy Dapp

David Taylor Research Center, Annapolis, MD 21402–5067

Viscoelastic IPNs have potential utility in
many noise and vibration damping applications. Three groups of simultaneous polyurethane (4,4 diphenyl methane diisocyanate,
polytetramethylene ether glycol, butanetane-
diol)/Epoxy (diglycidylether of bisphenol A,
boron trichloride amine complex) IPNs were
synthesized varying only the molecular
weight of polyurethane component polyol. The
polyols were selected at the molecular
weights of 650, 1000, and 2000. These IPNs
were characterized by density measurements,
dynamic thermal mechanical analysis, and
transmission electron microscopy. Results
show an increase in glass transition temperature (Tg) for the polyurethane component as
molecular weight of the polyol decreases,
and a significant broadening of Tg at or
around PU/EP composition of 70/30. This is
true for all three groups studied. At this
composition, storage modulus showed much
less steep variations with temperature during the transition from glassy to rubbery
state.

Viscoelastic Interpenetrating Polymer Networks (IPNs)
have potential utility in many noise and vibration damping applications. Interpenetrating Polymer Networks
(IPN) are a new class of materials consisting of multicomponent crosslinked polymer systems. IPNs are distinguished from their parent polymer materials by a general
characteristic that crosslinking occurs exclusively in,

but not between, the distinct polymer systems (1). Intimate component mixing during crosslinking results in
permanent physical entanglement of the polymer chains
and gives rise to unique physical and dynamic properties
unattainable in single component systems, graft copolymers or compatible polymer blends (2). Of particular
interest to the Navy is the enhanced noise and vibration
damping performance exhibited by certain IPN systems
over wide temperature ranges.

Even though the performance of any sound damping
material system is ultimately determined by its application, relative performance evaluation and ranking for
materials of interest can be obtained from dynamic-mechanical analysis techniques designed to provide viscoelastic (vibration damping) property information. Specifically, this technique provides storage modulus, E'
(elastic response); loss modulus, E" (viscous response);
and loss tangent, tan δ (E"/E'). All polymer systems
exhibit a maximum value for tan δ, and hence maximum
damping efficiency at their glass transition temperature, Tg. Two component (different Tg), partially miscible IPNs typically show characteristic "inward" shifts
of loss tangent where the component peaks are blurred
into a central region of relatively constant tan δ response. Chang, Thomas and Sperling (3) have subsequently shown that the area under the linear loss modulus
versus temperature curve obeys linear mixing rules for
certain sequential IPNs, essentially confirming that, in
general, IPNs distribute the same amount of damping
"efficiency" per unit volume of material over a wider
temperature range.

An early attempt to utilize the vibration absorbing
effect of an IPN mixture was made by Sperling et al (4),
who produced "Silent Paint", of which one layer was an
IPN. Hourston et al (5) illustrated typical IPN behavior in a 1:1 weight ratio Polyethylacrylate/Polyethylme-
thacrylate latex IPN. A continued need for similar types
of materials has prompted investigation of all polymeric
materials known to be effective energy absorbers.

EXPERIMENTAL

In this work, polyurethane (PU) and epoxy (EP) mixtures
were selected for investigation because they are known
to form partially miscible IPNs with broad glass transition temperatures. These were first prepared by Frisch
et al(6) using a simultaneous polymerization technique
in bulk. These materials showed the effects of crosslinking only one polymer component (pseudo-IPN) and in-
tentional grafting between the component polymers.
Klempner et al (2) also studied PU/EP IPNs for vibration
attenuation. The polyurethanes in this work were chain
extended and crosslinked with a 4:1 equivalent ratio of
butanediol (BD) and trimethylol propane (TMP).

In our work, PU/EP SINs were prepared from a polyurethane component consisting of 4,4 diphenyl methane
diisocyanate (MDI, Isonate 125M, Upjohn Chemicals) and
polytetramethylene ether glycol (Teracol, DuPont Chemical Co.) at different molecular weights of 650, 1000,
and 2000. Butanediol (BD) was the chain extender and
trimethylol propane (TMP) the cross linking agent in a
BD:TMP equivalent ratio of 7:1. An isocyanate index of
1.1 was used for all formulations. No catalyst was used
to maximize the pot life. The epoxy component consisted
of diglycidyl ether of bisphenol A (DER 332, Dow Chemical) cured with a boron trichloride amine complex (DY
9577, Ciba-Geigy) as the latent curing agent at a ratio
of 5 parts per hundred (phr). All materials were used as
received from the manufacturer except polyols which were
heated at reduced pressures to remove dissolved water.
Tables I,II,and III summarize the IPNs produced in Group
1 (polyol 650 Mw), Group 2 (polyol 1000 Mw), and Group 3
(polyol 2000 Mw).

Simultaneous PU/EP IPNs were prepared by a casting
technique. The polyol was heated to 100-110°C at reduced
pressure for one hour. It was cooled to 60°C and the
isocyanate and epoxy resin were added. The preheated,
premixed mixture of extender and crosslinking agents for
polyurethane and epoxy were then added and mixed thoroughly. The mixture was poured into a preheated mold,
allowed to gel, and cured for 16 hours at 110°C. Rods of
approximately 3/16" diameter X 12" were also produced
for dynamic mechanical testing.

Very early in this phase of work, it was recognized
that curing of epoxy resins with conventional curing
agents, such as primary and secondary diamines, did not
work well. As soon as the component A (consisting of
epoxy resin, polyol, isocyanates) was mixed with component B (consisting of catalyst, curing agent, and crosslinking agent), amines reacted with isocyanates to produce a non-castable solid mass that could not be easily
processed. Limited success was achieved by curing epoxy
resin with liquid anhydride (nadic methyl anhydride) but
this system produced some bubbles into the cast samples.

Best results were obtained when a catalytic curing
agent, such as boron trichloride amine complex, DY 9577,
was used. In this case, the curing agents are not co-
reactants, but serve to catalyze homopolymerization of
the epoxy resin. This complex of Lewis acid required
curing at elevated temperatures as it has negligible
reactivity at room temperature. These formulations produced satisfactory castable mixtures that could be processed easily by pouring into the mold or tube and curing
it in the oven. Samples were produced relatively free of
bubbles.

RESULTS AND DISCUSSIONS

The pure components, as well as IPNs produced at various ratios, were characterized by density measurements, dynamic thermal mechanical analysis, and electron microscopy. Results are presented and discussed below.

Table I. Group 1 IPNs - PTMEG 650 (PU)

PU/EP RATIO	SPECIFIC GRAVITY
100/0	1.1239
90/10	1.1434
80/20	1.1508
70/30	1.1596
60/40	1.1561
50/50	1.1768
0/100	1.1966

Table II. Group 2 IPNs - PTMEG 1000 (PU)

PU/EP RATIO	SPECIFIC GRAVITY
100/0	1.0965
90/10	1.1156
70/30	1.1298
50/50	1.1510
0/100	1.1966

Table III. Group 3 IPNs - PTMEG 2000 (PU)

PU/EP RATIO	SPECIFIC GRAVITY
100/0	1.0598
84/16	1.0824
72/28	1.0999
56/44	1.1262
0/100	1.1966

DENSITY. The specific gravity of all formulations was
measured by a displacement method in accordance with
ASTM D792. The density composition data for Groups
1,2,and 3 are shown in Figure 1. Straight lines were
drawn for each system based on a linear rule of mixtures.

The density composition curves show increased density over that expected using the linear rule of mixtures
for all IPN samples. IPN samples based on PTMEG 650
(Group 1) exhibited the highest increase in density
where as samples based on PTMEG 2000 (Group 3) showed
the least increase in density. As discussed later, Group
3 materials showed rather broad glass transition temperatures.

Observed IPN densities, higher than those predicted
by a linear rule of mixtures assumption, have been in-
terpreted to indicate the degree of molecular mixing (7)
and hence the extent of system interpenetration. Kim et
al (8) have explained the increased density of IPNs
qualitatively by means of chain entanglements at the
domain boundaries.

DYNAMIC THERMAL MECHANICAL ANALYSIS. As mentioned
previously, the purpose of this study is to synthesize
systems which produce broad glass transitions. Using
William, Landel, Ferry (WLF) frequency-time superpositioning techniques, it can be shown that broad material
tan δ values (over a given temperature range) trans-
late into similar tan δ performance over a much wider
frequency range. The usual accepted relationship is approximately one decade of frequency for every 5-8°C temperature increment. In an optimized IPN system, with
elevated tan δ values over a fairly wide temperature
range (50°C), WLF analysis would indicate effective
sound damping performance over 6-10 decades of frequency. This would be a significant improvement over state-
of-the-art damping materials.

Although the molecular basis of damping is not yet
completely understood, features such as flexible chains
rubbing over stiffer ones are thought to be important.
For obtaining a broad glass transition range, it is required that compositions be nearly, but not quite, miscible and that the domains of these phases be very small
in order to exhibit one very broad peak . Compositions
that are immiscible exhibit two glass transitions and
two tan δ-temperature peaks with relatively little damping in the area between them. Compositions with a homogeneous amorphous phase exhibit a single glass transition. A single glass transition yields good evidence of
molecular mixing, provided that the glass transitions of
the homopolymers are different and the glass transition
of the mixture is intermediate between the two.

The dynamic mechanical property data for Groups
1,2,and 3 materials were obtained from a Polymer Laboratory Model 983 Dynamic Mechanical Thermal Analyzer
(DMTA), and include log tan δ (loss factor), log
E'(storage modulus), and log E"(loss modulus). Frequency
was held constant at 10 Hz for all samples. The superposed results are shown for each group in Figures 2-10.
Samples produced in all three groups exhibited glass
transition peaks which were intermediate between the
glass transition temperature range of polyurethane and
epoxy at low concentrations of epoxy component. This
would indicate enhanced interpenetration with very little or no phase separation.
Around PU/EP ratio of 70/30, broadening of glass
transition is observed, and phase separation does become
evident as the concentration of epoxy component increases. Of particular importance was the fact that a
broadening of the glass transition temperature region is
also accompanied by flattening of modulus curve. Also,
broadening the glass transition temperature region for
the IPN system results in a lowering of the maximum tan
δ values. This is compensated for by the broadening of
tan δ curves to introduce lossiness in IPN materials in
the frequency bands not previously covered by the pure
components.

ELECTRON MICROSCOPY. Transmission electron microscopy
was done with a Philips 420 STEM on samples microtomed
at -78°C with an RMC ultramicrotome. Samples were
stained by placing the microtomed sections on electron
microscopy grids in ruthenium tetraoxide vapor for 15
minutes.
Electron micrographs for the samples belonging to
Group 2 are shown in Figures 11-15. The ruthenium tetraoxide vapor, which was used to stain the samples,
preferentially stains double bonds. Because of the extensive saturated hydrocarbon regions in the polyurethanes, they contain fewer double bonds than does epoxy.
Therefore epoxy is preferentially stained by ruthenium
tetraoxide, and is believed to account for the darker
features on the micrograph.
The pure materials showed no phase separation (Figures 11 & 12). The 50/50 PU/EP sample, shown in Figure
13, has features of two different sizes. There are elongated features approximately 20 nm in length and chain
like features approximately 5 nm wide. The dark epoxy
regions are larger than in the 70/30 PU/EP mixture shown
in Figure 14. For this sample, chains of polyurethane
approximately 5 nm in width appear to permeate and surround the epoxy domains which are about 2 nm in diameter. The 90/10 PU/EP mixture had a much more varied appearance than did the other two mixtures. Polyurethane

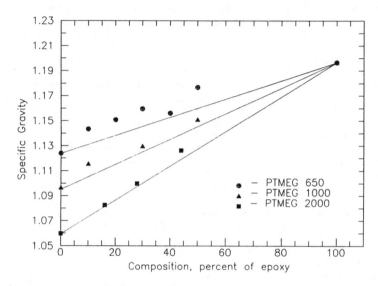

Figure 1. Specific Gravity Vs Composition for Groups
1,2, and 3 IPNs.

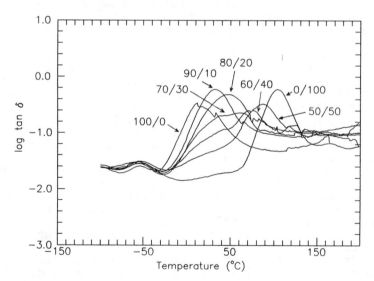

Figure 2. Composite log (loss factor) vs Temperature
for Group 1 IPNs, PTMEG 650(PU).

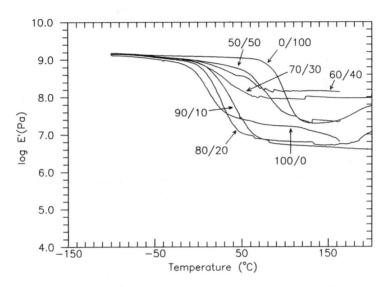

Figure 3. Composite log (E') vs Temperature for Group 1 IPNs, PTMEG 650(PU).

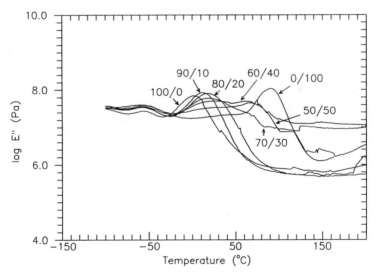

Figure 4. Composite log (E'') vs Temperature for Group 1 IPNs, PTMEG 650(PU).

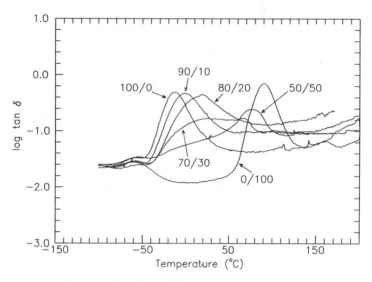

Figure 5. Composite log (loss factor) vs Temperature for Group 2 IPNs, PTMEG 1000(PU).

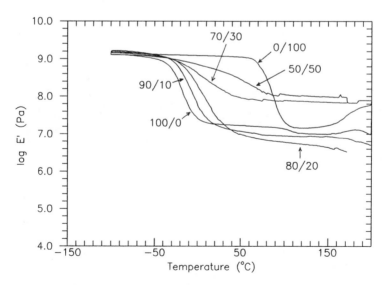

Figure 6. Composite log (E') vs Temperature for Group 2 IPNs, PTMEG 1000(PU).

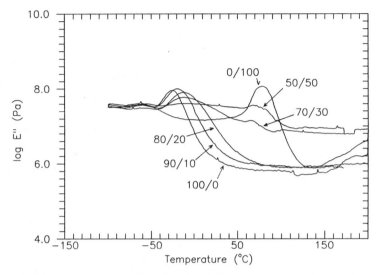

Figure 7. Composite log (E'') vs Temperature for Group 2 IPNs, PTMEG 1000(PU).

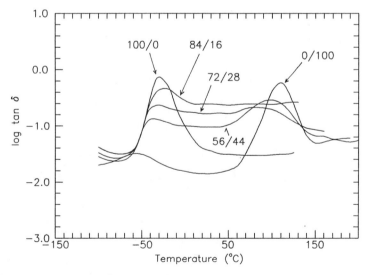

Figure 8. Composite log (loss factor) vs Temperature for Group 3 IPNs, PTMEG 2000(PU).

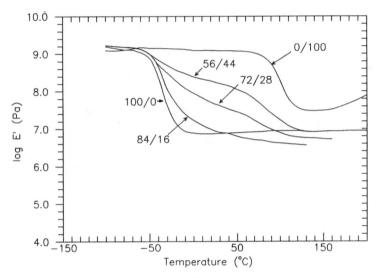

Figure 9. Composite log (E') vs Temperature for Group 3 IPNs, PTMEG 2000(PU).

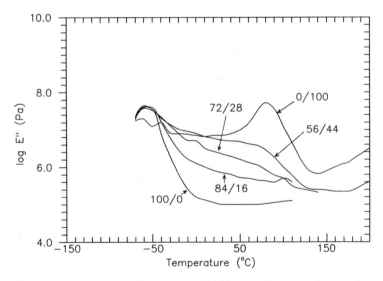

Figure 10. Composite log (E'') vs Temperature for Group 3 IPNs, PTMEG 2000(PU).

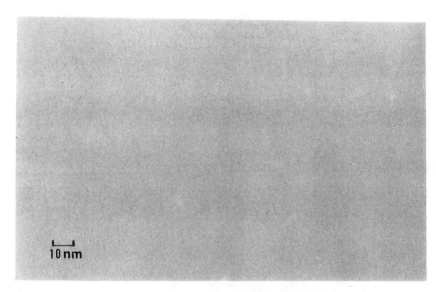

Figure 11. Electron Micrograph, 100% PU

Figure 12. Electron Micrograph, 100% EP

Figure 13. Electron Micrograph, 50/50 PU/EP

Figure 14. Electron Micrograph, 70/30 PU/EP

Figure 15. Electron Micrograph, 90/10 PU/EP

predominates, and the epoxy features are much more separated than in the other two samples. Five nm polyurethane features are also evident for this sample as shown in Figure 15.

CONCLUSIONS

Simultaneous PU/EP IPNs were synthesized and characterized by density measurements, dynamic thermal mechanical analysis, and transmission electron microscopy.
    Higher miscibility, increased density, and higher glass transition temperature for Group 1 materials was obtained using lower molecular weight polyol (mw 650) and producing shorter polymer segment lengths. Dynamic thermal mechanical analysis indicates significant broadening of glass transition temperature at or around PU/EP composition of 70/30. This is true for all three groups studied. At this composition, storage modulus showed much less steep variations with temperature during the transition from glassy to rubbery state.

ACKNOWLEDGMENTS

The authors wish to express their thanks to Dr. B. Howell and Mr. H. Telegadas, both of the David Taylor Research Center, for their technical assistance. Dr. B. Howell produced the electron micrographs.

REFERENCES

1. Sperling, L. H. Interpenetrating Polymer Networks and Related Materials, Plenum Press, New York, 1981.
2. Klempner, D.; Frisch, K. C.; Xiao, X. H.; Frisch, H. L. Two and Three Component Interpenetrating Polymer Networks, Polymer Engineering and Science, Vol. 25, No. 8, June 1985.
3. Chang,M.C.O.; Thomas, D.A.; Sperling L.H. J. Applied Polymer Sc., 34, pp.409-422, (1987).
4. Sperling, L.H.; Chin, T.W.; Giamlich, R.G; Thomas, D.A. Synthesis and Behavior of Prototype Silent Paint, Journal of Paint Technology, Vol. 46, No. 588, Jan. 1974.
5. Hourston, D.J.; R. Satgurunathan; H. Varma, J. Appl. Polymer Sc., 31, 1955 (1986).
6. Frisch, H.L.; Frisch, K.C.; Klempner, D. Polym. Eng. Sci., 14 (9), 646, 1974.
7. Frisch, H.L.; Frisch, K.C.; Klempner, D. Advances in Interpenetrating Polymer Networks, Pure & Appl. Chem., Vol.53, pp. 1557-1566, Printed in Great Britain.
8. Kim, S.C.; Klempner, D.; Frisch, K.C.; Radigan, W.; Frisch, H.L. Macromolecules, 9, 258 (1976).

RECEIVED January 24, 1990

Chapter 22

Synthesis of Composite Latex Particles and Their Role in Sound and Vibration Damping

D. A. Greenhill, D. J. Hourston, and J. A. Waters[1]

Polymer Centre, University of Lancaster, Lancaster, Lancashire, LA1 4YA, United Kingdom

A short review of the synthesis of composite latex particles is presented and their utility as damping materials addressed. The principal concern of this chapter is the control of morphology, and, hence, damping ability, in a series of semi-1 latex interpenetrating polymer networks. It was concluded that the relative hydrophilicities of the pair of polymers involved, the extent of crosslinking in the first-formed polymer and the state (glassy or rubbery) of this first polymer could be markedly influential in controlling the morphology of these composite latex particles.

A prolific variety of composite latex particles appears in both the open and patent literatures. The subject has been reviewed (1,2) by several authors. Composite implies the presence of at least two dissimilar components either of which could, in principle, constitute the major component by volume. Some features of composite particles, which retain colloidal stability during preparation and subsequent storage, that is where the product is a dispersion in which flocculation, aggregative, and coalescence processes are largely absent so long as the continuous phase remains, will be described here. There are alternative and important processes for preparing composite particles which give flocculated particles readily separated from the liquid diluent phase and dried for use as powder.

Particle morphology is partly determined by the method of particle preparation and by the ability or inability of the particles to

[1]Current address: Paints Division, ICI plc, Wexham Road, Slough, Berkshire, SL2 5DS, United Kingdom

undergo internal rearrangement to approach the thermodynamically preferred morphology.

The particles are heterogeneous by definition. As with single-polymer particles, heterogeneities in density can arise as a result of the mechanism of particle formation. Usually a polymer has a different density to the liquid monomer from which it is derived, and in most emulsion polymerisation processes to produce reasonably concentrated dispersions, propagation is dominated by arrival at the particle surface of oligomeric radicals which can lead to non-homogeneous shrinking within the particle.

Preparation by Sequential Polymerisation. Two-polymer composite latex particles may be prepared using either emulsion or dispersion polymerisation techniques. A dispersion (latex) of particles of a first polymer may be prepared in the usual manner; after complete conversion of monomer to polymer, a different monomer or monomer mixture is added and polymerised to provide the second polymer.

These processes must be monitored to confirm that the second polymer does not emerge as a separate particle type. Re-nucleation, producing a crop of new particles, may be detected by progressive determination of particle size and comparing actual with the theoretical size calculated on the basis of constant particle number. This is easier to do in those processes where the monomer for the second polymer is added slowly at a steady and known rate and samples can be taken at regular time intervals for particle-size determination by electron microscopy, photon correlation spectroscopy or by disc centrifuge photo-sediometry. For particles prepared by non-aqueous dispersion polymerisation using amphipathic graft copolymers as stabiliser, the conditions which promote re-nucleation have been elucidated (3). A major contributor is surplus stabiliser in the continuous phase. By careful control of the amount of stabiliser added, it has been possible to grow progressively larger particles from less than 0.1 micron to more than 3 microns without any re-nucleation (4) and without losing steric stabilisation of the growing particles, but this would be a slow and tedious process for preparing composite particles. In these systems re-nucleation was readily detected by electron microscopy because the new particles were of very fine size compared to the primary particles and usually with further growth the system became deficient in stabiliser and gave gross flocculation. With aqueous systems, using ionic initiators which furnish the polymer chains with ionic end-groups, renucleation is more difficult to control and detect because the new particles may be able to grow without losing colloidal stability. In the final dispersion the new particles may be indistinguishable from the primary particles by size characterisation alone.

Sequential polymerisation may be operated by producing particles of the first polymer by either emulsion or dispersion polymerisation and by adding the monomer for the second polymer together with free-radical initiator at a slow and controlled rate such that the rate of addition and the rate of polymerisation are equal. The amount of free monomer remains at a low level throughout the process. This 'monomer-starved' process (5) has been used with an aqueous continuous phase in many studies. Samples may be taken at regular intervals to monitor free monomer level and to check absence of re-nucleation. Usually the initiator and derived radicals are water-soluble, for example, potassium or ammonium persulphate are often used. The monomers have a finite, even if very low, solubility in water and oligomeric radicals of the second polymer are formed in the continuous phase and sweep-up of these at particle surfaces would be expected to be a dominant mechanism (6) with propagating polymers largely confined to the surface region. Because the diffusion of the propagating oligomer/polymer is vastly slower than that of the monomer molecules, the site of the propagating radical can only move into the particle as far as the new polymer chain grows. This restriction of growth to the particle surface is partly off-set by transfer of the radical to molecules which are able to diffuse into (and out of) the particles, especially transfer to monomer. The transfer rate to monomer varies: the transfer constant (7) of vinyl acetate is approximately ten times the value for methyl methacrylate. Transfer to molecules which are able to diffuse into the particles leads to some propagating sites in the interior of the particles. Ionic initiators give ionic radicals and ionic terminal groups on the polymer chains. This gives the polymer chains an element of surface activity because of the unfavourable energy change which accompanies the burying of solvated ionic groups within the particle and is a factor favouring retention of the polymer chains at the interface with water.

Composite particles which have been made using the monomer-starved process (8) include the poly(styrene-co-butadiene)-poly(ethyl acrylate-co-methacrylic acid) system.

In an alternative process (9), the monomer for the second polymer is added to a dispersion of particles of first polymer, allowed to swell the particles and is then polymerised. This has been called a 'monomer-flooded' process. It requires the monomer to be a good solvent for the first polymer. It is easy to identify systems which meet this criterion. The role of swelling appears to depend on the nature of the polymer and temperature and is unexpectedly slow (10) in some cases. Particle swelling can be promoted and controlled using selected agents as a precursor to forming large, monodisperse latex particles (11).

Non-Aqueous Processes.   Dispersions of composite particles in non-aqueous media (12) have been prepared. The particles were sterically stabilised to prevent flocculation and aggregation. This was achieved by physical absorption of amphipathic graft or block copolymer (13,14) or by covalent attachment of diluent-soluble oligomer or polymer chains (15) at the particle surface so that by definition different polymers were situated at the surface and in the bulk of the particles, even for 'single-polymer' particles. Composite particles were prepared by slow addition of the second monomer which was fully miscible with the diluent phase, obviating a monomer droplet phase; further monomer-soluble initiation and amphipathic graft stabiliser was included as appropriate so that the process comprised continued dispersion polymerisation with the monomer partitioned between particle and diluent phases and polymerisation largely confined to within the particles (16). This mechanism for particle growth should favour formation, initially, of inclusions of second polymer within the first rather than a core-shell morphology and examples of this have been reported (17), such as particles of polyethyl acrylate with inclusions of polymethyl methacrylate.

Processes Involving Precipitation of Polymer. Composite particles or micro-encapsulated particles may also be prepared by precipitating a (second) polymer onto the surface of pre-formed particles. For example, toner particles may be prepared using carbon black and styrene copolymers (18). The product required from these processes is a powder. Dispersions of colloidally stable composite particles are not usually sought or achieved.

Recently, a process has been described (19) in which two components are dissolved or dispersed in a common solvent and emulsified in water. Removal of the solvent leaves composite particles dispersed in water.

Importance of Interfacial Energies. The importance of interfacial energy (20-22) in determining the morphology of composite particles has been pointed to by a number of workers. Given two components in fixed proportions, a number of different particle morphologies may be envisaged for the composite particles. With the component volumes being the same in each case, differences in free energy arise predominantly from different interfacial areas and their associated interfacial energies.

For composite particles in water, the three extreme morphologies may be considered as core-shell, with the more hydrophilic component interfacing with water, inverted core-shell, with the more hydrophobic component interfacing with water, and separated particles.

Uses of Composite Particles. Composite particles are of substantial commercial importance because they provide new property balances for tougher plastics, films, and coatings. They are used to combine in a cost-effective manner functional groups or high performance components which are expensive with low-cost components and they provide opportunities to vary the optical properties of coatings. Micro-encapsulation interests are diverse, ranging from pharmaceuticals to dyes and toners and to magnetic particles.

Characterisation Techniques. Characterisation techniques are used to: (a) identify material on the particle surfaces, (b) confirm that particles comprise phase-separated material, (c) identify the locations of the different components, (d) check particle size against theoretical size, and (e) check that the components do not exist as separate particle types.

Scanning electron microscopy and transmission electron microscopy have been used extensively (23,24). Other methods in addition to dynamic mechanical analysis, include photon-correlation spectroscopy (25); differential scanning calorimetry (26); measurement of film-forming temperature (26); surfactant analysis (27); disc centrifuge photo-sediometry; and small-angle neutron (28) and light scattering (25).

Sound and Vibration Damping.

Much has already been written in other chapters in this book on the use of polymers in sound and vibration damping. Consequently, only a brief outline of this complex topic will be included here.

The principal characteristics of polymers which control their ability to dissipate energy are their stiffness, surface mass, and their inherent damping characteristics. In this chapter we are only concerned with the last of these three. Briefly, however, if lack of stiffness is a problem with a particular polymer in a damping system, this can be enhanced, if not completely obviated, by using the material in a constrained layer (29) mode. Surface mass refers to the mass of material behind a unit surface area. Clearly, for polymers which are all inherently low density materials, this can be increased by the incorporation of dense, particulate fillers such as lead, barytes etc.

To return to the principal topic, it is well-known that the region of greatest inherent damping is the glass transition region of amorphous polymers. The magnitude of this transition is a measure of the energy dissipated as heat when the material goes from its glassy to its rubbery state. Remember that polymers are viscoelastic materials (30), and, consequently, the position of the transition is frequency dependent. It follows, therefore, that what is desired for damping applications is a large, broad transition so that if the frequency of the sound or vibration to be damped changes substantially, the transition does not move out of range to either the low damping rubbery or glassy states. The frequency dependence of glass transitions is ca. $6\text{-}10^{\circ}C$ per decade of frequency.

A convenient index of damping peak width is the 'half-peak width' (the width at half-peak height). It is common for homopolymers to have relatively narrow Tg regions. The breadth of these transitions is controlled by the distribution of environments (31) available to the relaxing segments. For example, the Tg transition for a segmented polyether urethane (32) is relatively broad, having a half-peak width of $38^{\circ}C$, occasioned by the presence of both hard and soft segments which are partially phase separated yielding a much wider range of environments for the relaxing soft segments. This peak broadening in polyurethanes gives the clue that the desired transition breadth for polymers with potential damping applications would be better obtained by the use of polymer blends rather than with homopolymers and random copolymers.

What is necessary with a polymer blend in order to achieve the desired breadth of transition is partial miscibility. Complete immiscibility leads to two Tgs unshifted with respect to the Tgs of the components, and complete miscibility leads to the same relatively narrow transitions observed for homopolymers. Of course, with immiscible blends, it is possible to mix two or more polymers with relatively close Tgs and achieve broad damping transitions in that way. Hourston and Hughes (33) have reported broad transitions for polyether ester-polyvinyl chloride (PVC) blends where specific interactions occur between the ether oxygens and the chlorines in the PVC leading to partial miscibility.

Interpenetrating Polymer Networks.
      A sub-set of polymer blends is the group of materials known as interpenetrating polymer networks (IPN). An IPN may be defined as a material containing two polymers both in network form which has been prepared so that at least one of the networks has been synthesised and/or crosslinked in the intimate presence of the other. This form of chemical rather than mechanical blending of pairs of polymers has been reviewed by Sperling (34) who has pioneered this field. IPNs may be sub-divided into a number of categories (see ref. 34) including latex IPNs. Here a cross-linked latex polymer is formed by a conventional emulsion polymerisation route and then a second monomer and its crosslinker are added to the polymerisation vessel with new initiator, but no further surfactant, to allow the polymerisation of this second monomer to occur on or in the first-formed particles. Such latex IPNs are much more formable than any other class of IPN. They could be used conveniently as emulsions in paints or in grouts or mastics. Consequently, they are potentially interesting damping materials. Sperling (35) has reported on latex IPNs with very broad transition regions.

Experimental .
      The monomers, styrene, and methyl methacrylate (Aldrich) and ethyl methacrylate and n-butyl acrylate (Fluka) were purified prior to use. The crosslinking agent used in the synthesis of these semi-1-IPNs was tetraethyleneglycol dimethacrylate (Fluka). The emulsion polymerisation (90°C) was initiated with ammonium persulphate/sodium metabisulphite (both from BDH) and the resulting latex, which was prepared by a seed and feed procedure, was stabilised with sodium dioctyl sulfosuccinate (OT75 from Cyanamid).
      The excess heats of mixing experiments were performed using a modified McGlashan calorimeter. All the dynamic mechanical analyses were conducted using a Polymer Laboratories instrument at a frequency of 10 Hz and a heating rate of 2°C per minute.

Results and Discussion.
      Three factors, listed below, which might influence the morphology of composite latex particles will now be discussed.
      (a)   The degree of inherent miscibility of the polymer pair.
      (b)   The extent of crosslinking in the first-formed polymer.
      (c)   The state (glassy or rubbery) of the first-formed polymer when the second monomer is polymerised.

      The monomers chosen for these investigations are shown in Table I. They were selected so that a range of properties changed depending on the pair selected for the synthesis of these semi-1-IPNs (34). Factors of particular interest were the water solubility of the monomers, which could influence the locus of polymerisation and the relative hydrophilicities of the homopolymers as indexed by critical surface tension. In a semi-1-IPN the first-formed polymer is a network, but the second-formed polymer is linear.
      These four monomers allow the synthesis of the six pair-wise combinations shown in Table II. Of course, the order of synthesis can

be changed to give the so-called inverse semi-1-IPNs, but no comment will be made here about these systems.

Table I. Monomer Water Solubility and Polymer Critical Surface Tensions.

Monomer	Water solubility at $20^{\circ}C$	Critical surface tension (γ_c)[1] (mN/m)
styrene (S)	0.029	33
n-butyl acrylate (nBA)	0.30	28
ethyl methacrylate (EMA)	0.77	33
methyl methacrylate (MMA)	1.59	33-44

1. Values taken from Polymer Handbook. (Ref. 7).

Table II. The Systems Investigated

Polymer 1	Polymer 2
PS	PnBA
PS	PEMA
PEMA	PnBA
PMMA	PEMA
PMMA	PnBA
PS	PMMA

Extent of Miscibility. The extent to which a pair of polymers in a blend are mutually miscible varies from the situation of complete miscibility, as is the case for polystyrene with poly(vinyl methyl ether) (36), to the situation of complete immiscibility, which is by far the most common occurrence. However, there is a range of extent of miscibility which may in mechanical blending be influenced by such factors as temperature, shear rate, time of mixing and the molecular weights.

The ability to predict the magnitude of the thermodynamic driving force towards miscibility or immiscibility shown by a given pair of polymers is an active area of research in polymer science. No approach yet combines complete reliability with ease of use via the ready availability of the fundamental data required. The work to be reported here relies on the application of two methods of prediction. The first of these is the method developed by Krause (37) in which the values of the solubility parameters, δ , of the homopolymers are necessary. It is recognised that methods based on solubility parameters are only strictly applicable to non-polar materials. However, it was thought worthwhile to apply the method to these systems whose dielectric constants are at least relatively low (38).

The solubility parameters of the homopolymers may be calculated using

$$\delta = \frac{\ell}{M} \ell \Sigma F_i \qquad (1)$$

where ℓ is the density of the homopolymer and ΣF_i is the sum of the molar attraction constants for all the chemical groups in the repeat unit. M is the repeat unit molecular weight.

The Krause (37,39) method involves the comparison of the calculated values of χ_{12}, the interaction parameter between the two polymers and ($\chi_{12})_{cr}$, the interaction parameter at the critical point on a phase diagram for that particular binary system. Krause (37,39) stated that if $\chi_{12} > (\chi_{12})_{cr}$, the two polymers should be immiscible at some compositions. The greater the difference between these two values, the smaller will be the range of compositions over which the polymers will be compatible.

The χ_{12} and the $(\chi_{12})_{cr}$ values are calculated from equations (2) and (3), respectively.

$$\chi_{12} = \frac{Vr}{RT} (\delta_1 - \delta_2)^2 \qquad (2)$$

V_r is the reference volume which was taken to be the molar volume of the smaller polymer repeat unit. The temperature was taken to be 298 K.

$$(\chi_{12})_{cr} = \frac{1}{2} (\frac{1}{n_1^{\frac{1}{2}}} + \frac{1}{n_2^{\frac{1}{2}}})^2 \qquad (3)$$

n_1 and n_2 are the respective degrees of polymerisation.

In this study the molecular weight of the first component in the polymer system, M_1, was taken to be equal to 100,000 g/mol. The molecular weight of the second component M_2 was varied up to 100,000 g/mol and the corresponding ($\chi_{12})_{cr}$ values were calculated for comparison with χ_{12}. Krause (37,39) states that this scheme can only serve as a guide to polymer-polymer miscibility and that it is no substitute for experimentation.

By this method, the systems were ranked in the order given in Table II with the PS-PnBA combination being the most miscible and the PS-PMMA material being the least miscible.

In addition to these theoretical methods, direct measurement of the excess heats of mixing were determined for the systems using the hydrogenated monomers as model compounds. As the free energy of mixing is controlled by both the heat of mixing (ΔH_m) and the entropy of mixing, and given that the latter is very small for polymers of high molecular weight, the sign and magnitude of ΔH_m controls the free energy of mixing which must be negative for a system to be miscible. The results of this exercise are shown in Figure 1. Here it can be seen that the systems are ranked precisely in the same order as by the Krause method.

Figure 2 shows the tan δ versus temperature plot for the PS-PnBA semi-1-IPN. Clearly, the presence of two Tgs and the low inter-transition value of tan δ indicate, despite the predictions, that this is a quite immiscible combination, at least when combined in the already described manner as composite latex particles. The PnBA and the PS Tgs are, in fact, shifted towards each other by 2^O and 8^OC, respectively, indicating that some mixing of segments has occurred. In Figure 3, the plot of log E' versus temperature, the modulus drop associated with the PnBA Tg is much less than would be expected for this 1:1 composition. This can be explained if the PnBA is not present as a continuous phase. In other words, if the PnBA, the second-formed polymer, were on the outside of the latex particles as a shell layer, a larger modulus drop would be expected. It may, therefore, be postulated that in the as-formed latex particles the PnBA is present as discrete phases within the first-formed latex particles. As the PnBA is significantly more hydrophobic than PS (see Table I), this is believed to be the principal driving force (40) for the proposed particle morphology.

The tan δ versus temperature data for the PMMA-PnBA material, Figure 4, show evidence of a somewhat higher degree of mixing than was evident for the PS-PnBA semi-1-IPN. The total peak shift was again 10^OC, but the level of damping in the inter-transition region is increased and there is a transition at around 40^OC ascribable to a mixed phase. Again, the same morphology may be proposed as the PnBA transition is still suppressed.

For the PEMA-PnBA combination the tan δ versus temperature plot (Figure 5) shows the greatest extent of mixing with the PnBA Tg shifting by 5° and the PEMA Tg by 22^OC. There is also considerable damping in the inter-transition region, but, here again, the evidence is that the PnBA component is not present as a shell on the original latex particles, but is dispersed internally. The preceding body of dynamic mechanical analysis evidence is sufficient to suggest that the inherent miscibility ranking as predicted by the Krause and heats of mixing methods does not dominate the choice of morphology adopted by the particles. Indeed, it appears that the relative hydrophilicities of the pair of polymers chosen can be much more influential. Indeed, the importance of this factor is shown in Figure 6. Here the tan δ versus temperature curves are shown for the PEMA-PnBA semi-1-IPNs synthesised in water as the continuous phase and again in a water-ethanol mixture. When the polarity of the continuous phase in the polymerisation system is reduced, the results indicate a decreased degree of mixing.

Extent of Crosslinking.   Figure 7 shows plots of log E' versus temperature for three PEMA-PnBA semi-1-IPNs in which the extent of crosslinking of the first-formed polymer (PEMA) is altered from the uncrosslinked state to the situation where 5 mole% of crosslinking agent (tetraethyleneglycol dimethacrylate) has been incorporated. The

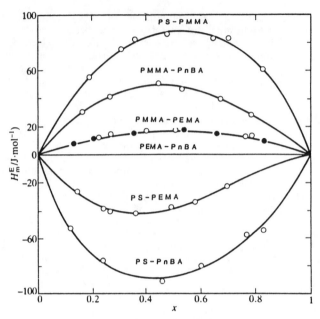

Fig. 1.   Excess heats of mixing versus composition for the
               six model   systems.

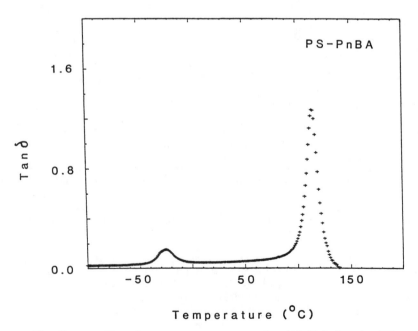

Fig. 2.   Tan δ versus temperature plot (10 Hz) for the PS-
               PnBA semi-1-IPN.

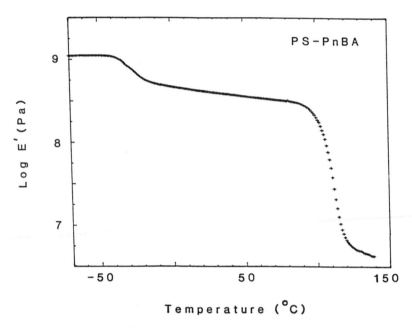

Fig. 3. Log E' versus temperature plot (10 Hz) for the PS-PnBA semi-1-IPN.

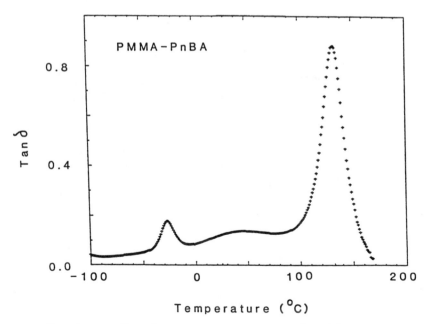

Fig. 4. Tan δ versus temperature plot (10 Hz) for the PMMA-PnBA semi-1-IPN.

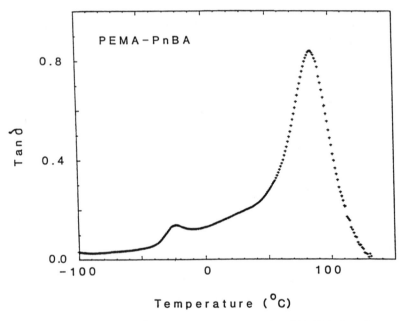

Fig. 5.   Tan δ versus temperature plot (10 Hz) for the
PEMA-PnBA semi-1-IPN.

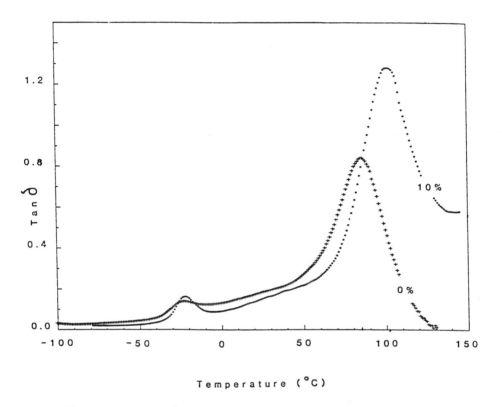

Fig. 6.   Tan δ versus temperature plots (10 Hz) for the
               PEMA-PnBA semi-1-IPN synthesised in water (0%)
               and in an ethanol water mixture (10%).

effect is both to shift and broaden the rubber Tg indicating an
enhancement of enforced mixing as a consequence of network
tightening. This observation for latex semi-IPNs is entirely in line with
what has been observed (41) for the generality of IPNs, where the
conclusion is that tightening the first-formed network can have a
significant effect on the extent of mixing, whilst the influence of
tighter second-formed networks is markedly less.

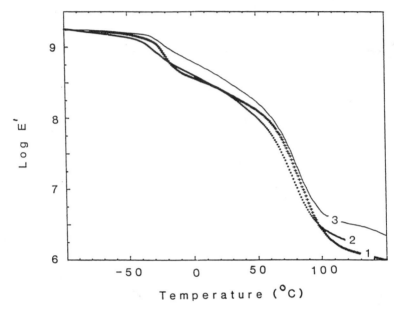

Fig. 7.   Log E' versus temperature plots (10 Hz) for the
               PEMA-PnBA semi-1-IPN system in which the PEMA
               contains zero (1), three (2) and five (3) mole
               percent of crosslinker.

Polymerisation on Glassy Core Particles.   Another way of influencing
particle morphology is to ensure that the first-formed latex is either
below its Tg when the second monomer is added and/or polymerised.
Figure 8 shows some of the tan δ-temperature data from a series of
experiments on a full IPN based on polyethyl acrylate (PEA) and PEMA
in equal parts by weight.   In these experiments the first network latex
was formed and then the vessel and contents were cooled to 0°C and
the second monomer added.   The polymerisation vessel was then held at
this temperature for periods of time ranging from zero (i.e. not cooled)
to 240 hours after which the temperature was raised to 60°C, initiator
added and the monomer polymerised.   At 0°C PEA is close to its Tg.
Both cases show by shifts in Tg and the high inter-transition damping
evidence of some degree of miscibility, but clearly the material allowed
240 hours to swell the glassy core particles is the more intimately
mixed.   It should be added that there was relatively little change in
properties (42) after about 100 hours.
        A similar experiment has been performed via another route.
Here a semi-1-IPN based on PEMA and PnBA was synthesised by both
use of a thermal initiator (ammonium persulphate) at 90°C and via a
redox initiator (ammonium persulphate-sodium metabisulphate) at 20°C,
which is well below the Tg of PEMA.   The data are shown in Figure 9.

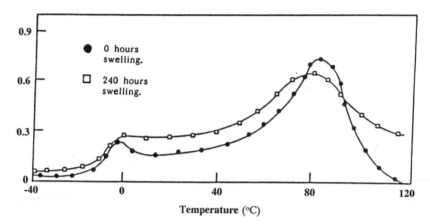

0.9

● 0 hours swelling.

□ 240 hours swelling.

0.6

0.3

0

-40   0   40   80   120

Temperature (°C)

Fig. 8.   Tan δ versus temperature plots (10 Hz) for the PEA-PEMA IPNs after zero (●) and 240 hours (□) of swelling.

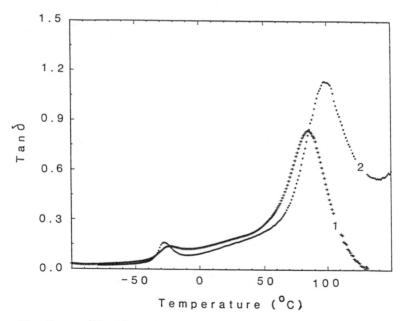

Fig. 9.   Tan δ versus temperature plots (10 Hz) for PEMA-PnBA semi-1-IPNs prepared using a thermal initiator (1) and a redox initiator (2) during the second-stage polymerisation.

Clearly, the material made by the redox-initiated second-stage polymerisation shows a lower degree of mixing as indexed by transition shifts and by inter-transitional values of tan δ. Thus, the state of the first-formed latex represents another way of influencing particle morphology.

Conclusions.
    In conclusion, it has been shown that the predicted order of miscibility in composite latex particle systems is not necessarily bourne out when the extent of miscibility is guaged by dynamic mechanical analysis, and, very recently, by the same authors using solid-state NMR spectroscopy. Control over particle morphology, and, hence, over damping behaviour can be exercised by the differences in hydrophilicity between the polymer pair in question, by the degree of crosslinking in the first network and by whether or not the first-formed polymer is above or below its Tg when the second monomer is polymerised.

Literature Cited.

1.   Eliseeva, V.I. *Prog. Org. Coat.* 1985, 13, 195.

2.   Daniel, J.C. *Makromol. Chem. Suppl.* 1985, 10/10, 359.

3.   Barrett, K.E.J.; Thomas, H.R. In *Dispersion Polymerisation in Organic Media;* J. Wiley and Sons: New York, 1975.

4.   Walbridge, D.J.; and Waters, J.A. *Discuss. Faraday Soc.* 1966, 42, 294.

5.   Sperling, L.H.; Chiu, T.W.; Hartman C.P.; Thomas, D.A. *Int. J. Polym. Mat.* 1972, 1, 331.

6.   Fitch, R.M. *J. Paint Technol.* 1965, 37 32.

7.   Brandrup, J.; Immergut, E. *Polymer Handbook;* J. Wiley and Sons: New York, 1975.

8.   Lee, D.I.; Ishikawa, T. *J. Polym. Sci.* 1983, 21, 147.

9.   Keusch, P.; Prince, J.; Williams, D.J. *J. Macromol. Sci.* 1973, A7, 623.

10.   Chung-li, Y.; Goodwin, J.W.; Ottewill, R.H. *Prog. Coll. Polym. Sci.* 1976, 60, 163.

11.   Ugelstad, J.; Kaggerud, K.H.; Hansen, F.K. Berge, A. *Macromol. Chem.* 1979, 180, 737.

12.   Backhouse, A.J. *J. Coat. Technol.* 1982, 54, 83.

13.   Osmond, D.W.J.; Thomson, H.H. British Patent 893 429, 1962.

14.   Waite, F.A. *J. Oil Col. Chem. Assoc.* 1971, 54, 342.

15.   Walbridge, D.J. In *Dispersion Polymerisation in Organic Media;* J. Wiley and Sons: New York, 1975.

16.   Barrett, K.E.J.; Thomas, H.R. In *Dispersion Polymerisation in Organic Media*; J. Wiley and Sons: New York, 1975.

17.   Thompson, M.W. In *Dispersion Polymerisation in Organic Media*; J. Wiley and Sons: New York, 1975.

18.   Wellman, R.E.; Brown, R.W. U.S. Patent 4 016 099, 1984.

19.   Sundberg, D. Private Communication.

20.   Berg, J.; Sundberg, D.; Kronberg, B. *Polym. Mat. Sci. Eng. ACS Sym. Series*. 1986, 54.

21.   Dimonie, V.L.; Chen, Y.C.; El Aasser, M.S.; Vanderhoff, J.W. *Dieuxieme Colloque Internationale sur les Copolymerisations et les Copolymers en Milieu Disperse.* Lyons, 1989.

22.   Waters, J.A. European Patent Application 89.3000 93.5.

23.   Bradford, E.B.; Morford, L.E. *Colloid Int. Sci., Proc. Int. Conf.* 1976, 4, 183.

24.   Narkis, M.; Talmon, Y.; Silverstein, M. *Polymer* 1985, 26, 1359.

25.   Ford, J.R.; Rowell, R.L.; Bassett, D.R. In *Emulsion Polymers and Emulsion Polymerisation. ACS Symp. Ser.* 1981, 165, 279.

26.   Erickson, J.R.; Seidewand, R.T. In *Emulsion Polymers and Emulsion Polymerisation. ACS Symp. Ser.* 1981, 165, 483.

27.   Okubo, M.; Yamada, A.; Matsumoto, T. *J. Polym. Sci.* 1980, 18, 3219.

28.   Bootle, G.A.; Lye, J.E.; Ottewill, R.H. In Press.

29.   Ball, G.L.; Salyer, I. *J. Acoust. Soc.* 1966, 39, 663.

30.   Aklonis, J.H.; MacKnight, W.J.; Shen. M. *Introduction to Polymer Viscoelasticity*; Wiley-Interscience: New York, 1972.

31.   Ferry, J.D. *Viscoelastic Properties of Polymers*; John Wiley and Sons: New York, 1961.

32.   Hourston, D.J.; Zia, Y. *J. Appl. Polym. Sci.* 1984, 29, 629.

33.   Hourston, D.J.; Hughes, I.D. *J. Appl. Polym. Sci.* 1981, 26, 3487.

34.   Sperling, L.H. *Interpenetrating Polymer Networks and Related Materials*; Plenum: New York, 1981.

35.   Sperling, L.H.; Chiu, T.W.; Gramlich, R.G.; Thomas, D.A. *J. Paint Technol.* 1974, 46, 47.

36.   MacMaster, L.P. *Macromolecules* 1973, 6, 760.

37. Krause, S.J. In *Polymer Blends*; Paul, D.R.; Newman, S., Eds.; Academic Press: New York, 1978.

38. Brandrup, J.; Immergut, E.   *Polymer Handbook*; Wiley-Interscience: New York, 1975.

39. Krause, S.J. *J. Macromol. Sci. Rev. Macromol. Chem.* 1972, 2, 251.

40. Lee, D.I.; Ishikawa, T. *J. Polym. Sci. Polym. Chem.* 1983, 21, 147.

41. Hourston, D.J.; McCluskey, J.A.; *J. Appl. Polym. Sci.* 1985, 30, 2157.

42. Hourston, D.J.; Satgurunathan, R.; Varma, H.C. *J. Appl. Polym. Sci.* 1987, 34, 901.

RECEIVED January 24, 1990

Chapter 23

Cross-Poly(vinyl methyl ether)-*Inter-Cross*-Polystyrene Interpenetrating Polymer Networks

Loss Modulus and Damping Behavior

J. J. Fay[1], C. J. Murphy[2], D. A. Thomas[3], and L. H. Sperling[3,4]

[1]Department of Chemistry; [3]Department of Material Science and Engineering; and [4]Department of Chemical Engineering, Lehigh University, Bethlehem, PA 18015
[2]Department of Chemistry, East Stroudsburg University, East Stroudsburg, PA 18301

Cross-poly(vinyl methyl ether)-inter-cross-polystyrene, PVME/PS, IPNs and their corresponding blends were synthesized and characterized by dynamic mechanical spectroscopy, differential scanning calorimetry, and other methods. Midrange compositions of the IPNs were observed to be phase separated at room temperature. The microheterogeneous phase separation in the IPNs appears to be dependent on composition and the degree of crosslinking. DSC shows that the IPNs have broader glass transitions than the corresponding miscible polymer blends, which suggests decreased miscibility in the IPNs. The development of a group contribution anaylsis to predict the integral of the linear loss modulus-temperature curves, the loss area, LA, is reviewed and utilized to compare with the LA's obtained for the PVME/PS IPNs and blends. The LA's for the IPNs were found to be equal to the blends, independent of the exact extent of phase separation, and following the group contribution analysis to a good approximation.

Polymers are useful sound and vibration damping materials near their glass transition temperature, T_g ([1],[2]). The onset of coordinated chain molecular motion in the T_g region permits the conversion of vibrational energy into heat, thereby reducing transmitted noise and vibration. Homopolymers and random copolymers constitute useful damping materials over a 20-30°C temperature range near

0097–6156/90/0424–0415$06.00/0
© 1990 American Chemical Society

T_g, corresponding to about three decades of frequency. Coincidentally, the acoustic spectrum spans a frequency range of 20 to 20,000 Hz, which is also three decades of frequency.

Immiscible polymer blends and grafts, which are phase separated, exhibit glass transitions corresponding to the transitions of the individual components and do not damp effectively outside these regions (3-5). With partially miscible systems, increased mixing of the components broadens the damping range. Comparatively, interpenetrating polymer networks, IPNs, which are defined as a combination of two polymers in network form, have been shown to exhibit damping over much broader ranges of temperature and frequency due to microheterogeneous morphologies (6,7). Figure 1 illustrates typical modulus and tan δ versus temperature curves for polymer blends with different degrees of miscibility. With a microheterogeneous morphology (case 3), in which exists an infinite number of phases of varying composition, damping may be obtained over broad temperature ranges. Since automobile, aircraft and machinery applications may require effective damping over broad frequency and temperature ranges, IPNs comprise a class of materials that can be utilized to obtain effective damping in these and other applications. Several reviews of the IPN literature have recently been published (8-9).

Damping Research at Lehigh

Sound and vibration damping research with IPNs began in the early 1970's and resulted in the formation of a constrained layer damping system with the inner damping layer a latex IPN paint (10). The constrained layer system results in a shearing effect within the IPN layer along with flexural and extensional motions as the composite panel vibrates. The added shear mechanism, not present in extensional applications, increases the amount of energy that is dissipated in each vibrational cycle.

Sound and vibration damping research at Lehigh University started with an accidental discovery, that certain IPN compositions exhibited substantially constant tan δ values over very broad temperature ranges. This behavior was first observed with IPNs based on poly(ethyl acrylate) and poly(methyl methacrylate), see Figure 2 (11). These two polymers are chemically isomeric, and hence the heat of mixing must be close to zero. It is not surprising, then, that in modern terminology these compositions exhibited a microheterogeneous morphology, and that they must lie close to the phase boundary between one and two phases.

Shortly thereafter, it was discovered that when the swelling of monomer mix II into polymer network I was incomplete, a sandwich structure resulted with the outside being stiff, and the interior being soft and flexible.

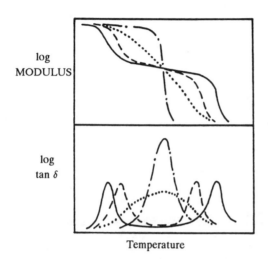

Figure 1.   Generalized mechanical loss (tan δ) and modulus behavior for different types of polymer blends. Case 1 (dashed-dotted line), miscible; case 2 (dashed line), limited miscibility; case 3 (dotted line), microheterogeneous; case 4 (solid line), heterogeneous. (Reproduced with permission from Ref. 5. Copyright 1979 Academic Press.)

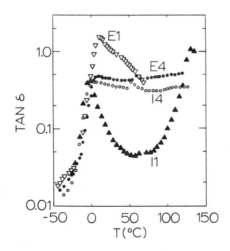

Figure 2.   Temperature dependence of tan δ for four IPN's: E1, 74 PEA/26 PS; E4, 72 PEA/28 PMMA; I1, 25 PS/75 PEA; I4, 27 PMMA/73 PEA.   The components of the IPN's were synthesized in the order as listed. (Reproduced from Ref. 11.   Copyright 1972 American Chemical Society.)

It was recognized at that time that this had the same
overall structure as a constrained layer system, leading
to a US Patent (12).

Development of "Silent Paint". A more elegant method of
applying such materials to vibrating substrates was through
the use of latex IPN materials, which were formulated in
a manner similar to ordinary house paint, that could be
brushed or sprayed onto a substrate. After drying, a
stiff, glassy polymer could be added, in a prepolymer form,
and polymerized *in situ*. Epoxy materials served this
latter purpose in an excellent manner. The advantages of
the new material included (a) a wide temperature and
frequency range of damping, (b) ease of application,
especially to curved or irregular surfaces, and (c)
relatively low price.
    The "Silent Paint" program studied latex-based IPNs
constrained by a stiff epoxy based layer for applications
in acoustic damping. The latex, based on acrylic/
methacrylic polymers exhibiting a microheterogeneous
morphology, was then thickened and several minor
ingredients were added to impart antioxidant and mildew
resistant properties. After coating the latex component
onto the substrate, the filled epoxy constraining layer
was overcoated to complete the system.

Damping Behavior of "Silent Paint". The damping behavior
of a typical "Silent Paint" composition, cross-poly(ethyl
methacrylate)-inter-cross-poly(n-butyl acrylate) 25/75,
constrained with epoxy resin, was then compared with
several commercial materials as a function of temperature,
Figure 3. The "Silent Paint" composition, curve A, damped
effectively over a broad temperature range whereas the
commercial materials, curves B and C [poly(vinyl acetate)],
and D, E, and F, of various compositions, damped
effectively over narrower temperature ranges.
    It is apparent from Figure 3 that the "Silent Paint"
formulation enables useful damping over a temperature
range from -20 to +50°C, as evidenced by a nearly constant
percent critical damping. Transitions in the epoxy
material immediately above the latex IPNs transitions
increase the effective damping range to to +90°C.

Group Contribution Analysis. The area under the linear
loss modulus vs. temperature curves was developed as a
quantitative measure of the damping behavior of
multicomponent polymer materials by Fradkin et al (13,14).
It was observed that the loss area, LA, the integral of
the linear loss modulus versus temperature curve, was
characteristic of the polymer chain and not affected by
decrosslinking or annealing. As a consequence, subsequent
research at Lehigh University centered about the manner
in which the polymer molecules actually absorb mechanical
energy in vibrating systems. The absorption of mechanical

energy by the chemical moieties in a vibrating system was considered analogous to the absorption of infrared energy in IR spectroscopy. The area under the absorption curve could then be related to the concentration of the functional groups in the material.

The basic theory for the group contribution analysis is predicated on the assumption that the structural groups within the repeat unit contribute to the total loss area on an additive weight fraction basis (15). In a quantitative manner (15):

$$LA = \sum_{i=1}^{n} \frac{(LA)_i M_i}{M} = \sum_{i=1}^{n} \frac{G_i}{M} \qquad (1)$$

where M_i is the molecular weight of the ith group in the repeat unit, M is the molecular weight of the whole mer, G_i is the molar loss constant for the ith group, $(LA)_i$ is the loss area contributed by the ith group and n represents the number of moieties in the mer. As a corollary to eq. (1), an additive mixing rule can be written for use with copolymers and multicomponent polymer materials,

$$LA = w_I (LA)_I + w_{II}(LA)_{II} + \cdots = \sum_{i=1}^{n} w_i (LA)_i \qquad (2)$$

where w_I, $w_I I$, ... represent the weight fractions of the components in the material.

A large number of homopolymers and statistical copolymers were evaluated in formulating the group contribution analysis. In a formal way there were n simultaneous equations to be solved, and n+1 unknowns. The n+1st unknown was the vinyl polymer backbone, see Table I, which was varied empirically until all the other values fell into place in a reliable, predictable manner. Table I summarizes the loss areas, $(LA)_i$, and molar loss constants, G_i, determined for each moiety (16). Application of the group contribution theory enables a predictive method for determining the loss area of a polymer or multicomponent polymer material based upon its chemical composition. Figure 4 illustrates the correlation between the calculated values of the loss area using the group contribution analysis and the values obtained experimentally for a series of nine statistical copolymers and 13 IPNs, all compositions based on mixes from Table I. The results are a better than expected fit, showing that the primary variables are the mer moieties. However, phase structure, continuity and inversion variables were not addressed.

While the LA is primarily a molecularly based quantity for pure materials, graphite filled IPNs and many other composite structures have been shown to exhibit greater LA's than the neat material (17). Morphological effects on the loss area are not considered by the group

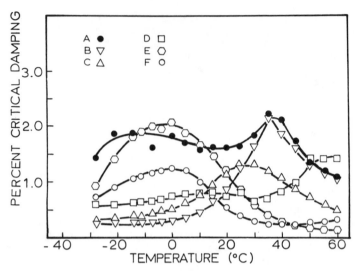

Figure 3.   The damping capability of "Silent Paint," composition A, and several commercial materials available in the early 1970's, B-F.   Composition B, based on poly(vinyl acetate), damps well at room temperature and slightly above, but fails at 0°C and below. (Reproduced with permission from Ref. 2. Copyright 1975 John Wiley and Sons.)

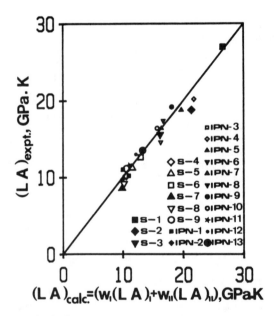

Figure 4.   Additivity of Homopolymer LA's corresponding to LA's of IPN's and statistical copolymers. (Reproduced from Ref. 16. Copyright 1988 John Wiley and Sons.)

Table I. Summary of Group Contributions to LA

Row	Group	Group Location[a]	$(LA)_i$ (GPaK)	G_i (GPaK g/mol)
1	H H | | —C—C— | | H	1	3.4	91.8
2	—O—	1	19.1	305.8*
3	O || —C—O—( )—H	2	20.8	936
4	O || —O—C—( )—H	2	20.1	905
5	O || —C—OH	2	20.8	936
6	—⟨benzene⟩	2	11.9	916
7	—CH_3	2	11.0	165
8	—OCH_3	2	21.7	674
9	—C≡N	2	23.2	603
10	—O—H	2	4.7	80
11	—Cl	2	9.2	327
12	—⟨cyclohexane⟩	3	3.5	287
13	—⟨benzene⟩	3	2.2	166
14	CH_3 | —CH_2—CH—CH_3	3	-1.7	-98
15	>CH—	3,4	0.5	7
16	—CH_2—	3	-3.0	-42
17	—C≡N	3	14.5	377
18	—Cl	3	15.7	556

[a]1: Backbone; 2: side group attached to backbone directly; 3: side group not attached to backbone; 4: value derived from isobutyl side group.

*T. Hur, J. A. Manson and L. H. Sperling, to be published.

SOURCE: Reference 16. Copyright 1988 American Chemical Society.

contribution approach.   Decrosslinking and annealing
effects on IPNs were studied by Fradkin et al (14) and
were found to affect the shape of the loss-modulus-
temperature curve, however, the area remained constant.
At present the contribution of morphology to the damping
behavior of a material is unknown. Similarly, there is
little experimental evidence as to whether or not an IPN,
based on a miscible polymer blend, is more or less miscible
than the corresponding blend. These questions form the
basis of current research at Lehigh University.

Current Research

The only single phase, miscible IPNs reported are
homopolymer IPNs, in which both networks are composed of
the same polymer, and IPNs based on poly(2,6-
dimethylphenylene oxide) (PPO) and polystyrene (PS) (18).
The corresponding blend of the latter system is miscible
and does not undergo thermally induced phase separation
below its degradation temperature (19).
     Miscible blends of poly(vinyl methyl ether) and
polystyrene exhibit phase separation at temperatures above
100°C as a result of a lower critical solution temperature
and have a well defined phase diagram (20). This system
has become a model blend for studying thermodynamics of
mixing, and phase separation kinetics and resultant
morphologies obtained by nucleation and growth and spinodal
decomposition mechanisms. As a result of its accessible
lower critical solution temperature, the PVME/PS system was
selected to examine the effects of phase separation and
morphology on the damping behavior of the blends and IPNs.

Experimental

Synthesis.   Styrene monomer, Polysciences Inc., and
divinylbenzene, DVB, Scientific Polymer Products, were
cleaned by a column chromatographic technique using neutral
alumina.   Poly(vinyl methyl ether), Scientific Polymer
Products, M_w = 1.3x10^5 g/mole, was supplied as a 50 wt%
toluene solution.   Dicumyl peroxide, Pfaltz and Bauer,
benzoin, Aldrich Chemical, and toluene, Fisher Chemical,
were used as received.
     Crosslinked PVME sheets were prepared in the following
manner. Dicumyl peroxide was added to the PVME by a
solution blending technique, in concentrations ranging from
1% to 10% After mixing, the solution was placed in a
Teflon mold and the toluene was removed under vacuum at
100°C. The resulting 2 mm thick sheet was then cured at
160°C for 60 mins. in a nitrogen atmosphere.
     Sequential IPNs were prepared from the crosslinked
PVME sheets by swelling in a mixture of styrene, DVB and
benzoin. A typical recipe was 100g styrene, 2.26g DVB and
0.41g benzoin. After swelling to equilibrium, the swollen
mass was sandwiched between Mylar film and glass plates,

the mold sealed, and photopolymerized in a UV box for 72 hours. Semi-I IPNs in which only the first component, PVME, is crosslinked were prepared in a similar fashion except DVB was omitted. Similarly, Semi-II IPNs, in which only the second component is crosslinked, were prepared by dissolving PVME in styrene, DVB and initiator. A chemical blend is made by dissolving PVME in styrene monomer and then polymerizing the latter. Physical blends were produced by synthesizing PS separately and then solution blending with PVME in toluene, followed by drying under vacuum.

Instrumental. Samples were characterized with a Mettler TA3000 DSC30 differential scanning calorimeter, DSC, to obtain heat capacity thermograms. An Autovibron Dynamic Viscoelastometer, Rheovibron DDV-III-C, coupled with a computer and plotter, assembled by Imass Inc., was used to obtain storage modulus, E', loss modulus, E'', and the loss tangent, tan δ, versus temperature curves at a frequency of 110 Hz and a heating rate of 1°C/min. The classical logarithmic loss modulus vs. temperature curves were converted to linear plots by computer programming. Cloud points were determined using a Zeiss optical microscope equipped with a LinKam microprocessor controlled hotstage at a heating rate of 2°C/min.

Results

The PVME/PS semi-I, semi-II and sequential IPNs are slightly hazy or turbid, indicating phase separation, whereas the corresponding blends are clear and miscible.

Optical Microscopy. The cloud point phase diagram determined for the PVME/PS blends at varying compositions is shown in Figure 5. The results are in accord with previously determined phase diagrams for the PVME/PS system, excepting molecular weight differences. Microscopic phase separation was observed with generally coarser morphologies being obtained at increased PVME content. With annealing at temperatures below the LCST and above the glass transition of the blend, phase separation was observed to be reversible and clear films were obtained.

Differential Scanning Calorimetry. Thermal analysis of PVME/PS blends and IPNs by DSC indicates only one glass transition temperature, which is located between the homopolymer T_g's. The position of T_g is dependent upon composition although it does not follow the prediction of the Fox equation (21), Table II. The breadth of the transition also increases significantly with increased PS content for the blends and IPNs. The broad transition might result from either the clustering of like mers near

the lower critical solution temperature or from a downward
shift of the lower critical solution temperature. Figure
6 compares the the glass transition region of a physical
blend, cast from toluene, and an IPN of the same
composition. Both exhibit broad transitions, however, the
IPN transition is significantly than that of the blend.

Table II. Characterization of PVME/PS IPNs and
Homopolymers

% PVME	Loss Area[a], GPa·K		Loss Peak °C	Tg, °C		
	exper.	calc.[b]		DSC	Fox Eqn.	
100	8.9	13.2	-15	-24	-15
70	10.8	12.1	-4	-16	14
50	12.4	11.5	16	-8	38
30	11.6	10.7	(20) 67	20	65
0	9.4	9.7	117	93	117

[a]Crosslinkers: 7.5 wt% dicumyl peroxide (PVME),
1 mol % DVB (PS). M_c (g/mol): 18,000 (PVME), 7,700 (PS)
[b]based on E'= 2.85×10^{10} dyn/cm^2

Dynamic Mechanical Spectroscopy. Dynamic mechanical
spectra of a series of PVME/PS IPNs are shown in Figure
7. Both homopolymers show narrow, well defined loss peaks.
The 70/30 PVME/PS IPN loss modulus curve is slightly
broader than the PVME homopolymer peak and may indicate
clustering. However, there is a dramatic broadening of the
loss curve with increasing PS weight fraction as shown for
the 50/50 and 70/30 IPNs. The loss modulus peak extends
over a range of 100°C and in the 30/70 sample there is an
observable shoulder on the low temperature end of the peak.
These last results are best explained by phase separation.
Recently, Bauer et al. (22) found that phase separation
occured as the crosslink density of the polystyrene network
was increased for semi-II IPNs. The IPNs synthesized at
Lehigh University have PS crosslinking levels corresponding
to those which resulted in phase separation of the
semi-IPNs in the work of Bauer et al. (22). Similarly,
phase separated semi-II PVME/PS IPNs, in which the PS was
crosslinked with maleic acid anhydride, have also been
reported by Felisberti et al. (23).
    The loss areas, LA's, obtained for the IPNs and
homopolymers are comparable with the values predicted
from equation (2) and Table I. Except for the crosslinked
PVME homopolymer, the LA values obtained experimentally
by DMS are within 10% of the values predicted by the group

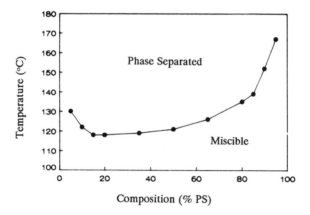

Figure 5. Cloud point phase diagram for physical blends of PVME and PS, showing lower critical solution temperature, LCST, behavior.

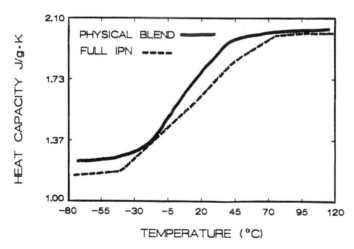

Figure 6. Differential scanning calorimetry of a PVME/PS physical blend and full-IPN. 50 wt % PS in each sample. Heating rate 10°C/min.

ectorectorectorectorectorectorectorectorectorectorectorectorectorectorectorectorectorectorectorectorectorectorectorectorectorectorectorectorectorectorectorectorectorectorectorectorectorectorectorectorectorectorяectorErrorector

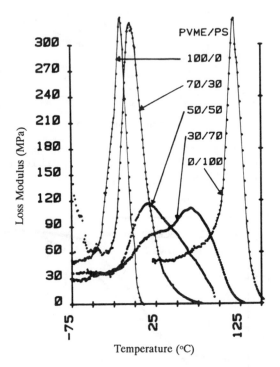

Figure 7. Linear loss modulus versus temperature curves for PVME/PS full-IPN's and homopolymer networks. Frequency 110Hz.

contribution analysis. The large deviation in the LA value obtained for the crosslinked PVME homopolymer, Table II, may be attributed to oxidation during sample preparation, as evidenced by sample discoloration which was not observed in the PVME used for the IPN samples.

To examine the effect of phase separation on the loss modulus, a sample of a 50/50 chemical blend was phase separated by annealing at 130°C and then evaluated by DMS, Figure 8. Note that the unannealed 50/50 blend shows a broad transition similar to that obtained by DSC and by Hourston and Hughes (24). As compared to the unannealed blend, the annealed PVME/PS blend shows a broader transition. Similarly, the IPN has a still broader transition than the blend. However, the LA's for the three samples are relatively constant, ±5%, agreeing also with theory, Table II. Note that overall experimental error is ±10%. Also, it was observed that the phase separated blend has a milky white appearance whereas the IPN is slightly hazy. This indicates that the size of the phase separated domains in the IPN are smaller.

Discussion

While examining the glass transition regions reported for PPO/PS and PVME/PS blends, it is of interest to note that the former has narrow glass transition regions over the entire range of composition (5), the location of which can be predicted on the basis of the Fox equation (21), whereas the latter exhibits much broader transitions for blends at mid to high polystyrene content. The broad transitions are indicative of molecular clustering which occurs in the vicinity of the LCST (24,25). The broader transitions associated with the PVME/PS blends as compared to the PPO/PS blends strongly suggest differences in the degree of molecular mixing for the two blends.

Microheterogeneous morphologies on the order of 100Å which were first observed by Huelck et al. (26) result in large interfacial areas, or interphase regions with pure phase regions being small or excluded. Since each region then has a slightly different composition and a correspondingly different T_g, a spectrum of glass transitions is obtained resulting in the observed broadening of the glass transition (27). Similarly, for small compositional fluctuations in PVME/PS blends, broadening of the glass transition would also be expected.

It must be noted that the crosslinks themselves pose multiple problems relative to the corresponding blends in determining the actual miscibility of the systems. (1) The chemical nature of the crosslinks (both in PVME and PS) must always be different from that of the main chain, hence the heats of interaction and Flory's X_1 value will be changed. While this can be minimized, it cannot be made zero. Thus, while the blends of PVME and PS teter on the edge of miscibility, chemical alterations such as

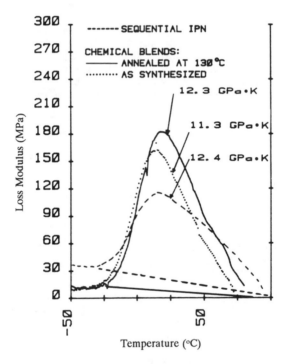

Figure 8.   Linear loss modulus curves for two 50/50
PVME/PS chemical blends, and for a 50/50 full-IPN.
Baseline corrections are also shown.

those imposed herein probably drive the system toward immiscibility. (2) The networks act to reduce phase domain size, since phase separation takes place after gelation. Consequently, while the blends go from clear to quite turbid on raising the temperature, the corresponding IPN's are only translucent by comparison. (3) It is thought that crosslinks in such systems should behave in a manner similar to block copolymer junctions, in that they tend to hold the two polymers together. Following Krause's zeroth approximation theory (28), the crosslinks might, from an entropic point of view, increase miscibility.

Conclusions

The PVME/PS IPNs synthesized appear to be less miscible than the corresponding chemical and physical blends. Dynamic mechanical spectroscopy indicates that midrange IPN compositions are phase separated. Broad glass transitions observed for PVME/PS blends and IPNs may indicate molecular level clustering. Surprisingly, LA values for the PVME/PS blends and IPNs are identical within experimental error, Figure 8, independent of morphology. The group contribution analysis method seems to be followed for all of these compositions, to a good approximation.

Acknowledgments

The authors are pleased to thank the Office of Naval Research for support and also the Plastics Institute of America for a supplemental Fellowship to JJF.

Literature Cited

1.   Nielsen, L. E. Mechanical Properties of Polymers and Composites; Marcel Dekker: New York, 1983; Vol. 1.
2.   Grates, J. A.; Thomas, D. A.; Hickey E. C.; Sperling, L. H. J. Appl. Polym. Sci. 1975, 19, 1731.
3.   Molau, G. E. Colloidal and Morphological Behavior of Block and Graft Copolymers; Plenum Press: New York, 1971.
4.   Platzer, N. A. Multicomponent Polymer Systems; Adv. Chem. Ser. No. 99, American Chemical Society: Washington, DC, 1971.
5.   Olabisi, O.; Robeson, L. M.; Shaw, M. T. Polymer-Polymer Miscibility; Academic Press: New York, 1979.
6.   Sperling, L. H.; Chiu, T. -W.; Thomas, D. A. J. Appl. Polym. Sci. 1973, 17, 2443.
7.   Sperling, L. H. Interpenetrating Polymer Networks and Related Materials; Plenum Press: New York, 1981.
8.   Sperling, L. H. In Comprehensive Polymer Science; Allen, G., Ed.; Pergamon Press: New York, 1989; Vol. 6.

9. Klempner, D.; Berkowski, L. In _Encyclopedia of Polymer Science and Engineering_; Mark, H. F.; Bikales, N. M.; Overberger, C. G.; Menges, G., Eds.; John Wiley and Sons: New York, 1987; Vol. 8.

10. Sperling, L. H.; Chiu, T. -W.; Gramlich, R. G.; Thomas, D. A. _J. Paint Technol_. 1974, _46_, 4.

11. Huelck, V.; Thomas, D. A.; Sperling, L. H. _Macromolecules_ 1972, _5_, 348.

12. Sperling, L. H.; Thomas, D. A. U.S. Patent 3 833 404, 1974.

13. Fradkin, D. G.; Foster, J. N.; Sperling, L. H.; Thomas, D. A. _Rubber Chem. Technol_. 1986, _59_, 255.

14. Fradkin, D. G.; Foster, J. N.; Thomas, D. A.; Sperling, L. H. _Polym. Eng. Sci_. 1986, _26(11)_, 730.

15. Chang, M. C. O.; Thomas, D. A.; Sperling, L. H. _J. Appl. Polym. Sci_. 1987, _34_, 409.

16. Chang, M. C. O.; Thomas, D. A.; Sperling, L. H. _J. Appl. Polym. Sci., Part B, Polym. Phys_. 1988, _26_, 1627.

17. Foster, J. N.; Sperling, L. H.; Thomas, D. A. _J. Appl. Polym Sci_. 1987, _33_, 2637.

18. Frisch, H. L.; Klempner, D.; Yoon, H. K.; Frisch, K. C. _Macromolecules_ 1980, _13_, 1016.

19. Macknight, W. J.; Karasz, F. E.; Fried, J. R. In _Polymer Blends_; Paul, D. R.; Newman, S., Eds.; Academic Press: New York, 1978.

20. Nishi, T.; Kwei, T. K. _Polymer_ 1975, _16_, 285.

21. Fox, T. G. _Bull. Am. Phys. Soc_. 1956, _1_, 123.

22. Bauer, B. J.; Briber, R. M.; Han, C. C. _Macromolecules_ 1989, _22_, 940.

23. Felisberti, M. I.; Abetz, V.; Kriegler, V.; Stadler, R.; Cantow, H-J. _Polym. Prepr_. 1989, _30(1)_, 111.

24. Hourston, D. J.; Hughes, I. D. _Polymer_ 1978, _19_, 1181.

25. Gunton, J. D.; San Miguel, M.; Sahni, P. S. In _Phase Transitions and Critical Phenomena_; Domb, C.; Lebowitz, J. L., Eds.; Academic Press: London, 1983.

26. Huelck, V.; Thomas, D. A.; Sperling, L. H. _Macromolecules_ 1972, _5_, 340.

27. Couchman, P. R.; Karasz, F. E. _J. Polym. Sci., Polym. Symp_. 1978, _63_, 271.

28. Krause, S. _Macromolecules_ 1970, _3_, 84.

RECEIVED January 24, 1990

Chapter 24

Vinyl Compound and Phenolic Interpenetrating Polymer Networks

Synthesis and Properties

K. Yamamoto and A. Takahashi

Hitachi Chemical Company, Ltd., Shimodate Research Laboratory, 1500 Ogawa, Shimodate, Ibaraki-ken 308, Japan

Interpenetrating polymer networks (IPNs) composed of vinyl compounds and phenolics were synthesized by combining simultaneous radical polymerization and phenolic curing reactions. To investigate the structure of products, two different types of IPNs were prepared. One, IPN 1, was formed through simultaneous reactions; the other, IPN 2, was formed through sequential reactions. Dynamic mechanical analysis, SEM observations, and tensile tests of these IPNs were performed. IPN 1 showed only one loss maximum, while IPN 2 showed two distinct loss maxima. These results indicate that vinyl polymer chains and cured phenolic chains in IPN 1 were well entangled with each other, while limited intermixing occurred in IPN 2. Compared with common vibration-damping materials, such as chloroprene rubber, the loss maxima of IPNs were at higher temperatures. Judging from the loss tangent values, IPNs are effective damping materials at elevated temperatures.

Phenol-formaldehyde resins were first manufactured commercially in 1907. (1-2) Even now they are still widely used in many fields because of their features of temperature resistance and good mechanical and electrical properties.

However, they are basically brittle materials, and many types of additives and modifiers have been developed to improve some of their properties. While working to develop special-purpose phenolic materials, the authors found that the compounds being produced had excellent vibration-damping properties (high loss tangent values), and that these properties were preserved to higher temperatures than elastomers generally used for the

purpose. It is possible that they may be suitable for specific
damping applications.
   These compounds have the structure of an interpenetrating
polymer network (IPN) of phenolic resin with materials having
elastomeric properties. (3-8) Except for early work by Aylsworth
in 1914, there has been no work on full IPN phenolic materials.
(9) This paper describes the preparation and testing of full IPN
phenolic systems with vinyl compounds.
   At first, methyl methacrylate (MMA) polymerization was used
as a model reaction system. However, phenolic derivatives are
well known as antioxidants and inhibitors for radical
polymerization (10), so such a reaction would not be expected to
occur. However, by proper choice of vinyl compounds and
initiators, it was found that such polymerization is possible, and
that a variety of IPNs can be produced.
   This paper reports on radical polymerization of MMA in
phenolic resol and confirmation of the structure by measurement of
dynamic mechanical properties, scanning electron microscopy, and
tensile tests, then the damping ability of these vinyl
compound/phenolic IPNs is evaluated.

Experimental Methods

Model Reactions

Phenolic resol was synthesized by the usual procedure (See Scheme
1.); 1 mol of phenol, 1.2 mol of formaldehyde, and 0.04 mol of
ammonia were heated at 70°C with vigorous stirring for 3 hours,
then dehydration was carried out under vacuum.
   MMA polymerization was conducted at 70°C in phenolic resol,
as shown in Scheme 2. The reaction product was poured into a
large volume of methanol to precipitate the poly (methyl
methacrylate) (PMMA). This was separated and washed several times
with methanol, then dried at reduced pressure.
   The phenolic resol curing reaction and dimethacrylate
polymerization were conducted at 170°C for 90 minutes, yielding
vinyl compound/phenolic IPN. (See Scheme 3.)
   The structure of the polymer obtained was determined from its
IR spectrum; the extent of conversion by polymerization and
condensation (phenolic curing reaction) were also calculated from
IR spectra.

Synthesis of Resol (molar ratio)

$$\left.\begin{array}{ll} \langle\bigcirc\rangle\text{— OH} & 1.0 \\ \text{HCHO} & 1.2 \\ \text{NH}_3 & 0.04 \end{array}\right\} \xrightarrow[]{70°C,\ 3h} 0 \xrightarrow[\text{Dehydration}]{70°C,\ i.\ vac.} \text{Resol} \qquad \text{Scheme 1}$$

Polymerization of MMA (mass ratio)

$$\left.\begin{array}{ll} \text{Resol} & 0.7 \\ \text{MMA} & 0.3 \\ \text{Initiator} & 0.01 \end{array}\right\} \xrightarrow[]{70°C} 0 \xrightarrow[\text{Wash}]{\text{MeOH}} \text{Polymer} \qquad \text{Scheme 2}$$

Synthesis of IPN (mass ratio)

```
Resol          0.7 ⎤
Dimethacrylate 0.3 ⎥ ── 170°C, 90min ──→ IPN        Scheme 3
Initiator      0.01⎦
```

Structural Analyses

In order to investigate the structure, vinyl compound/phenolic
IPNs were synthesized as follows (See Schemes 4, 5.):
Method 1: Poly (ethylene glycol) dimethacrylate (23G), and
phenolic resol (PR) were mixed at 50°C for 10 minutes and dicumyl
peroxide (DCP) was added. The polymerization and phenolic curing
reactions then took place simultaneously at 170°C for 90 minutes
in 200mm x 10mm x 1.5mm stainless steel mold. This product was
called IPN 1.
Method 2: First, 23G was polymerized at 70°C for 60 minutes, then
the product, poly 23G, was ground finely. Poly 23G particle size
was about 7.9×10^{-7}m in diameter. PR and the finely-powdered
poly 23G were mixed at 50°C for 10 minutes and suspended in
phenolic resol, then cured at 170°C for 90 minutes. This product
was called IPN 2.

Synthesis of IPN (mass ratio)

```
Method 1:
  PR   0.7 ⎤
            │  50°C, 10min        170°C, 90min
  23G  0.3 ─┤ ────────────→ 0 ────────────────→ IPN 1   Scheme 4
            │    Mixing        ↑ Polymerization
  DCP  0.01─┘                    Curing Reaction
```

```
Method 2:
  23G  0.7 ⎤
            │  70°C, 60min
  AIBN 0.01─┤ ──────────────→ Poly 23G ────────────→ 0
            │  Polymerization              Grinding    ↑
  PR   0.3 ─┘

              50°C, 10min        170°C, 90min
            ────────────→ 0 ──────────────────→ IPN 2   Scheme 5
               Mixing          Curing Reaction
```

Dynamic mechanical property data were obtained using Du Pont
DMA 982 instrument for structural analyses and Rheology DVE
instrument for measurement of damping ability. Scanning electron
microscopy was performed on samples etched with strong chromic
acid. The mechanical properties were measured at 20°C by tensile
test.

Damping Ability

IPNs consisting of vinyl compounds, phenolic novolacs, and epoxies
were synthesized for evaluation of damping ability. The test
specimens were prepared by curing the raw materials to IPN on
0.8mm thick steel sheets. The thickness of these IPNs were about
3.0mm.

Damping properties were measured by the impulse hammer technique at resonant frequency, and the logarithmic decrement was calculated by the half-power-width method.

Results and Discussion

Reaction Dynamics

Figure 1 shows time-conversion curves of MMA polymerization in phenolic resol by using various initiators.

When BPO was used, MMA did not polymerize, and a long induction period resulted when DCP was used. However, even in phenolic resol, MMA polymerization proceeded when certain kinds of initiator were used, such as AIBN.

Figure 2 shows time-conversion curves of MMA polymerization in phenolic resol compared with that in toluene. In phenolic resol, MMA polymerization proceeded more rapidly. However, there was a short induction period. These results suggest that by selecting the appropriate initiator, vinyl compound/phenolic IPNs can be expected.

Next, polymerization and curing reactions were conducted simultaneously. Table I shows the results of the reactions. By using various vinyl compounds and initiators, polymerization of over 90% and condensation conversion over 80% was obtained (as measured by IR spectrum of the cured materials). Figure 3 shows typical IR spectra of dimethacrylate and phenolic resol IPN. Reaction conversion was calculated from the peak of C-C double bond, methylol, and carbonyl groups.

Structural analyses

Dynamic mechanical properties of IPN 1, IPN 2, poly 23G and cured phenolic resol (CPR) were examined, in order to investigate IPN structure.

Figure 4 shows loss tangent-temperature curves. Each IPN contained 30% of 23G and 70% of phenolic resol. Poly 23G had its loss tangent peak in the narrow region of -60°C to -20°C. In cured phenolic resol, no peak was found, but the loss tangent increased gradually at temperatures over 200°C. IPN 1 had a broad peak in the region of 30°C to 160°C, and a very small one at -40°C, just like poly 23G. On the other hand, two distinct peaks were found in IPN 2, a small one at -50°C, and another at 0°C, and the loss tangent increased gradually at temperatures over 200°C. The curve for IPN 2 indicated a phase separation: the peak in the lower temperature region was similar to that for poly 23G, and the curve over 200°C was just as for cured phenolic resol. The peak at 0°C was attributed to IPN structure. This peak is discussed again later. Compared with IPN 2, the method of making IPN 1 introduced a high level of compatibility between the components being reacted.

Polymerization of 23G proceeds in a way different from the curing reaction of phenolic resol. A cross-linking reaction between 23G and phenolic resol cannot be considered because there

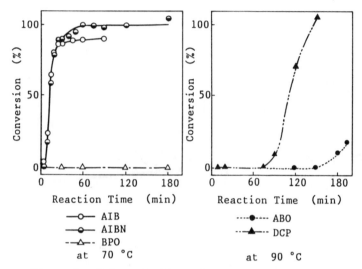

Figure 1.   Polymerization of MMA in phenolic resol.

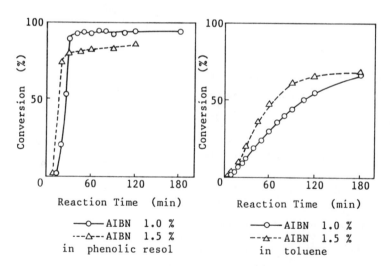

Figure 2.   Time–conversion curve of MMA polymerization.
(Reproduced from ref. 11.   Copyright 1986 American
Chemical Society.)

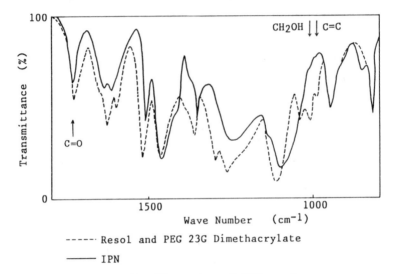

----- Resol and PEG 23G Dimethacrylate

——— IPN

Figure 3.   IR spectra of IPN samples.

Table I   Results of IPN Reactions

No.	Oligomer	Initiator	Temp. (°C)	Conversion (%) Polymerization	Condensation
1	Unsaturated Polyester	AIBN[a)]	170	93	81
2	"	ACHN[b)]	"	84	81
3	"	ABO[c)]	"	89	88
4	"	DCP[d)]	"	89	82
5	"	ACHN.ABO	"	90	78
6	PEG 23 Dimethacrylate	ACHN.ABO	"	100	96
7	PEG 9 Dimethacrylate	"	"	100	95
8	Polyurethane Acrylate	"	"	91	91
9	PEG 23 Dimethacrylate	AIBN.ACHN	"	98	82
10	"	AIBN.ABO	"	98	76

a) AIBN 2,2'-Azobisisobutyronitrile
b) ACHN 1,1'-Azobis (1-cyclohexanecarbonitrile)
c) ABO 2,2'-Azobis (2,4,4-trimethylpentane)
d) DCP Dicumyl peroxide

are no active points where reaction can occur. These results indicate that when 23G polymerization and the phenolic resol curing reaction take place simultaneously, the polymer chains are well entangled with each other and have a strong interaction. On the other hand, when each reaction proceeds sequentially, the polymer chains have poor entanglement, and that permits a phase separation.

Figure 5 shows the influence of IPN component ratio on T_g. Each sample was of IPN 1 type, prepared by the simultaneous method. Increasing the amount of 23G lowered the Tg of the IPN. This result also suggests that these IPNs show a good ability to mix. The Tg of IPN containing 70% of 23G was shown to be about 0°C, just like IPN 2 synthesized by the sequential method. The structure of 70%-23G-IPN can be considered to be cured phenolic resol in a continuous poly 23G phase. So, the loss tangent peak of IPN 2 at 0°C was attributed to the structure of 23G rich IPN such as 70%-23G-IPN.

Scanning electron microscopy of IPN 1 and IPN 2 was conducted next. Figure 6 shows micrographs of the surface etched with strong chromic acid, which selectively dissolves the 23G phase. IPN 1 had a smooth surface, indicating that the 23G particles were very small and evenly distributed, while these of IPN 2 were rough, indicating that the 23G particles were much larger and less well distributed, leaving a larger pit when removed. IPN 1 showed a high level of compatibility between the two components, much more than that of IPN 2.

Finally, mechanical properties were measured at 20°C by a tensile test. The tensile strength, elongation and modulus are shown in Table II.

Table II   Mechanical Properties

Sample	Tensile Strength (MPa)	Elongation (%)	Tensile Modulus (MPa)
IPN 1	34	1.5	2650
IPN 2	8	0.6	1270
Poly 23G	1	11.0	10
CPR	11	0.3	4310

Experiments were carried out at 20°C.

Cured phenolic resol showed low strength, small elongation, and high modulus. This is a typical brittle material. Poly 23G showed low strength, large elongation and very low modulus. This material is very weak in mechanical strength. IPN 2 showed low strength, elongation and modulus. This material is also weak. However, IPN 1 showed high strength, relatively large elongation, and moderately high modulus. This is a tough material compared with the others. These results also suggest that poly 23G chains

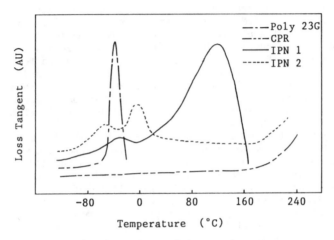

Figure 4.   Loss tangent-temperature curve.

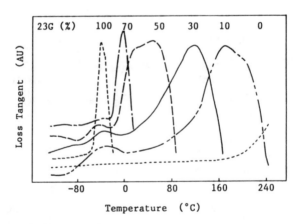

Figure 5.   Influence of IPN component ratio on T_g.

and cured phenolic resol chains in IPN 1 are well entangled with each other, and that this entanglement makes it tough.

Damping Ability

From the results of the previous section, it is considered that IPN 1 has properties of both poly 23G and cured phenolic resol. If a rubber-like polymer is used as the vinyl polymer, this IPN will show good damping properties at elevated temperatures. So, butyl acrylate, ethylene glycol dimethacrylate, phenolic novolac, and bisphenol A type epoxies were used as IPN components. The dynamic mechanical properties of these IPNs were examined first, because the loss tangent is very important to damping properties. Then the damping properties of IPN and commercial chloroprene rubber were measured at various temperatures.

Table III shows the composition of IPNs prepared for dynamic mechanical analysis.

Table III   Composition of IPNs

No.	Acrylates BA (g)	EGD (g)	Phenolics PN (g)	EP (g)	Initiator (g)	Acrylates (wt%)	EGD/BA (molar ratio)	EP/PN (equiv. ratio)
1	19	1	30	50	0.1	20	0.04	1.0
2	28.5	1.5	26	44	0.1	30	0.04	1.0
3	38	2	22.5	37.5	0.1	40	0.04	1.0
4	47.5	2.5	19	31	0.1	50	0.04	1.0
5	28	2	26	44	0.1	30	0.08	1.0
6	23	7	26	44	0.1	30	0.20	1.0
7	17	13	26	44	0.1	30	0.50	1.0
8	45	5	19	31	0.1	30	0.08	1.0
9	45	5	19	31	0.5	50	0.08	1.0
10	45	5	19	31	1.0	50	0.08	1.0
11	47.5	2.5	27.5	22.5	0.1	50	0.04	0.5
12	47.5	2.5	22.5	27.5	0.1	50	0.04	0.75

BA Butyl Acrylate,   EGD Ethylene Glycol Dimethacrylate,
PN Phenolic Novolac,   EP   Epoxy,
0.5 wt% of curing accelerator was used.

The relationship between the loss tangent or T_g and acrylate content, the molar ratio of dimethacrylate to acrylate, the initiator concentration, or the equivalent ratio of epoxy to phenolic are examined. The loss tangent was not affected by the acrylate content, but T_g was slightly lowered with an increase of acrylate content. (See Figure 7.) Figure 8 shows the influence of the molar ratio of dimethacrylate to acrylate, but this parameter had no effect on the loss tangent or T_g. The loss tangent was also not affected by the initiator concentration, but T_g was raised as it increased. (See Figure 9.) Figure 10 shows the

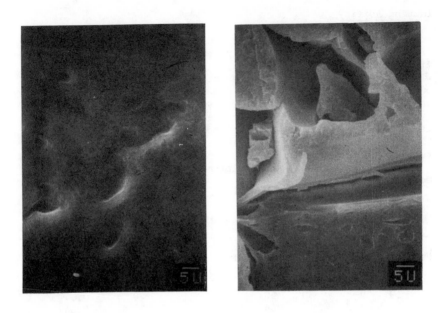

IPN 1   IPN 2

Figure 6.   SEM micrograph of etched surface of IPN 1 and IPN 2.

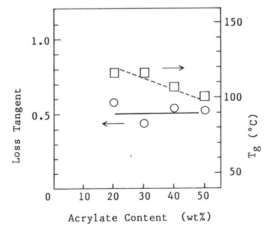

Figure 7.   Influence of acrylate content on loss tangent and T_g.

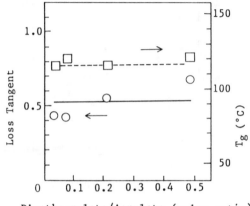

Figure 8.   Influence of dimethacrylate/acrylate on loss tangent and T_g.

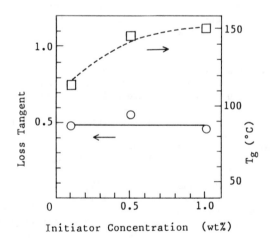

Figure 9.   Influence of initiator concentration on loss tangent and T_g.

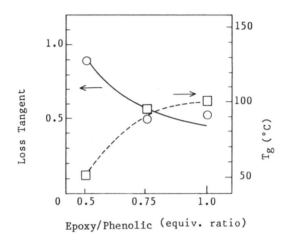

Figure 10.   Influence of epoxy/phenolic on loss tangent and T_g.

influence of the equivalent ratio of epoxy to phenolic.   The loss
tangent was decreased, and T_g was raised with an increase in the
equivalent ratio of epoxy to phenolic (which controlled the density
of cross-linking in the cured phenolic).   An IPN with an epoxy to
phenolic ratio of 0.5 had a relatively large loss tangent, about 0.9
at 50°C.
        These results lead to the conclusion that the parameter most
effective in controlling the loss tangent and T_g is the equivalent
ratio of epoxy to phenolic.   In this system, phenolic components
mainly determine the dynamic mechanical properties rather than
acrylates.
        An IPN with a ratio of epoxy to phenolic of 0.5 had a
relatively large loss tangent, so it was expected to show good
damping properties.   Figure 11 shows the loss tangent and the
logarithmic decrement of this IPN, along with chloroprene rubber
used as a commercial damping material.   In the case of IPN, the
maximum logarithmic decrement was at 40°C, which was near the T_g
of this material.   On the other hand, chloroprene rubber, which has
T_g of -30°C, showed no damping effect above room temperature.
However, the logarithmic decrement increased with decrease of
temperature below -5°C.
        These results support the proposal that the loss tangent is an
index of damping properties.   Judging from the loss tangent, IPNs
are useful damping materials at elevated temperatures.
        Figure 12 shows the temperature dependence of the logarithmic
decrement of neat IPN and filled IPN.   The logarithmic decrement of
neat IPN showed relative high attenuation, while filled IPN prepared
by adding platelet fillers showed an even higher attenuation over a
wide temperature range.

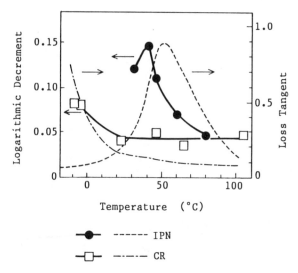

Figure 11. Influence of temperature on logarithmic decrement and loss tangent.

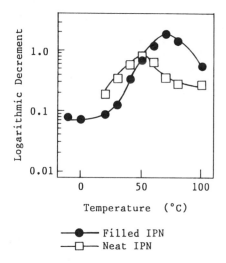

Figure 12. Temperature dependence on logarithmic decrement of IPN damping material.

Conclusions

Properties of vinyl compound/phenolic IPN were discussed and the following conclusions drawn. MMA radical polymerization proceeded rapidly in the presence of phenolic resol. Poly (ethylene glycol) dimethacrylate, 23G, and phenolic resol IPNs were synthesized by simultaneous radical polymerization and phenolic resol curing reaction. These IPNs had a structure of poly 23G chains and cured phenolic resol chains so well entangled with each other that the whole acted as a single phase. This type of IPN is considered to

have properties of both vinyl polymers and phenolics. By choosing
the IPN component materials and their ratio, the T_g of IPNs can be
designed to be in the region of 50°C to 150°C.

Judging from the results of dynamic mechanical analyses, IPNs
showed more effective damping properties than commercial chloroprene
rubber at elevated temperatures. In addition, filled IPNs prepared
by adding platelet fillers showed even higher attenuation
(logarithmic decrement).

Literature Cited

 1. Bakeland, L. H. U.S. Patent 939 966, 1909.
 2. Bakeland, L. H. U.S. Patent 942 852, 1909.
 3. Azlak, R. G.; Joesten, B. L.; Hale, W. F. Polym. Sci. Technol.
    1975, 9A, 233.
 4. Demmer, C. G.; Garnish, E. W.; Massy, D. J. R. Br. Polym. J.
    1983, 15, 76.
 5. Bachman, A.; Müller, K. Plaste Kautshuk 1977, 24, 158.
 6. Kosfeld, R.; Borowitz, J. Prog. Colloid Polym. Sci. 1979, 66,
    253.
 7. Cahill, R. W. Polym. Eng. Sci. 1981, 21, 1228.
 8. Borowits, J.; Kosfeld, R. Angew. Makromol. Chem. 1981, 100,
    23.
 9. Aylsworth, J. W. U.S. Patent 1 111 284, 1914.
10. Bailey, A. E. In Encyclopedia of Chemical Technology; Kirk, R.
    E.; Othmer, D. F., Ed.; The Interscience Encyclopedia, Inc.
    New York, 1948; Vol. 2, p.71.
11. Yamamoto, K.; Kumakura, T.; Yoshimura, Y. Polymer Preprints,
    Japan, 1986, Vol. 35, p 802.

RECEIVED January 24, 1990

Chapter 25

Dynamic Mechanical Response of In Situ Polyurethane–Poly(methyl methacrylate) Interpenetrating Polymer Networks

Influence of Kinetics of Formation

M. T. Tabka, J. M. Widmaier, and G. C. Meyer

Institut Charles Sadron (CRM–EAHP), Ecole d'Application des Hauts Polymères, 4, rue Boussingault, 67000 Strasbourg, France

A rather new synthesis principle yielding the so-called *in situ* simultaneous interpenetrating polymer networks (IPNs) has been elaborated: by selecting an adequate synthesis parameter, the kinetics of formation of both networks, followed by Fourier transform infra-red spectroscopy, interfer more or less, and a not negligeable change in the final structure results without having to change the composition. Thus, in a series of polyurethane/poly-(methyl methacrylate) IPNs, the damping behavior differs according to the amount of the polyurethane catalyst added, showing either two transitions, or, more interestingly, one broad transition extending over a large temperature scale.

The morphology of multicomponent systems is primarily related to the miscibility between polymers and to the type of blending. Usually, mechanical blends exhibit large phase domains of various types. Smaller, and well-defined morphologies are obtained by forming block and graft copolymers. Interpenetrating polymer networks (IPNs), in which the polymers are held together in their network form by permanent entanglements, have shown to lead to smaller domains than other polymer mixtures (1). In such materials, any further evolution of the morphology is impeded once both constituents have been crosslinked, and therefore no subsequent change in properties has to be feared.

In our laboratory, much attention has been devoted to the investigation of *in situ* sequential polyurethane/poly(methyl methacrylate) interpenetrating polymer networks (SEQ PUR/PAc IPNs) (2-5) in which the elastomeric polyurethane network is completely formed in the presence of the methacrylic monomers before the onset of the radical copolymerization which leads to the second network. To each polymerization process corresponds a typical kinetics, which however is not completely independent from each other (6-8). The results obtained with such SEQ IPNs show that the properties do in

0097–6156/90/0424–0445$06.00/0

fact not change significantly when the various synthesis parameters
are changed. Different properties can be obtained only by varying
the composition of the components.

A more selective approach consists in trying to influence the
kinetics of formation of at least one network; in this case, the two
networks are formed more or less simultaneously, and the resulting
morphology and properties can be expected to vary to some extent
without changing the overall composition. The same system as
previously studied, PUR/PAc, has been utilized in order to prepare a
series of *in situ* simultaneous IPNs (SIM IPNs), by acting
essentially on two synthesis parameters: the temperature of the
reaction medium and the amount of the polyurethane catalyst. Note
that the term simultaneous refers to the onset of the reactions and
not necessarily to the process. The kinetics of the two reactions
are followed by Fourier transform infra-red (FTIR) spectroscopy as
described earlier (7,8). In this contribution, the dynamic
mechanical properties, especially the loss tangent behavior, have
been examined with the aim to correlate the preceding synthesis
parameters to the shape and temperature of the transitions of the
IPNs.

E X P E R I M E N T A L

MATERIALS

PLURIISOCYANATE. Desmodur L75, provided by Bayer AG, is a 1,1,1-
trimethylol propane/toluene diisocyanate adduct containing 25% ethyl
acetate by weight. Density: 1.17 g/ml; NCO equivalent weight per kg:
3.06 (by standard titration with di-n-butylamine). Desmodur L75 was
used as received. Gel permeation chromatography has shown that this
product contains in fact four species of different molecular weights
and functionalities (9).

POLY(OXYPROPYLENE GLYCOL). The diol used in this work was poly-
(oxypropylene glycol) (POPG) supplied by Arco Chemical Co. under the
trade name Arcol 1020. It has a number-average molecular weight of
1890 g/mol with a polydispersity index of 1.5. Hydroxyl content:
1.06 mol/kg; density: 1.0 g/ml; viscosity at 25°C: 370 cP. Before
use, POPG was dried at least for three weeks over molecular sieves,
and the water content checked by a Karl Fischer titration.

CATALYST. Stannous octoate (OcSn) from Goldschmidt was stored under
nitrogen at low temperature and was used without further
purification. Tin content: 29.1% by weight; density: 1.25g/ml.

METHACRYLIC SYSTEM. Methyl methacrylate purchased from Merck and
1,1,1-trimethylol propane trimethacrylate (TRIM) supplied by
Degussa, used as crosslinker, were dried over molecular sieves but
not otherwise purified, so that they still contained 15 ppm and 100
ppm methylethylhydroquinone, respectively. The initiator of radical
polymerization was azobisisobutyronitrile (AIBN).

SYNTHESIS

The synthesis of polyurethane/poly(methyl methacrylate) IPNs was performed by the following general procedure (2). The reagents were mixed together and stirred under dry nitrogen, stannous octoate being added last, as its catalytic action begins immediately upon contact with the polyurethane precursors. The end of mixing was set as the origin of the reaction times. The mixture was poured into a glass mold and allowed to react. The IPNs were annealed at 75°C for one night and further cured at 120°C for 3 hours. For a given composition, the following values have been taken : [NCO] / [OH] = 1.07 , TRIM = 5 wt-% and AIBN = 1 wt-%.

KINETICS

The kinetics of IPN formation were followed by Fourier transform infra-red (FTIR) spectroscopy. For FTIR experiments, the mixture was injected in a cell formed by two NaCl windows separated by a 20 μm thick gasket. The spectra were obtained on a Nicolet 60SX spectrophotometer equipped with a Specac heating chamber with automatic temperature controller. Reaction conversion was calculated from the change of the normalized absorbance of the isocyanate peak at 2275 cm^{-1} and C=C peak at 1639 cm^{-1}.

DYNAMIC MECHANICAL MEASUREMENTS

Dynamic mechanical measurements were performed with a Rheometrics model RMS 7200 mechanical spectrometer at a fixed frequency of 1 rad/s through a temperature range from -100°C to 150°C under dry nitrogen. The test specimens were prepared in rectangular shape about 60 mm in length, 11 mm in width, and 4 mm in thickness. The applied strain was 1%.

R E S U L T S   A N D   D I S C U S S I O N

Figure 1 shows conversion versus time curves in the case of an *in situ* sequential synthesis. Due to differing synthesis modes (polyaddition and radical copolymerization), these curves are not superposable, and may therefore only cross each other when moved along the temperature scale or when their slopes change.

By setting the temperature of the reaction medium at 60°C from the beginning of the IPN formation, the PUR synthesis is accelerated, and that of the methacrylic system begins after the usual inhibition period. The competition between the two processes can still favour the complete formation of PUR before appreciable radical copolymerisation may have taken place, though the kinetic curves may change or even cross. For this reason, a second factor, the content of PUR catalyst, is varied too : with less stannous octoate, the formation of the first network is more or less delayed, even at 60°C, and counterbalances to some extent the effect of temperature. In such a case, the conversion of the methacrylic phase may proceed further before higher or even post-gel conversions are reached for polyurethane. Thus, IPNs in which both networks have been formed more or less simultaneously, are obtained by this procedure.

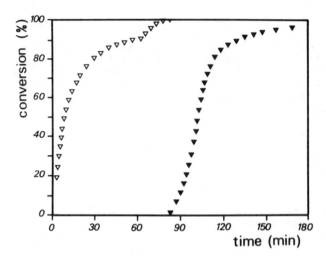

Figure 1.   Conversion profiles for a 25/75 *in situ* SEQ PUR/PAc IPN.
(▽): polyurethane; (▼): poly(methyl methacrylate)

For a series of 25/75 SIM PUR/PAc IPNs, the OcSn content was
varied from 1.0 to 0.1 wt-%; the first amount is also used in the *in
situ* sequential IPNs (*7*) and allows therefore comparison between the
both types of materials. Conversion curves obtained by FTIR are
given in Figures 2 to 4. Curves corresponding to the formation of

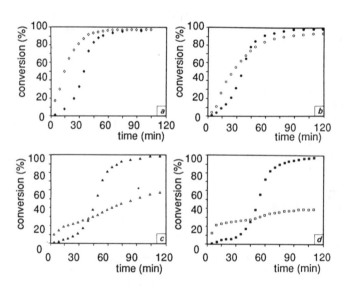

Figure 2.   Conversion profiles for PUR formation (open symbols) and
PAc formation (full symbols) in 25/75 SIM PUR/PAc IPNs as a function
of catalyst concentration: (a): 1.0% ; (b): 0.5% ; (c): 0.25% ; (d):
0.1% .

the PUR phase are displayed in Figure 3 and that of the PAc phase in Figure 4 : for both components, though mainly for polyurethane, they differ largely with the catalyst content, indicating that the two parameters chosen influence adequately the kinetics. Arrows on the polyurethane conversion curves show the actual crossing by the corresponding methacrylic curves. For 1% OcSn, PUR is completely formed before crossing occurs, and only about 10% acrylic double-bonds have been converted at its gelation point (i.e. 70% NCO conversion). Such an experimental situation is in fact close to that realized in an *in situ* sequential IPN (6), though a slight difference in methacrylic conversion exists.

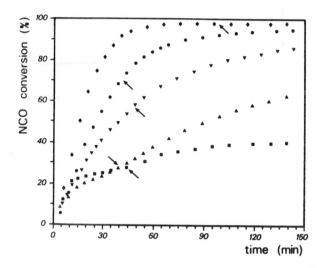

Figure 3. Conversion profile for PUR formation in 25/75 SIM PUR/PAc IPNs as a function of catalyst concentration: (♦): 1.0%; (●): 0.5%; (▼): 0.35%; (▲): 0.25%; (■): 0.1%. Arrows indicate location of crossing with the corresponding PAc conversion curve.

As the catalyst content decreases, crossing occurs at earlier NCO conversion. Above 0.5% OcSn, polyurethane still reaches gelation before crossing; below that content, its conversion is uncomplete even after 150 minutes, a rather long time as compared to the usual 30 minutes required to reach gelation. For 0.1% OcSn, the conversion is stopped at around only 30% of the available NCO functions. If a decrease in conversion rate is normal with decreasing the amount of catalyst, more detailed studies have shown that at least two other factors are involved in both slow down and low level of final NCO disappearence. First, an interaction exists between OcSn and AIBN, which forms an 1:1 complex (Tabka, M. T.; Widmaier, J. M.; Meyer, G. C., to be published), and the part of OcSn implied in this complex is no more available as catalyst. On the other hand, the partially concomitant development of a rigid methacrylic phase, may perturb the formation of polyurethane.

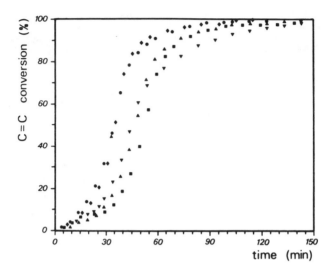

Figure 4. Conversion profile for PAc formation in 25/75 SIM PUR/PAC IPNs as a function of catalyst concentration: (♦): 1.0%; (●): 0.5%; (▼): 0.35%; (▲): 0.25%; (■): 0.1%.

Once the synthesis is completed, the IPNs are annealed. In the materials with incomplete NCO conversion, the unreacted functions disappear within a week at 60°C, even in the absence of water. However, the parallel occurrence of new urethane, urea... groups could not be detected accurately by FTIR spectroscopy or other methods. It remains that these IPNs contain no more reactive groups when utilized for testing.

Figure 4 shows that the formation of the methacrylic network is also somewhat influenced by the amount of OcSn, though a 100% conversion is always reached after about two hours. This observation may be related to the viscosity of the reaction medium : its increase accelerates the radical copolymerization. The already mentioned OcSn-AIBN interaction has a similar effect, by inducing a faster decomposition of AIBN into radicals (Tabka, M. T.; Widmaier, J. M.; Meyer, G. C., to be published).

The above results demonstrate that a rather minor change in some parameters is able to influence largely the kinetics; this in turn signifies that materials of various morphologies may be obtained at constant IPN composition; whether this change in morphology will induce appreciable differences in the final properties, has still to be demonstrated, and depends on the polymers involved and the property wanted.

For this purpose, the dynamic mechanical properties of a series of 25/75 in situ SIM IPNs have been investigated (Figure 5). With 1% OcSn, the tan δ vs temperature curve shows a classical shape, as for in situ SEQ IPNs and corroborates the kinetic results : two separated transitions, broadened and damped, exist (10). The lower transition corresponding to the polyurethane phase, is shifted

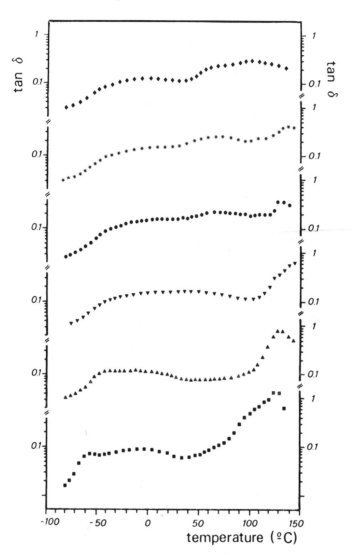

Figure 5. Loss tangent versus temperature curves for 25/75 SIM PUR/PAC IPNs as a function of catalyst concentration: (♦): 1.0%, (∗): 0.75%, (●): 0.5%; (▼): 0.35%; (▲): 0.25%; (■): 0.1%.

towards a higher temperature; in some cases, it overlaps with the β peak of PMMA in these systems, so that their respective contribution cannot be properly separated. The upper transition is lower than for the neat methacrylic network, and indicates some mixing with the elastomeric component. At lower catalyst contents, a third

transition   appears   in   the   75°C   temperature   range.   Such   a
supplementary transition has been reported by other authors (11,12).
As it had not been observed in our previous work, it may well
correspond to a new phase, i.e. a special state of interpenetration
of the two components due to the specific experimental conditions
used. Simultaneously, the peak corresponding to the methacrylic
phase is shifted towards higher temperatures and becomes again
sharper, a sign that larger domains of PMMA, containing no or little
PUR, exist in such materials. For 0.35%, one very broad transition
extends from   -50°C to   100°C, an indication of the existence of
one phase of continuously varying composition (13,14). The PMMA peak
is shifted to an unexpected 150°C, perhaps due to a very high
crosslink density made possible by a more complete crosslinking
reaction in a nearly pure methacrylic medium, than when PUR is
present. For still lower amounts of catalyst, two more and more
individualized peaks indicate an increasing separation of the
phases : due to the incomplete formation of the first network, an
irreversible and intimate interpenetration of the two components is
not possible. In this 25/75 series, the SIM IPN with 0.1% OcSn
contains the most uncompletely formed PUR, and the lower peak is
shifted to -56°C (Tg_{PUR} = -40°C) due to free or pendant diol chains.

Table I. Storage modulus and loss tangent values at room temperature
and at 100°C for a series of 25/75 SIM PUR/PAc IPNs
obtained with various amounts of stannous octoate

[OcSn] (%)	Temperature			
	24°C		100°C	
	G' (MPa)	tan δ	G' (MPa)	tan δ
0.10	730	0.07	70	0.45
0.25	550	0.08	195	0.10
0.35	440	0.14	105	0.10
0.50	405	0.15	25	0.18
0.75	445	0.14	30	0.19
1.00	555	0.11	25	0.28
1.00[*]	650	0.10	20	0.33
PAc	1360	0.08	390	0.19

[*] : *in situ* SEQ IPN

The values of tan δ, given in Table I, vary also with the
concentration of OcSn, by a factor of 2 to 4 depending on
temperature. So, for both the position on the temperature scale and
the loss value, quite different loss behaviors can be obtained by
influencing the kinetics of formation of the networks. The variation
of the storage moduli G' with temperature, shown in Figure 6,
complete these results. They differ most at higher temperatures,
upwards 50°C/60°C. At low catalyst contents, they are highest, due

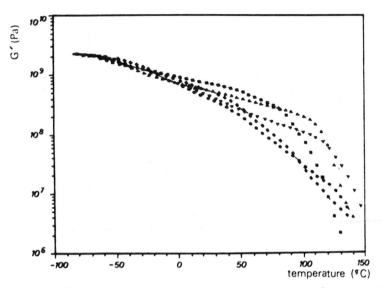

Figure 6. Shear storage modulus versus temperature for a series of 25/75 SIM PUR/PAc IPNs obtained with various amounts of stannous octoate: (◆): 1.0%; (●): 0.5%; (▼): 0.35%; (▲): 0.25%; (■): 0.1%.

to the well individualized rigid phase (15); when the latter disappears progressively with increasing concentration of OcSn, the storage modulus of the IPNs at 100°C decreases by a factor of ten (Table I).

The loss properties of SIM IPNs with other PUR contents, from 15 to 50 wt-%, have also been examined. Obviously, the amount of elastomer influences the overall properties by itself, but the specific behavior observed for the 25/75 series is roughly found again for the other compositions. However, in the 15/85 SIM IPNs, the loss curves are not quite different from each other, and present all a distinct methacrylic transition. For such a composition, the low polyurethane content does not contribute much to the shape of the curve, and the networks have also many defects (9), so that, at the same time, the lower transition is weak, and not precisely located on the temperature scale, whereas the upper transition is sharper and well defined. The 50/50 SIM IPNs, on the other hand, show (Figure 7) a most regular and broad transition for the higher OcSn contents, [OcSn] ≥ 0.50%, and also higher loss values at low temperatures, due to their elastomer content. The extension to the other compositions gives therefore the possibility to combine both appropriate loss behavior and elastomeric properties for an IPN, depending on the specified end-use. More generally, the composition is no more the only parameter available for changing the properties of such materials.

However, a full quantitative interpretation of the preceding results cannot yet be given, due to the very complex morphology of IPNs (1). It seems normal that allowing the PUR, or any other host network, to form more or less completely, will induce morphological changes in the resulting IPNs. In so far, obtaining either rather

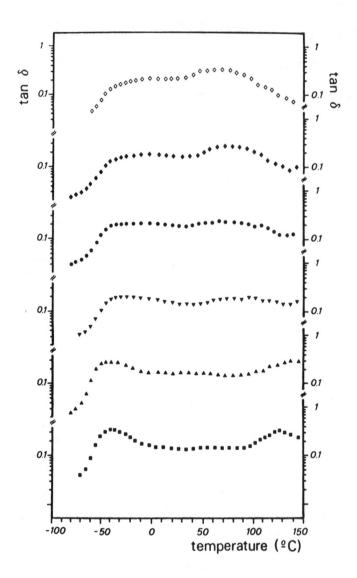

Figure 7. Loss tangent versus temperature curves for 50/50 SEQ (open symbols) and SIM (full symbols) PUR/PAc IPNs as a function of catalyst concentration: (◇,◆): 1.0%; (●): 0.5%; (▼): 0.35%; (▲): 0.2%; (■): 0.1%.

individualized phases or on the contrary interpenetrated phases can be explained by the experimental procedure utilized. But no theoretical model allows to foresee a third transition (16) or one very broad transition (17,18). One must admit that interphases resulting from specific synthesis conditions are formed, which correspond to combinations of PUR and PAc with well defined composition and degree of interpenetration. Accordingly, a phase enriched in PUR with regard to the initial composition, would coexist with a (perhaps pure) methacrylic phase.

Transmission electron microscopy and small-angle X-ray scattering experiments, still at their beginning, corroborate at least qualitatively these findings and interpretations, so that more accurate synthesis-properties relationships may be established soon, despite the complexity of the problem.

C O N C L U S I O N

Two conclusions emerge from this investigation : an adequately selected synthesis parameter is able to change the kinetics of formation of one or both networks to a not negligeable extent; the subsequent evolution in kinetics allows a stepwise modification of properties of IPNs, without the necessity to change their composition. This approach, which yields the so called "*in situ* simultaneous*" class of IPNs, offers new possibilities for these materials. According to their final destination, it becomes possible to select, after chemical nature and composition, a synthesis parameter which modifies the spectra of properties in the expected way. Especially, in this example, materials with a broad temperature range of damping were obtained by this means.

L I T E R A T U R E   C I T E D

1. Sperling, L. H. Interpenetrating Polymer Networks and Related Materials; Plenum Press: New York, 1981.
2. Djomo, H.; Morin, A.; Damyanidu, M.; Meyer, G. C. Polymer 1983, 24, 65 .
3. Djomo, H.; Widmaier, J. M.; Meyer, G. C. Polymer 1983, 24, 1415.
4. Morin, A.; Djomo, H.; Meyer, G. C. Polym. Eng. Sci. 1983, 23, 94.
5. Hermant, I.; Damyanidu, M.; Meyer, G. C. Polymer 1983, 24, 1419.
6. Jin, S. R.; Meyer, G. C. Polymer 1986, 27, 592.
7. Jin, S. R.; Widmaier, J. M.; Meyer, G. C. Polym. Comm. 1988, 29, 26.
8. Jin, S. R.; Widmaier, J. M.; Meyer, G. C. Polymer 1988, 29, 346.
9. Tabka, M. T.; Widmaier, J. M.; Meyer, G. C. Macromolecules 1989, 22, 1826.
10. Curtius, A. J.; Covitch, M. J.; Thomas, D. A.; Sperling, L. H. Polym. Eng. Sci. 1972, 12, 101.
11. Lipatov, Y. S.; Chramova, T. S.; Sergeeva, L. M.; Karabanova, L.V. J. Polym. Sci., Polym. Chem. Ed. 1977, 15, 427.
12. Frisch, H. L.; Frisch, K. C.; Klempner, D. Pure Appl. Chem. 1981, 53, 1557.
13. Lee, D. S.; Kim, S. C. Macromolecules 1984, 17, 268.
14. Akay, M.; Rollins, S. N.; Riordan, E. Polymer 1988, 29, 37.
15. Fox, R. B.; Bitner, J. L.; Hinckley, J. A.; Carter, W. Polym. Eng. Sci. 1985, 25, 157.
16. Wang, K. J.; Hsu, T. J.; Lee, L. J. Polym. Eng. Sci. 1989, 29, 397.

17. Jordhamo, G. M.; Manson, J. A.; Sperling, L. H. Polym. Eng. Sci. 1986, 26, 517.
18. Rosovizky, V. F.; Ilavsky, M.; Hrouz, J.; Dušek, K.; Lipatov, Y. S. J. Appl. Polym. Sci. 1979, 24, 1007.

RECEIVED January 24, 1990

Author Index

Affiliation Index

Subject Index

Production: Peggy D. Smith
Indexing: Deborah H. Steiner
Acquisition: Cheryl Shanks

Elements typeset by Hot Type Ltd., Washington, DC
Printed and bound by Maple Press, York, PA

Paper meets minimum requirements of American National Standard
for Information Sciences—Permanence of Paper for Printed Library
Materials, ANSI Z39.48–1984 ∞

Other ACS Books

Chemical Structure Software for Personal Computers
Edited by Daniel E. Meyer, Wendy A. Warr, and Richard A. Love
ACS Professional Reference Book; 107 pp;
clothbound, ISBN 0–8412–1538–3; paperback, ISBN 0–8412–1539–1

Personal Computers for Scientists: A Byte at a Time
By Glenn I. Ouchi
276 pp; clothbound, ISBN 0–8412–1000–4; paperback, ISBN 0–8412–1001–2

Biotechnology and Materials Science: Chemistry for the Future
Edited by Mary L. Good
160 pp; clothbound, ISBN 0–8412–1472–7; paperback, ISBN 0–8412–1473–5

Polymeric Materials: Chemistry for the Future
By Joseph Alper and Gordon L. Nelson
110 pp; clothbound, ISBN 0–8412–1622–3; paperback, ISBN 0–8412–1613–4

The Language of Biotechnology: A Dictionary of Terms
By John M. Walker and Michael Cox
ACS Professional Reference Book; 256 pp;
clothbound, ISBN 0–8412–1489–1; paperback, ISBN 0–8412–1490–5

Cancer: The Outlaw Cell, Second Edition
Edited by Richard E. LaFond
274 pp; clothbound, ISBN 0–8412–1419–0; paperback, ISBN 0–8412–1420–4

Practical Statistics for the Physical Sciences
By Larry L. Havlicek
ACS Professional Reference Book; 198 pp; clothbound; ISBN 0–8412–1453–0

The Basics of Technical Communicating
By B. Edward Cain
ACS Professional Reference Book; 198 pp;
clothbound, ISBN 0–8412–1451–4; paperback, ISBN 0–8412–1452–2

The ACS Style Guide: A Manual for Authors and Editors
Edited by Janet S. Dodd
264 pp; clothbound, ISBN 0–8412–0917–0; paperback, ISBN 0–8412–0943–X

Chemistry and Crime: From Sherlock Holmes to Today's Courtroom
Edited by Samuel M. Gerber
135 pp; clothbound, ISBN 0–8412–0784–4; paperback, ISBN 0–8412–0785–2

For further information and a free catalog of ACS books, contact:
American Chemical Society
Distribution Office, Department 225
1155 16th Street, NW, Washington, DC 20036
Telephone 800–227–5558